Lecture Notes in Electrical Engineering

Volume 850

The book series *Lecture Notes in Electrical Engineering* (LNEE) publishes the latest developments in Electrical Engineering - quickly, informally and in high quality. While original research reported in proceedings and monographs has traditionally formed the core of LNEE, we also encourage authors to submit books devoted to supporting student education and professional training in the various fields and applications areas of electrical engineering. The series cover classical and emerging topics concerning:

- Communication Engineering, Information Theory and Networks
- Electronics Engineering and Microelectronics
- Signal, Image and Speech Processing
- Wireless and Mobile Communication
- Circuits and Systems
- Energy Systems, Power Electronics and Electrical Machines
- Electro-optical Engineering
- Instrumentation Engineering
- Avionics Engineering
- Control Systems
- Internet-of-Things and Cybersecurity
- Biomedical Devices, MEMS and NEMS

For general information about this book series, comments or suggestions, please contact leontina.dicecco@springer.com.

To submit a proposal or request further information, please contact the Publishing Editor in your country:

China

Jasmine Dou, Editor (jasmine.dou@springer.com)

India, Japan, Rest of Asia

Swati Meherishi, Editorial Director (Swati.Meherishi@springer.com)

Southeast Asia, Australia, New Zealand

Ramesh Nath Premnath, Editor (ramesh.premnath@springernature.com)

USA, Canada:

Michael Luby, Senior Editor (michael.luby@springer.com)

All other Countries:

Leontina Di Cecco, Senior Editor (leontina.dicecco@springer.com)

** This series is indexed by EI Compendex and Scopus databases. **

More information about this series at https://link.springer.com/bookseries/7818

Rupam Goswami · Rajesh Saha

Editors

Contemporary Trends in Semiconductor Devices

Theory, Experiment and Applications

 Springer

Editors
Rupam Goswami
Department of Electronics
and Communication Engineering
Tezpur University
Napaam, Assam, India

Rajesh Saha
Department of Electronics
and Communication Engineering
Malaviya National Institute of Technology
Jaipur, Rajasthan, India

ISSN 1876-1100   ISSN 1876-1119 (electronic)
Lecture Notes in Electrical Engineering
ISBN 978-981-16-9126-3   ISBN 978-981-16-9124-9 (eBook)
https://doi.org/10.1007/978-981-16-9124-9

This Springer imprint is published by the registered company Springer Nature Singapore Pte Ltd.
The registered company address is: 152 Beach Road, #21-01/04 Gateway East, Singapore 189721,
Singapore

Preface

The evolutionary principles of the semiconductor device industry can be perceived through technological as well as philosophical viewpoints. On the more transparent technological standpoint, Moore's law initiated the roadmap for the industry, correlating the economic and technical aspects, which fueled the lines of great inventions and discoveries. Equally important is the philosophical standpoint which has motivated the upgradation of devices in an incremental fashion, moving from interesting theories to experiments, and eventually, applications. The exponential growth in the device industry is a result of the amalgamation of the two overlapping viewpoints.

With the roadmaps being consistently followed, and remodeled over time, the domain of semiconductor devices has led to the formation of many sub-domains, including overlaps with multiple areas. From materials, and physics to applications, semiconductor devices have been continuously evolving, and in the present era where feasibility of application of these devices has increased considerably, there is a need that the current trends in research are collected and reported.

This book is a collection of curated chapters on diverse areas of research in semiconductor devices in current times. The primary objective of this book is to present a few important reports in theory, experiment, and applications of semiconductor devices in order to acquaint the readers with the different dimensions of the domain. Most importantly, the categorization of the chapters into three sections (theory, experiment, and applications) is done on the basis of appropriateness of the coverage of the chapters and not particularly on the basis of a single perspective. The book is a collection of holistic reviews, experimental methods including fabrication techniques and simulation frameworks, and device-to-circuit applications.

The book is expected to act as a good reference for postgraduate as well as doctoral researchers who are in the lookout for interesting problems and solutions. For undergraduate scholars, the book can be helpful in knowing the different areas in the field of semiconductor devices and understanding the fundamental theories of novel devices or applications.

It is perfectly admissible that it is difficult, and all the more, futile, to include plenty of ideas into one book, and that too, in matters of contemporary research. Therefore, this book has selected chapters in the three sections, authored by researchers working in the corresponding areas. It is believed that the book is able to offer knowledge, and ideas to the readers, and orientation toward research in semiconductor devices in the present and future.

Napaam, India   Rupam Goswami
Jaipur, India   Rajesh Saha

Acknowledgements The editors would like to acknowledge the sponsorship of the following funded projects for motivating to create this collection of articles related to semiconductor devices.

- File No. SRG/2019/000660, funded by Science and Engineering Research Board, Government of India.
- File No. SRG/2019/000628 funded by Science and Engineering Research Board, Government of India.

Contents

Editors and Contributors

About the Editors

Rupam Goswami obtained his M.Tech. in 2014 and Ph.D. in 2018 from National Institute of Technology Silchar, India. Currently, he is Assistant Professor at the Department of Electronics and Communication Engineering, School of Engineering, Tezpur University, India. Before joining Tezpur University, he worked as Assistant Professor at Birla Institute of Technology and Science Pilani, Rajasthan, India. His research interests include simulation and modeling of TFETs, TFTs, FinFETs, and memristors.

His research works have appeared in three books, 22 international peer-reviewed journals, and 13 international peer-reviewed conferences. He is an editor of a book on carbon nanomaterial electronics.

Rajesh Saha has received B.E. with honours in Electronics and Telecommunication Engineering from Assam Engineering College, Assam, in 2012 and M.Tech. from NIT Arunachal Pradesh, Yupia, Arunachal Pradesh, in 2015. He has received Ph.D. from NIT Silchar, Assam, in 2018. He has worked as Junior Research Fellow in IIT Guwahati from September 2012 to April 2013. Currently, he is working as Assistant Professor at the Department of ECE in MNIT Jaipur. Before joining MNIT Jaipur, he has worked as Assistant Professor in School of Electronics Engineering, VIT AP University, Amaravati. His research interest includes modeling and simulation of nanoelectronics devices, biosensors, and MEMS. He has published his research work in 25 peer-reviewed journals and 6 international peer-reviewed conferences.

Contributors

Srimanta Baishya National Institute of Technology Silchar, Silchar, Assam, India

Kuheli Roy Barman National Institute of Technology Silchar, Silchar, Assam, India

Bhabana Baruah Department of Electronics & Communication Engineering, Tezpur University, Assam, India

Brinda Bhowmick Department of Electronics and Communication Engineering, National Institute of Technology Silchar, Silchar, Assam, India

Basab Das Department of ECE, GIMT Guwahati, Guwahati, Assam, India

Souradeep De School of Advanced Materials Green Energy and Sensor Systems, Indian Institute of Engineering Science and Technology, Shibpur, Howrah, India

Arighna Deb School of Electronics Engineering, KIIT Deemed To Be University, Bhubaneswar, India

Abhigyan Ganguly Department of Electronic Engineering, Howard College, University of KwaZulu Natal, Durban, South Africa

Puja Ghosh Department of Electronics and Communication Engineering, Indian Institute of Information Technology Ranchi, Ranchi, India

Rupam Goswami Department of Electronics and Communication Engineering, Tezpur University, Assam, India

Saral K. Gupta Department of Physical Sciences, Banasthali Vidyapith, Rajasthan, India

Chinmayee Hazarika Assam Energy Institute Sivasagar, Centre of Rajiv Gandhi Institute of Petroleum Technology, Jais, Amethi, India

Kavindra Kandpal Department of Electronics and Communication Engineering, Indian Institute of Information Technology, Uttar Pradesh, Allahabad, Prayagraj, India

Deeksha Kharkwal Department of Physical Sciences, Banasthali Vidyapith, Rajasthan, India

Santanu Maity School of Advanced Materials Green Energy and Sensor Systems, Indian Institute of Engineering Science and Technology, Shibpur, Howrah, India

Chandra Mohan Singh Negi Department of Physical Sciences, Banasthali Vidyapith, Rajasthan, India

Sujan Neroula Department of Electronics and Communication Engineering, Tezpur University, Tezpur, India

K. P. Pradhan Indian Institute of Information Technology Design and Manufacturing Kancheepuram, Chennai, Tamil Nadu, India

Soumyaranjan Routray SRM Institute of Science and Technology, Chennai, Tamil Nadu, India

Rajesh Saha Department of Electronics and Communication Engineering, MNIT Jaipur, Jaipur, India

Argha Sarkar Vishnu Institute of Technology, Bhimavaram, India

Nidhi Sharma Department of Physical Sciences, Banasthali Vidyapith, Rajasthan, India

Santanu Sharma Department of Electronics and Communication Engineering, Tezpur University, Tezpur, India

Biraj Shougaijam Department of Electronics and Communication Engineering, Manipur Technical University, Imphal, India

Aryamick Singh Department of Electrical and Electronics Engineering, Birla Institute of Technology and Science Pilani, Rajasthan, India

Salam Surjit Singh Department of Electronics and Communication Engineering, Manipur Technical University, Imphal, India

Akriti Srivastava Department of Electronics and Communication Engineering, Indian Institute of Information Technology, Uttar Pradesh, Allahabad, Prayagraj, India

Viranjay M. Srivastava Department of Electronic Engineering, Howard College, University of KwaZulu Natal, Durban, South Africa

K. Vanlalawmpuia Department of Electronics and Communication Engineering, National Institute of Technology Silchar, Assam, India

Introduction

Rupam Goswami and Rajesh Saha

Abstract This chapter introduces the theme of the book on the theory, experiment, and applications in semiconductors and semiconductor-based devices currently in research and development. The world of semiconductors has progressed immensely with the evolution of novel materials and device architectures. The International Roadmap for Devices and Systems (IRDS), formerly known as International Technology Roadmap for Semiconductors (ITRS), has been instrumental in offering detailed prediction of the electrical parameters, materials, devices, architectures, and applications. The downscaling of device dimensions to sustain the economics of the industry, and functionality per unit chip area, is the prime reason behind the emergence of new devices and systems. From metal oxide semiconductor field effect transistors (MOSFETs), and FinFETs to interconnects and solar cells, this book picks up some of the thrust areas in the light of its title and arranges them for better readability by researchers and individuals interested in the domain of semiconductor devices.

Keywords Semiconductor devices · FinFET · MOSFET · VLSI · TFT · Opto-electro-mechanical device · Solar cell · Gas sensors

1 Semiconductor Devices: A Brief Overview

The semiconductor device industry saw a paradigm shift since the invention of the monolithic integrated circuits (ICs) by Jack S Kilby and Robert Noyce [1][2]. While the transition of the industry from bipolar junction transistors (BJTs) to MOSFETs created new avenues, the complementary MOS (CMOS) technology finally achieved

R. Goswami (✉)
Department of Electronics and Communication Engineering, Tezpur University, Assam 784028, India

R. Saha
Department of Electronics and Communication Engineering, MNIT Jaipur, Jaipur, India
e-mail: rajesh.ece@mnit.ac.in

R. Goswami and R. Saha (eds.), *Contemporary Trends in Semiconductor Devices*,
Lecture Notes in Electrical Engineering 850,
https://doi.org/10.1007/978-981-16-9124-9_1

the objective of low power applications of devices [3][4]. Silicon became the *perfect* semiconducting material for a significant fraction of the devices due to its cost-effectiveness, stability, and ability to form its own oxide [5][6].

1.1 International Technology Roadmap for Semiconductors (ITRS)

The Moore's law and Dennard's scaling law can be said to have inspired the formation of the International Technology Roadmap for Semiconductors (ITRS) [7] comprising of industries and researchers working in the domain of semiconductors and their applications. The objective was to set goals and predictions based on viability of semiconductor materials and devices in research and production. The ITRS became the *guide* for researchers around the world, who worked in the area of semiconductors. From low power requirements to high performance computing, the ITRS has laid out a data-oriented roadmap after regular intervals of time. The ITRS is now known as International Roadmap for Devices and Systems (IRDS).

1.2 From MOSFETs to Novel FETs

The MOSFETs were in use until the advent of the fin-shaped FETs [8][9]. As device dimensions started reducing and entered the nanoscale regime, the short channel effects (SCEs) became prominent in MOSFETs [10–12]. Low power applications suffered a huge setback due to increased off-state leakage current and poor subthreshold swing. To counter these effects, the effective gate control on the channel was targeted to be improved, and hence, different solutions were proposed to re-design MOSFETs. In the process, double gate [13] and gate-all-around architectures [14] were proposed.

However, researchers also started searching for novel materials and device phenomena to solve the said problems in MOSFETs. Tunnel field effect transistors (TFETs) [15–17], fin-shaped field effect transistors (FinFETs) [8][9], carbon nanotube field effect transistors (CNT FETs) [18], ferroelectric FETs [19], thin film transistors (TFTs) [20][21], and spin FETs [22][23] emerged in the research spectrum. The objective was to mitigate the SCEs in conventional MOSFETs and primarily look for alternatives in principle of operation. While TFETs operated via quantum mechanical tunnelling, FinFETs operated on the principle of enhanced gate control [8]. CNT FETs utilized the low dimensionality of the nanotubes to have enhanced mobility [18], and ferroelectric FETs worked on the idea of negative capacitance to reduce the subthreshold swing [19]. TFTs operate by trapping-release mechanism on grain boundaries of amorphous polysilicon or oxide semiconductors

[20], whereas spin FETs work on the principle of conduction through spin-polarized electrons by tuning their spin–orbit interaction via modulation of the electric field [22].

1.3   Opto-Electronic Devices

The use of novel transistor designs as cited above may not always be sufficient in case of applications involving ultra-high-speed operations, minimum area, low switching time, and low operating power, especially required in domains of deep learning networks. The advances in silicon photonics, however, have enabled technologists to ponder over the use of optical circuits for extremely fast operations using smaller chip occupancy footprint [24]. Different Boolean functions may be represented using the optical logic circuits which contribute to the high-speed functions.

1.4   Solar Cells and Photovoltaics

On the other hand, the area of solar cells and photovoltaics has progressed at a significant pace [25][26]. The demand for clean and green energy has become one of the most important sustainable goals on the planet, considering the severe impacts of climate change on environment. The research aims of enhancing the efficiency and fabrication methods have led to experiments and reports on using organic materials and low dimensional materials for photovoltaic applications.

1.5   Other Sensing Applications

Apart from photovoltaic applications, semiconductors and semiconductor devices are instrumental in gas sensing and biosensing applications. Thin films of semiconductors, inorganic and organic, which include ZnO, graphene, and reduced graphene oxide are used in gas sensors [27][28]. On the other hand, the MOSFET can be modified on its gate structure to facilitate it for sensing purposes (biomolecule sensing, explosives, others). Such devices, known as ion-sensing field effect transistors (ISFETs) [29][30], are usually used to detect the target molecules through a carrier solution containing the target molecules with the help of a suspended or a free-end reference electrode.

1.6 Computational Tools

There have been revolutionary developments in computational tools used for designing and visualizing semiconductor devices from their geometrical as well as physics-based perspectives. Technology computer aided design (TCAD) tools for devices ranging from bulk geometry to atomistic analyses have occupied an important position in the predictive analysis and modelling of semiconductor devices in the current era [31][32]. Current semiconductor simulation tools have a wide range of functionalities and options for multiple physics-based models, allowing the designer to predict the characteristics of a device with more convenience. The availability of user-friendly resources has further contributed to the accessibility of these tools for in-depth analyses starting from calibration with experimental results and simulation of non-ideal device phenomena.

2 Organization of the Book

This book has been organized, keeping in view the three technical sub-areas of semiconductor devices, namely theory, experiment, and applications. However, it is important to point out to the reader that a chapter may qualify for more than one sub-area based on its content, but the categorization has been done for the sub-area for which it has a greater degree of inclination. A total of 14 chapters comprises this book out of which 13 chapters are categorized into the technical sub-sections.

2.1 Theory of Semiconductor Devices

This section consists of a total of five chapters. The first chapter presents an alternate three-dimensional non-planar geometry to MOSFET, named as the vertical super-thin body (VSTB) MOSFET. Using 3D TCAD tool, the chapter reports the performance of the device in terms of DC and RF analyses, response to stress, strain, and velocity saturation, and response to interface traps and noise sources.

The second chapter discusses ferroelectric TFETs, starting from the theory of ferroelectric materials, and TFETs and later extending it into analyses of temperature and noise sources through a well-defined TCAD framework. The authors discuss Landau's theory, laying out the motivation for negative capacitance FETs, and finally, the ferro-TFET.

The third chapter is a detailed review on organic photovoltaic applications. The chapter lucidly mentions the importance of renewable sources of energy in current times through data on the current status and finally delves into the geometry, materials, and fabrication techniques of organic solar cells. Different deposition techniques as well as material specifications are discussed in this chapter.

The fourth chapter in this section presents a review on dye-sensitized solar cells. The authors discuss about this technology which uses natural dyes as photochemical agents through the description of the principle of operation and recent developments in counter electrodes, electrolytes, and sensitizers. Not only do the authors discuss the material advances in detail, but also the prospects in flexible electronics.

The final chapter in this section is based on the theory of nanostructured kesterite solar cells with CZTS and CZTSe as absorber layers, where the authors discuss the different parameters through appropriate mathematical modelling. Using the Schrodinger wave equation, and current continuity and current density equations, the different parameters are discussed. Comments are provided for the consideration of traps in the analyses. The efficiency of the absorber material and the solar cell in general is reported in the chapter.

2.2 *Experiment in Semiconductor Devices*

The second section comprises of a total of four chapters. The first chapter in this sub-area is a detailed report on fabrication and experiments in quantum dot-based solar cells. The authors fruitfully discuss the methods and techniques to improve quantum efficiency in these solar cells. Through appropriate results and discussion, the chapter finds itself as a good reference in the area of nanomaterial sensitized solar cells.

The second chapter presents one of the most non-ideal cases in ion-implanted doping profiles of TFETs, the lateral straggle. The principle of lateral straggle is discussed along with a prior discussion on TFETs. Through calibrated TCAD simulations, the authors take up the case of a heterostacked-source geometry of TFET and discuss the impact of lateral straggle on the performance of such a device for low power applications.

The third chapter deals with fabrication methods of heterostructures for gas sensing applications. The chapter outlines the fabrication process flow for ZnO-based heterostructure gas sensors and discusses the optimization steps for fabrication of such devices. Reduced graphene oxide is used for two thin films, one in solo state and the other with Mg: ZnO for analysis. Furthermore, experimental observations are discussed with the help of characterization methods before discussing the performance of the gas sensors.

The final chapter in this section discusses on the experimental aspect of optimal positioning of reference electrode in ion-sensing field effect transistor (ISFET). An experimental set-up is designed to validate the results from a physics-based model regarding the appropriate distance of separation between the sensing layer and the reference electrode. The chapter discusses the model in detail using site-binding approach and Poisson equation. The microcontroller-based experimental set-up is pictorially represented and explained.

2.3 Applications of Semiconductor Devices

The final section of the book has four chapters in total. The first chapter in this section presents a detailed description of VLSI interconnects made of composite materials involving graphene, carbon nanotubes, copper, and others. The chapter not only delivers the facts and figures for VLSI interconnects, but also, highlights the modelling aspects of interconnects.

The second chapter in the section is about voltage-programming techniques used in TFT circuits for driving OLEDs. TFTs are convenient to manufacture and are quite promising candidates for flexible electronic applications. The authors discuss the voltage-programming techniques through phase-wise explanations and lucid examples.

The third chapter in the section introduces and discusses an emerging area of application of opto-electro-mechanical device in optical logic circuits. The author clearly outlines the motivation for optical logic circuits in high-speed applications and presents a survey on existing techniques and alternatives. The chapter further delves into the entire structure of the optical multiplexer, including the mapping to Boolean functions. Experimental results are included for a comprehensive understanding of the techniques.

The final chapter in the section and this book is based on the comparative analyses of electrical characteristics of a selective buried oxide (SELBOX)-based TFET and a dual tunnel diode TFET. The authors discuss about the simulation framework and extend the comparison for different parameters for DC as well as RF/analog applications.

3 Prospects and Outlook

It must be mentioned that the area of semiconductor devices is vast and extensive. The chapters edited and included in this book are selected based on the demands of the researchers in the current era, and the prospects for future applications. Therefore, the chapters are diverse in terms of areas which are on the path towards possible commercialization, or which can be derived as possible solutions to specific problems.

Acknowledgements The authors acknowledge the technical sponsorship for the funded projects: File No. SRG/2019/000660, and File No. SRG/2019/000628 by SERB, Government of India.

References

1. Kilby JS (2000) The integrated circuit's early history. Proc IEEE 88(1):109–111
2. Noyce RN (1964) Integrated circuits in military equipment. IEEE Spectr 1(6):71–72
3. Chandrakasan P et al (1992) Low-power CMOS digital design. IEEE J Solid-State Circuits 27(4):473–484
4. Veendrick HJM (1984) Short-circuit dissipation of static CMOS circuitry and its impact on the design of buffer circuits. IEEE J Solid-State Circuits SC 19:468–473
5. Anderson K (1993) Silicon and Germanium. MRS Bull 18(3):96–100
6. Weber L, Gmelin E (1991) Transport properties of silicon. Appl Phys A 53:136–140
7. The International Technology Roadmap for Semiconductors 2.0 (2015) Itrpv, 2015 [Online]. Available: http://www.itrs2.net/
8. Das R, Goswami R, Baishya S (2016) Superlattices Microstruct 91:51–61
9. Saha R et al (2020) Dependence of RF/analog and linearity figure of merits on temperature in ferroelectric FinFET: a simulation study. IEEE Trans Ultrason Ferroelectr Freq Control 3010(c):1–6.
10. Veeraraghavan S, Fossum JG (1989) Short-channel effects in SOI MOSFETs. IEEE Trans Electron Devices 36(3):522–528
11. Jaiswal N, Kranti A (2018) Modeling short-channel effects in asymmetric junctionless mosfets with underlap. IEEE Trans Electron Devices 65(9):3669–3675
12. Pourghaderi MA et al (2018) Universality of short-channel effects on ultrascaled MOSFET performance. IEEE Electron Device Lett 39(2):168–171
13. Reyboz M et al (2009) Continuous model for independent double gate MOSFET. Solid-State Electron 53(5):504–513
14. Rewari S et al (2019) Novel design to improve band to band tunneling and gate induced drain leakages (GIDL) in cylindrical gate all around (GAA) MOSFET. Microsyst Technol 25:1537–1546
15. Sahu SA et al (2020) Characteristic enhancement of hetero dielectric DG TFET using SiGe pocket at source/channel interface: proposal and investigation. SILICON 12:513–520
16. Manocha P et al (2021) Selection of low dimensional material alternatives to silicon for next generation tunnel field effect transistors. SILICON 13:707–717. https://doi.org/10.1007/s12 633-020-00452-y
17. Shukla S, Goswami R (2020) Perspective—performance assessment of TFETs for low power applications: challenges and prospects. ECS J Solid State Sci Technol 9(10):109001
18. Tamersit K (2021) A novel band-to-band tunneling junctionless carbon nanotube field-effect transistor with lightly doped pocket: Proposal, assessment, and quantum transport analysis. Physica E: Low-Dimensional Sys Nanostruct 128:114609
19. Ghosh P et al (2019) Optimization of ferroelectric tunnel junction TFET in presence of temperature and its RF analysis. Microelectron J 92:104618
20. Desai MS et al (2021) A Multiple-trapping-and-release transport based threshold voltage model for oxide thin film transistors. Journal of Elec Mater 50:4050–4057
21. Kandpal K, Gupta N (2017) Study of structural and electrical properties of ZnO thin film for Thin Film Transistor (TFT) applications. J Mater Sci: Mater Electron 28:16013–16020
22. Ringer S et al (2018) Spin field-effect transistor action via tunable polarization of the spin injection in a Co/MgO/graphene contact. Appl Phys Lett 113:132403
23. Malik GFA et al (2020) Performance analysis of indium phosphide channel based sub-10 nm double gate spin field effect transistor. Phys Lett A 384(19):126498
24. Midolo L et al (2018) Nano-opto-electro-mechanical systems. Nature Nanotech 13:11–18
25. Gul M et al (2016) Review on recent trend of solar photovoltaic technology. Energy Explor Exploit 34(4):485–526
26. Luceño-Sánchez JA et al (2019) Materials for photovoltaics: state of art and recent developments. Int J Mol Sci 20(4):976
27. Gupta SK et al (2010) Development of gas sensors using ZnO nanostructures. J Chem Sci 122:57–62

28. Varghese SS et al (2015) Recent advances in graphene based gas sensors. Sensors Actuators B Chem 218(SC):160–183

29. Lee C-S et al (2009) Ion-sensitive field-effect transistor for biological sensing. Sensors 9(9):7111–7131

30. Sharon E et al (2009) Detection of explosives using field-effect transistors. Electroanalysis 21:2185–2189

31. Demchenko O et al (2016) (2016) Research possibilities of Silvaco TCAD for physical simulation of gallium nitride power transistor. AIP Conf Proc 1772:060007. https://doi.org/10.1063/1.4964587

32. Synopsys SENTAURUS TCAD I-2013.12. [http://www.synopsys.com]

Theory

A Brief Insight into the Vertical Super-Thin Body (VSTB) MOSFET

Kuheli Roy Barman and Srimanta Baishya

Abstract Investigation of different attributes of a vertical super-thin body (VSTB) MOSFET has been performed through the 3-D Sentaurus TCAD platform. Same-scale FinFET and VSTB FET performance are also compared in terms of their performance. The proposed doping outline 3 (DO$_3$) decreases off-state leakage current (I_{off}) by 99.4% over DO$_1$. The gate overlap used on source/drain of 15 nm (OL$_3$) improves on/off current ratio (I_{on}/I_{off}) and subthreshold swing (SS) over OL_1 by about 83.6% and 6.47%, respectively. The explanation for such enhancement is addressed by examining carrier properties like density, mobility, and velocity in off/on state. Furthermore, with a supply voltage (V_D) sweep = (0.05–1) V, I_{on}/I_{off}, SS, and threshold voltage (V_{th}) show sufficient performance needed for ultra-low-power (ULP) and high performance (HP) nodes. For channel length (L_{ch}) = (15–100 nm), the device shows a negligible variation in I_{on}/I_{off}, SS, and DIBL, thus maintaining electrostatic integrity. For L_{ch} = 25 nm, the maximum $g_m/C_{gg}/C_{gd}/f_T/GBP$ at V_D = 0.45 V is 0.00155 μS/4.268 fF/0.277 fF/102.09 GHz/177.57 GHz, respectively. The device works well with stress/strain/velocity saturation effects too. Various noise (diffusion/generation-recombination/flicker) power spectral densities are reported at f = 1 MHz and 10 GHz. Lastly, the influence of uniform and Gaussian trap distribution with densities of 10^{11} and 10^{13} cm^{-2} existing at Si/Si$_3$N$_4$ interface on I_{off}, I_{on}, and SS is analyzed.

Keywords Vertical super-thin body (VSTB) · Overlap (OL) · Uniform trap · Gaussian trap

1 Introduction

So far, the semiconductor industries have greatly satisfied the huge demand for multi-functional electronic appliances across the globe. The key technique behind such incredible achievement of the industries is the continual scaling of device size. A

K. R. Barman (✉) · S. Baishya
National Institute of Technology Silchar, Silchar, Assam 788010, India

smaller device size enables various desired features such as reduced capacitance, increased functionalities in a single chip, reduced cost per function, and increased portability. However, devices with shorter channel lengths also suffer from various short channel effects (*SCEs*) such as off-state leakage current (I_{off}), drain-induced barrier lowering (*DIBL*), mobility degradation, and threshold voltage roll-off. To achieve efficient performance of a short channel device, the degree of these SCEs must be maintained within the considerable limit as outlined by International Technology Roadmap for Semiconductors (*ITRS*, now known as International Roadmap for Devices and Systems) [1]. At a point in time, researchers realized that only scaling of the basic MOSFET structure cannot lead to persistent improvement of performance parameters due to the unavoidable existence of SCEs [2]. Therefore, various new device structures such as SOI FETs, multi-gate FETs, and FinFETs came into existence to drive the technology node forward [3–10]. Although such novel devices have improved the problem of SCEs to a great extent, various new challenges have arisen out of the exercise of repetitive shrinkage of these devices. Despite providing significant enhancement of speed and power consumption, SOI FETs have never been largely produced by the microelectronic industries due to the high cost of the sophisticated fabrication process needed for such devices [3–6]. Furthermore, under the strong influence of oppositely directed electric fields in low-thickness multi-gate devices, the carrier mobility gets degraded due to the SOI-thickness-fluctuation-induced scattering and hence, the driving current (I_{on}) gets reduced [7]. Therefore, downscaling of multi-gate FETs for high performance (HP) applications is obstructed by such poor I_{on} issues. In FinFETs, the aspect ratio (height-to-width ratio) of the fins gets squeezed with every advancement of the technology node, and designers experience the challenge of sustainability of the thin fins which stand absolutely without any support. It has been reported that for lower channel lengths, the *fin* part of a FinFET may be *broken* anytime; thus the durability and reliability of FinFET based applications get hampered [8–10]. Therefore, such issues demand further modification of the FET structure. In this regard, a new structure called the vertical super-thin body (VSTB) FET was invented by Koldyaev and Pirogova in 2014 [10]. This novel architecture consisting of a vertical thin fin/body supported by one or more STI dielectric walls is a single gate structure. Hence, the mobility of the charge carriers in a low-thickness channel gets improved due to the use of only one gate contact in such a device. Moreover, the STI support to the vertical thin fin adds mechanical strength to it and as a result, the downscaling of the channel length becomes more reliable. The technological assessment of such a new device must be initiated with the analysis of its fundamental electrostatics. Also, various aspects of device performance analysis are needed to be explored widely for VSTB FET.

As this device is the outcome of improvisation over FinFET, it shall be scientifically correct to compare the performance of a FinFET and a VSTB FET on similar scales. Various research works have reported that the doping strategy in the MOSFET plays an important role in improving its device performance [11–13]. Hence, the doping profile assessment can be performed for VSTB FET to evaluate and achieve improved performance. Moreover, the use of gate overlap over the source or drain impacts the device electrostatics, and hence, various figures of merit (*FOM*) [14–17]. Thus, it will be interesting to investigate the effect of gate overlap on the performance as a part of the exploration of the device. The demand for ultra-low-power (*ULP*) devices is increasing day by day because the power is consumed in these devices in an effective manner [18–20]. The VSTB performance with low supply voltages must be explored to predict its applicability as a *ULP* device. The journey of technological advancement with any novel device must go through the initial walk of analysis of basic electrostatics of the same. Therefore, study of the nature of subthreshold swing (SS), *DIBL*, on/off current ratio (I_{on}/I_{off}), and electrostatic integrity of VSTB FET in terms of its channel length variation is extremely important in order to investigate its scalability [21–23]. Besides, as the applicability of FET in the radio frequency (RF) spectrum is crucial, the RF performance investigation should be considered as an integral part of the complete study of this new device [24–27]. The channel of the FET device may undergo stress or strain exerted by the adjacent materials used in the device [28]. Stress and strain effects play a significant role in determining carrier mobility by modifying the band structure of semiconductor crystals in short channel devices. The thick shallow-trench isolation (STI) walls surrounding the channel region in the VSTB FET structure may exert a significant amount of stress/strain on it. Hence, the study of I_D variation against gate voltage (V_G) by considering stress/strain will estimate the impact of STI on the channel current. The velocity of carriers is limited by various short channel effects for lower gate lengths [29–31]. So, the study of the drain current in presence of velocity saturation will assist one in studying the impact of the high field on the dynamics of the channel carriers. Moreover, a MOS device encounters different noise phenomena like diffusion, generation-recombination, and flicker noises. If the noise level is beyond the tolerable band, it may hamper device operation fully or partially. Therefore, the power of different noise sources must be analyzed to estimate the application of the device in the circuit level analysis [32–34]. Different studies show that the traps existing at the interface of channel and gate oxide interface substantially modify the device's electrical behavior [24, 35–38]. Since with the advancement of technology, the Si/SiO_2 interface is being replaced by Si/other high-k oxide materials, the trap density at the interface has become a more significant parameter in performance prediction strategy. Thus, studying the effect of traps on I_D will help us to acquire a deeper knowledge of the device.

In this chapter, we compare the performance of a FinFET and a VSTB FET on similar scales. The impact of doping strategy and gate overlap on source/drain on the device is presented in terms of I_{off}, I_{on}/I_{off}, and SS. Various parameters like I_{on}/I_{off}, SS, and threshold voltage (V_{th}) are presented for a supply voltage (V_D) sweep of (0.05 – 1) V to realize the *ULP* and *HP* applicability of the device. A concise investigation of the key performance parameters such as SS, *DIBL*, I_{on}/I_{off}, and other parameters of a gate-overlapped device was performed to develop a basic overview of the device's efficacy. Again, various RF FoMs such as input capacitance (C_{gg}), gate-drain-capacitance (C_{gd}), transconductance (g_m), unit-gain-cut-off-frequency (f_T), and gain-bandwidth-product (GBP) were thoroughly evaluated to develop the application perspective of this new device. Moreover, the effect of stress/strain/velocity saturation on the device performance was investigated. The noise power spectral densities (PSDs) for different noises are also studied for $f = 1$ MHz and 10 GHz. The influence of traps for uniform and Gaussian distributions introduced at the interface of Si/Si_3N_4 is also presented in terms of I_{off}, I_{on}, and SS.

2 Device Structure and Simulation Set up

The 3-D and 2-D conventional diagrams of n-type Si VSTB FET are presented in Fig. 1a, b. Highly doped source and drain are used, and the complete doping outline (*DO*) is explained in the next section. Different dopants used are as follows: source/drain: Arsenic, channel/substrate: Phosphorus. Both substrate and channel doping are kept fixed at 10^{15} cm^{-3}. Si_3N_4 as gate oxide and Ti as gate metal are used to improve the gate tunneling current [39, 40]. Various physical parameters of the device are as follows: gate thickness $(t_g) = 5$ nm, channel length $(L_{ch}) = 25$ nm, gate oxide thickness $(t_{ox}) = 1.5$ nm, body thickness $(t_{ch}) = 5$ nm, substrate height $(h_s) = 25$ nm, body height $(h_b) = 40$ nm, and SiO_2 isolation insertion between different contacts (gate/source/drain) and substrate $(h_i) = 10$ nm. The overlap of the gate used over source/drain is denoted as x, the values for which are considered as 0, 10, and 15 nm.

The analysis is performed with the help of 3-D Sentaurus TCAD simulation tool [41]. Various device phenomena are considered in the simulation by activating different physics models. Due to the presence of the highly doped source/drain regions, the Fermi–Dirac statistics model is used [41]. Different mobility models such as doping-dependent Masetti model (for mobility dependency on device doping profile), high-field saturation model (for velocity saturation), and the enhanced Lombardi model with high-k degradation (for high-k dielectric oxide) are included to capture the effective mobility [41]. To address the stress/strain effects of the STI wall into the vertical thin body, the stress-induced electron mobility model, stress-dependent saturation velocity model, and deformation potential model are turned on [28, 41]. Quantum effects are taken into account by considering the quantum density gradient model and thin-layer mobility model [41]. Finally, the doping-dependent

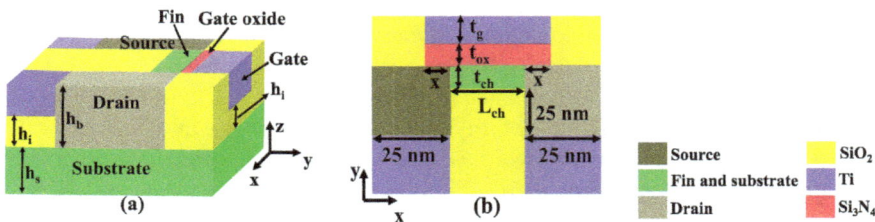

Fig. 1 Vertical super-thin body (VSTB) MOSFET **a** 3-D view, **b** 2-D view

Shockley-Hall-Read (*SRH*) recombination model was activated to address the generation and recombination process of carriers [41]. Regarding the trap analysis, two types of acceptor trap distributions are considered at the Si/Si$_3$N$_4$ interface: uniform trap distribution (UTD), and Gaussian trap distribution (GTD). The densities of both the distributions can be presented by the following equations [24, 35, 41]:

$$d_U = N_0 \text{ for } E_0 - 0.5E_S < E < E_0 + 0.5\,E_S \tag{1}$$

$$d_G = N_0 \exp(-(E - E_0))^2/2E_S^2 \tag{2}$$

In the above equations, d_U and d_G signify UTD and GTD, respectively. N_0 is the maximum trap density (uniform trap density) in *GTD* (*UTD*). Two values of 10^{11} and 10^{14} eV^{-1} cm^{-2} for N_0 are considered for both the types of distributions. E_0 and E_S are the energy-mid point and energy straggle parameters, respectively, for *GTD*. We have considered $E_0 = 0$ eV and $E_S = 0.1$ eV.

3 Results and Discussion

A gradual approach is followed to explore the different aspects of the device. Initially, a comparison based on *SS*, I_{off}, and $I_{\text{on}}/I_{\text{off}}$, of a same-scale FinFET and VSTB FET is explored in Sect. 3.1. The effect of doping on the I_{off} and *SS* is studied to address the suitable doping profile for better performance accomplishment in Sect. 3.2. The improvement in the SS and $I_{\text{on}}/I_{\text{off}}$ by the gate overlap on the source/drain region is discussed in the same section. In Sect. 3.3, the influence of V_D on the basic FoM is addressed. The variation of the key parameters concerning the change of channel length is presented in Sect. 3.4. Then, the RF performance of the device is evaluated in Sect. 3.5. Next to that, Sect. 3.6 describes the influence of stress/strain/velocity saturation effect on device performance. Different noise PSDs are explained in Sect. 3.7. Finally, the trap effects on the drain current are discussed in Sect. 3.8.

3.1 Comparison of FinFET and VSTB FET Performance

This section presents the comparison of the performance of the FinFET and the VSTB FET on similar scales in with the same doping level outlined as: source: 10^{20} cm^{-3}, channel: 10^{15} cm^{-3}, drain: 10^{20} cm^{-3}. The drain current (I_D)-gate voltage (V_G) plot shown in Fig. 2 expresses a huge improvement of more than two orders in I_{off} and about 47.27% in *SS* in case of VSTB FET as compared to FinFET. Such improvement originates from the pseudo-SOI type isolation provided by vertical *STI* walls between source and drain in VSTB FET [10]. On the other hand, this type of source and drain

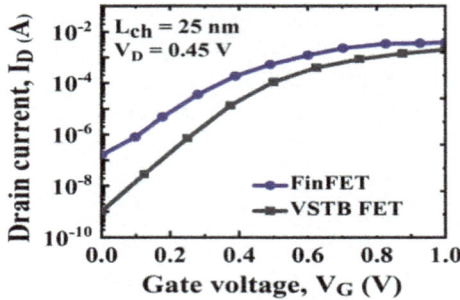

Fig. 2 $I_D - V_G$ plot comparison for FinFET and VSTB FET

Table 1 Relative performance of FinFET and VSTB FET

Device	SS (mV/dec)	I_{off} (nA)	I_{on}/I_{off}
FinFET	144.58	163.65	2.64×10^4
VSTB FET	76.23	1.026	1.72×10^6

isolation is not present in FinFET, which degrades SS and I_{off}. The assessment for various fundamental parameters of FinFET and VSTB FET is mentioned in Table 1.

3.2 Study of Doping Effect and Gate Overlap on Device Performance

The optimization of the source/channel/drain doping profile is performed to improve I_{off}, I_{on}/I_{off}, and SS. Three different doping outlines (DO$_1$, DO$_2$, and DO$_3$) are described below out of which the last one (DO$_3$) is observed to enhance the device performance significantly.

DO$_1$ (cm^{-3}) = source: channel: drain = 10^{20}: 10^{15}: 10^{20}.
DO$_2$ (cm^{-3}) = source: channel: drain = 10^{20}: 10^{15}: 10^{19}.
DO$_3$ (cm^{-3}) = source: channel: drain = 10^{19}: 10^{15}: 10^{19}.

Figure 3a, b represent doping dependencies of I_D-V_G plot and I_{off}, respectively. It is clear from Fig. 3b that low doping concentration used in the drain/source in DO_2 and DO_3 significantly improves the transfer characteristics concerning DO_1. A decrease in the drain doping level in DO_2 increases the depletion width in the drain region; hence, the strength of the drain electric field responsible to attract electrons from the source at $V_G = 0$ V weakens. Again in DO_3, I_D, and hence I_{off}, is slightly reduced by lowering the source doping from 10^{20} cm^{-3} to 10^{19} cm^{-3}. A large fall of I_{off} (about 99.4%) is observed in Fig. 3b while changing the doping profile from DO_1 to DO_3.

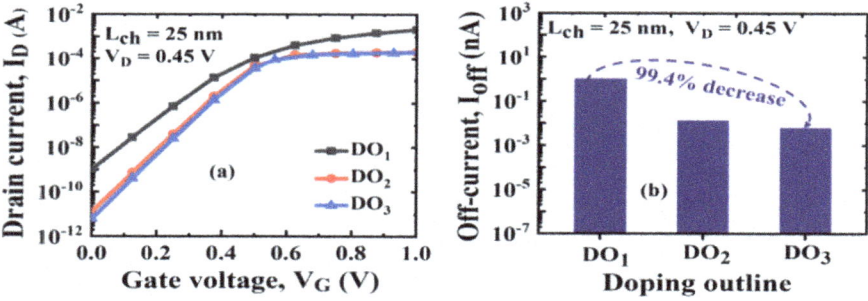

Fig. 3 a I_D-V_G plot for different doping dependencies, b I_{off} for doping outlines FET

Figure 4a shows the effect of three different degrees (OL_1, OL_2, and OL_3) of gate overlap over the source/drain on I_D-V_G characteristics. OL_1, OL_2, and OL_3 represent gate overlap length (x) on both, source and drain, of 0 nm, 10 nm, and 15 nm, respectively. It is observed that with the increasing degree of overlapping the transfer plot gets improved (Fig. 4a). Figure 4b, c show that OL_3 enhances I_{on}/I_{off} and SS by 83.6% and 6.47%, respectively, compared to OL_1. The key reason for such improvement can be understood by considering electron density, mobility, and velocity in off and on state. Figures 5a, b compare electron density and mobility of OL_1 and OL_3 across the channel length at $V_G = 0$ V and $V_D = 0.45$ V (off state). The electron density in off state for OL_3 is seen to be lesser than OL_1 (Fig. 5a); however, the mobility of the carriers shows the opposite nature (Fig. 5b), as expected. But since in its off state, instead of mobility, the density of carriers is the more vital character for deciding I_{off}, OL_3 exhibits low I_D at $V_G = 0$. On the other hand, due to the presence of gate fields in on state, the electron densities for OL_1 and OL_3 are almost the same (Fig. 5c), but interestingly the velocity of the carriers is higher for the overlapped device (Fig. 5d). This happens as a consequence of the enhanced channel horizontal field in OL_3 by distancing the gate fringing field from the channel.

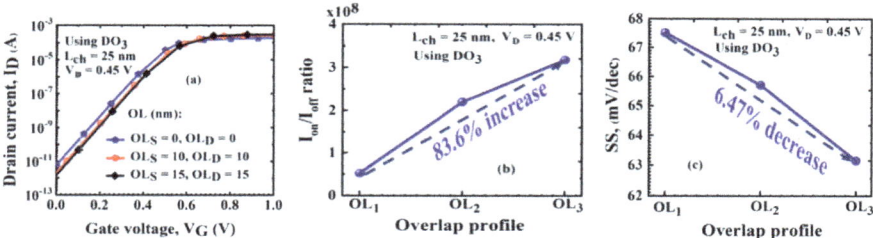

Fig. 4 Different OL conditions a I_D-V_G plot, b I_{on}/I_{off} ratio, c SS

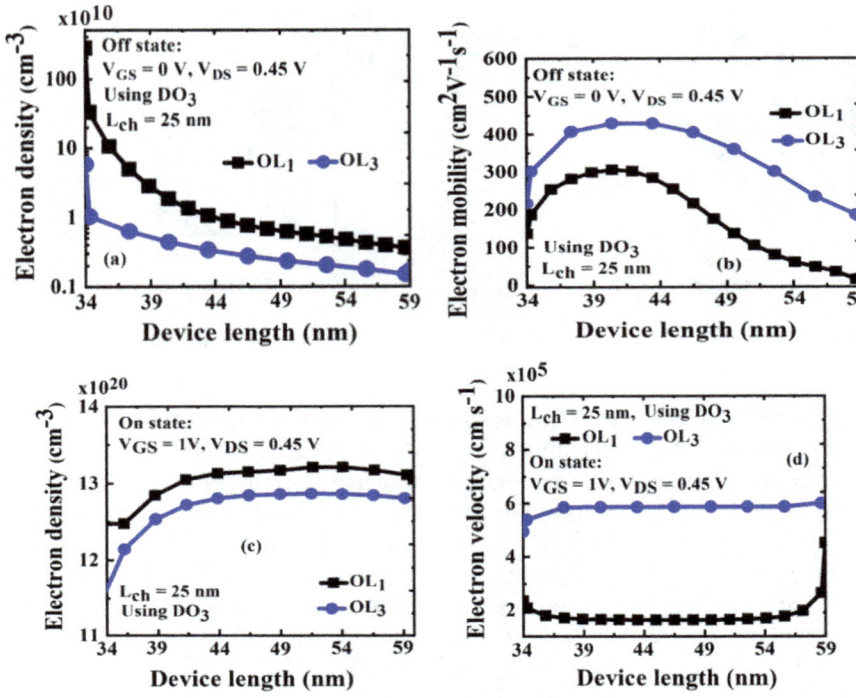

Fig. 5 **a** Electron density in off state, **b** electron mobility in off state, **c** electron density in on state, and **d** electron velocity in on state

3.3 Study of the Effect of Supply Voltage Variation on Various Device of Merit

This section presents variations of different key device parameters concerning the change of V_D. Figures 6a–c show the nature of I_{on}/I_{off}, SS, and V_{Th} for V_D variation,

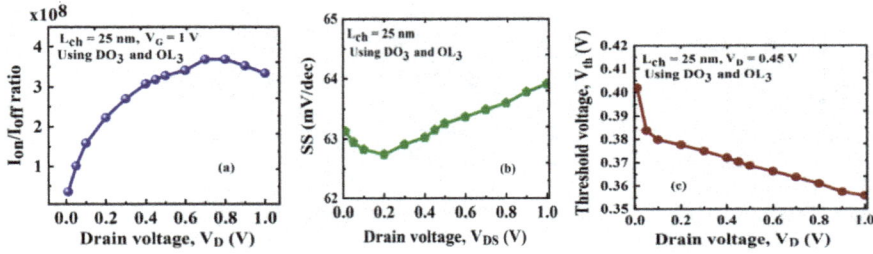

Fig. 6 Variation of different device parameters as a function of drain voltage (V_D), **a** I_{on}/I_{off} ratio, **b** SS, and **c** threshold voltage

respectively. We have estimated I_{on} at $V_G = 1$ V. It is clear from Fig. 6a that for lower drain voltages too (as low as $V_D = 0.05$ V to $V_D = 0.3$ V), I_{on}/I_{off} satisfies the ITRS 2015 predictions by showing high orders of ratio (in the order of 10^7 to 10^8) [1]. Moreover, with increasing V_D, I_{on}/I_{off} shows an increasing trend up to $V_D = 0.7$ V; however, for $V_D > 0.7$ V it decreases slowly. Higher V_D results in higher I_{on}, but simultaneously I_{off} increases. When V_D crosses 0.7 V, the increase in I_{off} dominates over the increase in I_{on}, and I_{on}/I_{off} shows a decreasing trend.

The SS versus V_D plot shown in Fig. 6b reveals that for $V_D \leq 0.2$ V, the value of SS decreases; however, the opposite trend can be noticed for $V_D > 0.2$ V. As SS signifies the change in gate voltage needed to change the drain current by one decade, it can be inferred that for a fixed value of V_G, when V_D increases from 0 to 0.2 V, the horizontal drain field also enhances and acts effectively to increase the velocity of channel electrons and thereby, I_D increases. Thus such increase in V_D helps to lower the SS by improving I_D. But, any increase in V_D beyond 0.2 V exerts a higher field in the channel, and consequently, scattering between charge carriers becomes prominent, which prevents further increase in velocity of carriers. As a result, SS deteriorates for higher V_D. However, overall the SS value is observed to be less than 64 mV/dec in Fig. 6b, which implies the efficient control of the gate in VSTB FET.

The V_{Th} curve presented for various values of V_D in Fig. 6c was estimated for a constant I_D of 10^{-7} A. With the rise of V_D, the value of V_{Th} is observed to be reduced. This is expected because a higher V_D contributes to higher I_D, and thus, the V_G value needed to reach any fixed I_D decreases with the increasing V_D.

3.4   Study of the Effect of Channel Length Variation on Various Device of Merit

Figure 7a–d show, respectively, the change of I_{on}/I_{off}, SS, SS-shift (ΔSS), and $DIBL$ for change in L_{ch}. An increase in L_{ch} causes an increase in I_{on}/I_{off} (Fig. 7a). This is caused since a longer L_{ch} helps to reduce I_{off} by minimizing the effect of $DIBL$. Also, at lower L_{ch}, the values of I_{on}/I_{off} are significantly high. This indicates the effective scalability of the device.

Further, SS shows gradual improvement with the extension of L_{ch} (Fig. 7b). This is quite expected because various SCEs drop with lengthening the L_{ch}. Besides, the values of SS for all L_{ch} also show a very satisfactory response by maintaining entirely low magnitudes.

The electrostatic integrity of the device was estimated by calculating the shift of SS (ΔSS) at a particular L_{ch} for the SS of $L_{ch} = 100$ nm. Figure 7c demonstrates that for a wide variation of L_{ch} from 15 to 95 nm, SS shows negligible shifts. This illustrates the fact that the downscaling of the device does not impact much on its electrostatics [42]. Such a property is highly desired for the shrinkage of the device size.

DIBL was estimated by the following relation [17, 43]:

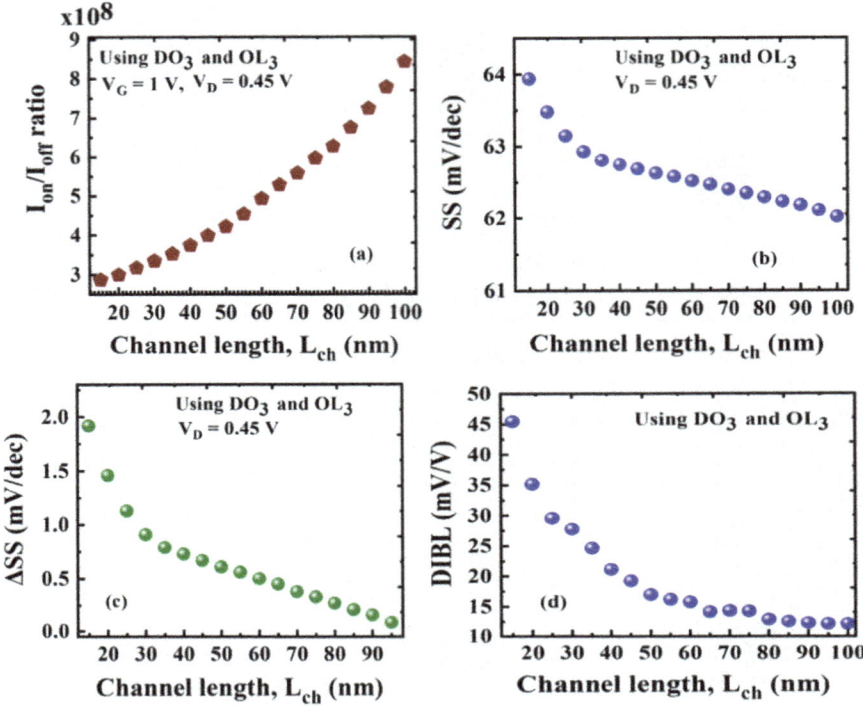

Fig. 7 Variation of different device parameters as a function of channel length (L_{ch}), **a** I_{on}/I_{off} ratio, **b** SS, **c** ΔSS, and **d** DIBL

$$\text{DIBL} = \left(V_{\text{Th,lin}} - V_{\text{Th,sat}}\right) / \left(V_{\text{Dsat}} - V_{\text{Dlin}}\right) \tag{3}$$

where $V_{\text{Th,lin}}$, and $V_{\text{Th,sat}}$ are, respectively, gate voltage magnitudes for reaching $I_D = 10^{-7}$ A at $V_D = V_{\text{Dsat}}$, and $V_D = V_{\text{Dlin}}$. Here, $V_{\text{Dsat}} = 1$ V and $V_{\text{Dlin}} = 0.05$ V. The *DIBL* values also exhibit very low amplitudes; thus ensures efficient gate control of the device [17, 42].

3.5 Study of the RF Performance of the Device

Various RF parameters are demonstrated in this section.

Figure 8a presents transconductance (g_m) versus V_G plot, in which g_m initially increases; but after V_G crosses a certain value (~0.67 V), g_m starts to saturate slowly (Fig. 8a). g_m is related to the change in I_D with respect to the change in V_G. Now, since the initial increase in V_G increases I_D by accumulating more electrons in the channel, the respective incremental change can also be observed in g_m. However, a further

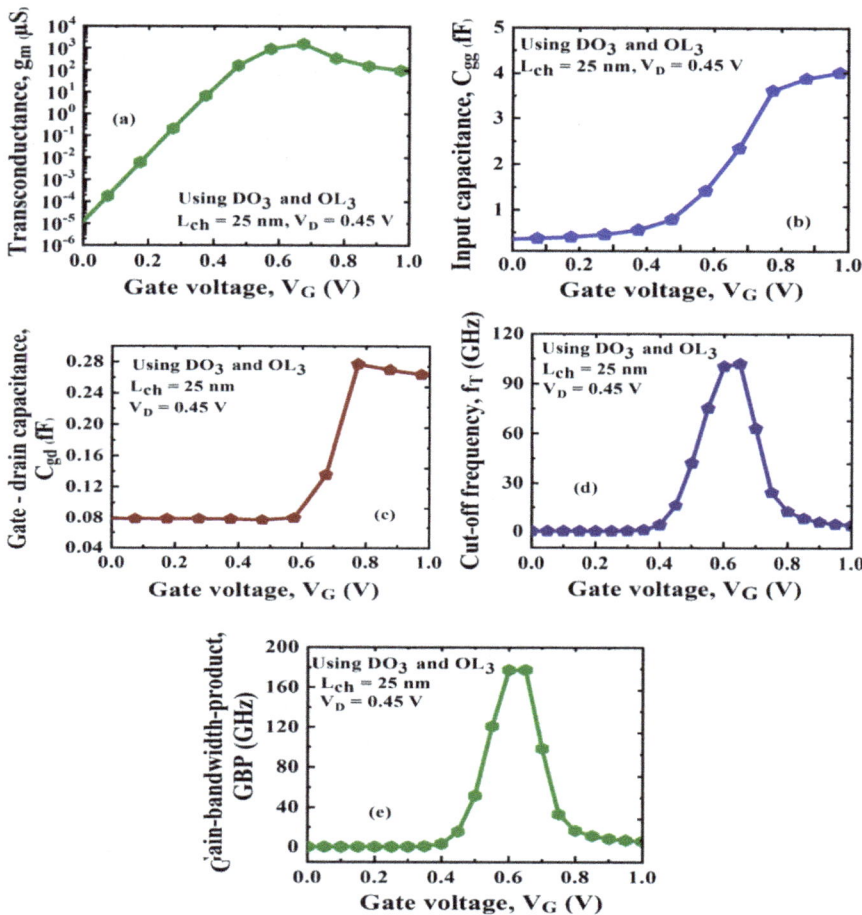

Fig. 8 Variation of different RF parameters as a function of gate voltage (V_G), **a** g_m, **b** C_{gg}, **c** C_{gd}, **d** f_T, and **e** GBP

increase in V_G results in high channel electron concentration due to which carrier-carrier scattering increases, and therefore, I_D starts to deteriorate. Consequently, g_m does not rise anymore at such high V_G. The maximum value of g_m estimated at V_G = 0.67 V exhibits a high value of 1557 μS. Such a property is helpful in design of high-gain amplifiers [17].

In Fig. 8b, the input capacitance (C_{gg}) versus V_G curve rises for an initial rise in V_G and slowly approaches saturation at high V_G. This is reasonable because charge density increases initially for the increase in the longitudinal field. However, after the formation of the inversion layer at a certain value of V_G, C_{gg} approaches oxide capacitance for further gate field rise. The variation of gate-drain capacitance (C_{gd}) concerning V_G depicted in Fig. 8c also shows the same trend as C_{gg}. It is evident

Table 2 RF parameters for
$L_{ch} = 25$ nm

RF parameters	values
Maximum g_m	0.00155 μS (at $V_G = 0.674$ V)
Maximum C_{gg}	4.268 fF (at $V_G = 1$ V)
Maximum C_{gd}	0.277 fF (at $V_G = 0.775$ V)
Maximum f_T	102.09 GHz (at $V_G = 0.65$ V)
Maximum GBP	177.57 GHz (at $V_G = 0.65$ V)

from Figs. 8b, c that both the capacitances, C_{gg} and C_{gd}, are in the range of fF, which ensures the enhanced speed of the device.

The effects of V_G variation on unit-gain cut-off frequency (f_T) and gain-bandwidth-product (GBP) are shown in Figs. 8d, e, respectively. f_T and GBP are calculated as follows [24, 26]:

$$f_T = g_m / 2\pi C_{gg} \tag{4}$$

$$GBP = g_m / 20\pi C_{gd} \tag{5}$$

The f_T versus g_m (Fig. 8d) plot rises, at first, for an increase in V_G, gradually attains a peak value, and thereafter falls for higher V_G. f_T is directly proportional to g_m and inversely related to C_{gg}. After V_G crosses a certain value, g_m (C_{gg}) decreases (becomes saturated); so f_T decreases at higher V_G. The plot of GBP versus V_G (Fig. 8e) can also be explained by considering the trend of g_m versus V_G and C_{gd} versus V_G plots. Additionally, the frequencies in GHz obtained for both, maximum f_T and GBP, ensure the high-frequency applicability of the device. The RF response for $L_{ch} = 25$ nm is listed in Table 2.

3.6 Study of the Stress, Strain, and Velocity Saturation Effect on the Device

Stress and strain effect plays a significant role in determining carrier mobility by modifying the band structure of semiconductor crystals in short channel devices. The thick STI walls surrounding the channel region in the VSTB FET architecture may exert a significant amount of stress/strain on it. Hence, the study of I_D variation against V_G by considering stress/strain helps to estimate the STI effect on channel current. For that, we have activated the stress values of two magnitudes across three stress tensor components: XX, YY, and ZZ. The stress values considered are 5×10^8 and 5×10^{10} Pa. Figure 9a–c present the I_D-V_G plots for the exertion of the different magnitude of stresses in XX, YY, and ZZ components, respectively. All of the graphs (Fig. 9a–c) demonstrate that the curve characteristics of drain current hold its integrity for different stress conditions. Also, the deformation potential (DP)

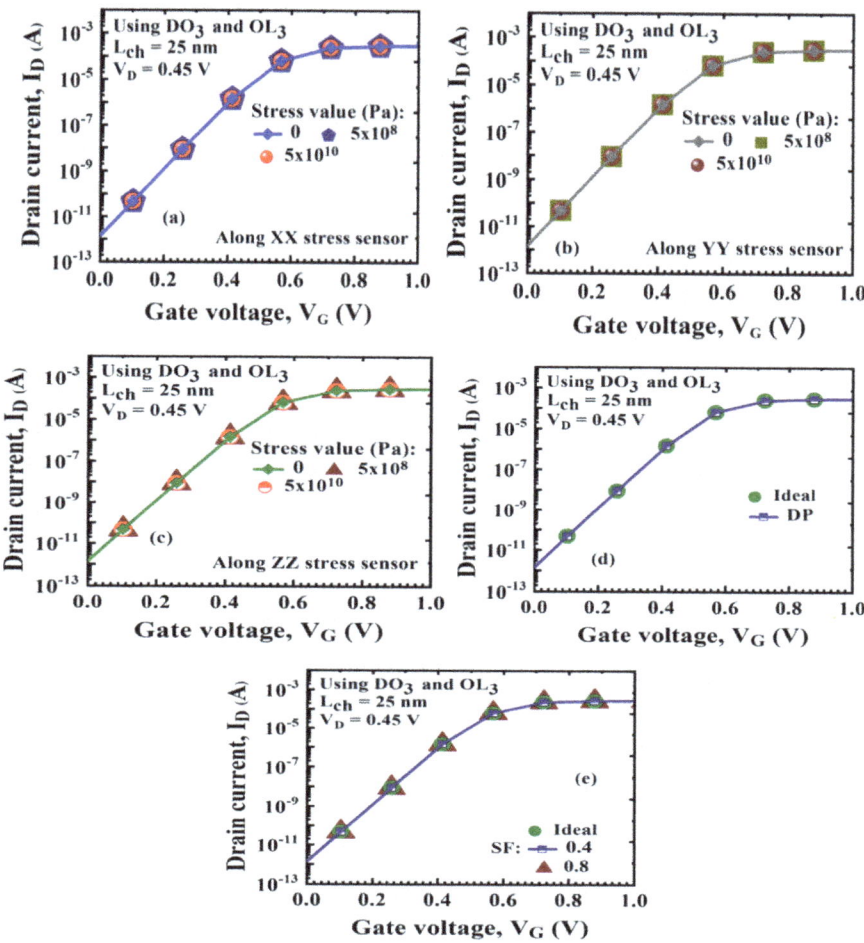

Fig. 9 I_D versus V_G for application of **a** stress along XX, **b** stress along YY, **c** stress along ZZ, **d** strain, and **e** stress-dependent saturation model

model was activated at the interface of Si/SiO$_2$ for studying the strain effect; however, Fig. 9d, showing the I_D variation over V_G where the strain effect included, reveals an almost intact curve of I_D-V_G. Furthermore, as the carrier velocity saturation may happen by the exertion of stress, the transfer plot has been analyzed by considering two saturation factors (SFs): 0.8 and 0.4. Both the SFs shows ineffectiveness over I_D (Fig. 9e).

3.7   Study of Impact of Noise on Device Performance

The impact of different noise sources (diffusion/generation-recombination (*GR*)/flicker) is studied in this section. The noise originating in gate voltage (S_{vg}) or drain current (S_{id}) was investigated in terms of power spectral density (*PSD*) form. As the frequency (*f*) plays a crucial role in noise management, this study is based on noise *PSD* study at two frequencies of 1 MHz and 10 GHz.

Figures 10a, b depict *PSD*s of gate voltage noise (S_{vg}) versus V_G, respectively at $f = 1$ MHz and $f = 10$ GHz. Also, the individual contribution of noise sources like diffusion noise (*DN*), GR noise (*GRN*), and flicker noise (*FN*) are addressed in the same plots. It is evident that the *PSD*s show a huge improvement at $f = 10$ GHz (Fig. 10b) compared to $f = 1$ MHz (Fig. 10a). Such a feature ensures robust high-frequency operation of the device. Furthermore, at $f = 1$ MHz, for the entire V_G range, the curve of the collective *PSD* of S_{vg} for various noise sources follows the same as *GR* noise; however, diffusion noise is followed by the same at $f = 10$ GHz [17, 24]. At low *f*, fluctuation in carrier number is mostly caused by the *SRH* and defect centers presented in the channel. Therefore, such phenomena contribute to the high dominance of *GRN* over collective noise *PSD* [32]. However, at higher *f*, the channel carrier constitution changes rapidly with fast variation of V_G. Thus, the events of generation and recombination decrease and hence the *GRN*. But, in such conditions, the diffusion rate of carriers changes a lot and as a consequence, the carrier concentration undergoes severe fluctuations. As the unsteadiness in the

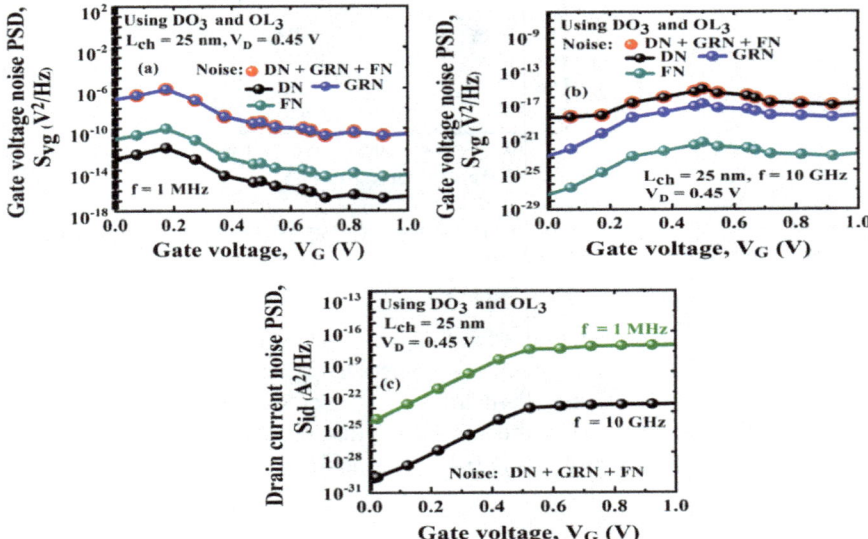

Fig. 10 Gate voltage noise PSD (S_{vg}) as function of V_G at **a** f = 1 MHz, **b** f = 10 GHz; **c** drain current noise PSD (S_{id}) as a function of V_G

carrier density is caused by the fast varying diffusion rate at high f, it is termed as diffusion noise, the high prominence of which is visible in the overall noise *PSD*. On the other hand, the *FN PSD* which is mostly caused by the conduction variation of a channel influenced by contaminants decreases greatly at high f [34, 44, 45].

Again, a gradual fall in S_{vg} value can be observed when V_G changes from 0.1 to 1 V at $f = 1$ MHz (Fig. 10a). Mainly at a lower sweep of V_G (<0.4 V), all noise *PSDs* exhibit high values; but for $V_G > 0.4$ V, a significant decrease in S_{vg} is observed. This is caused since at the lower gate field the channel starts to weakly invert itself by accumulating a light concentration of electrons at the surface; hence any fluctuation in the number of electrons caused by any type of noise is significantly sensed by the lightly dense channel charges and consequently show high S_{vg}. However, when V_G increases beyond 0.4 V, a greater gate field attracts a large number of electrons at the channel, and thus a highly inverted layer of electrons is created near the surface. Now, such a high population of electrons weakly sense any fluctuation in their number and due to this, the overall *PSD* decreases at high V_G. Though at $f = 10$ GHz, the *PSD* pattern shows a more random tendency for highly randomized charge density caused by fast changing gate field.

S_{id} vs V_G plot is shown in Fig. 10c, which illustrates that high frequency decreases the noise *PSD* to a significant extent. The reason is the same as explained for S_{vg}. Moreover, it is observed that for $V_G < 0.5$ V, S_{id} increases gradually; however, for $V_G > 0.5$ V, S_{id} gets saturated. As S_{id} signifies the noise level related to drain current, the change in I_D with the change in V_G directly impacts upon S_{id} change. Now it is a well-known fact that I_D increases gradually at lower V_G and saturates for higher gate field. Therefore, S_{id} also follows the same trend.

3.8 Study of Trap Impact on Device Performance

Figure 11a, b show the I_D-V_G plots for UTD and GTD, respectively. It is quite evident that the transfer plot gets shifted from its ideal nature with an increase in trap density for both the case. The I_D shifts in both the graphs (Fig. 11a, b) for all the trap

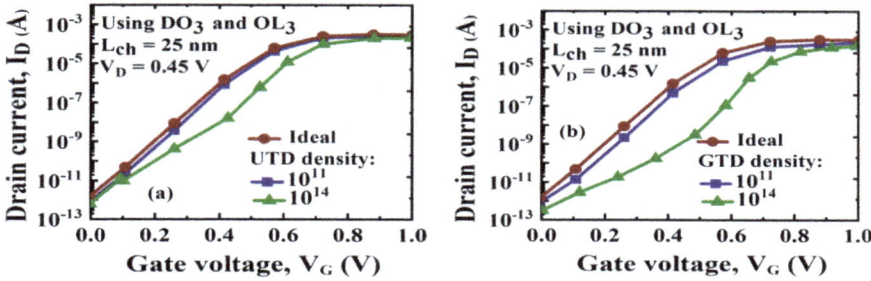

Fig. 11 I_D-V_G for different densities of **a** uniform trap and **b** Gaussian trap

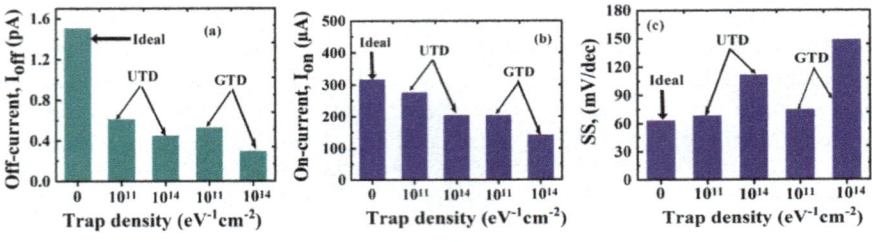

Fig. 12 Effect of various trap densities on **a** off-current (I_{off}), **b** on-current (I_{on}), and **c** SS

densities considered in this work are occurred in the linear region, mainly. The reason for such a nature can be elaborated by examining the trap activities in a MOS device. The unsatisfied atomic valances in the interface of Si/Si_3N_4 create additional energy states which trap electrons moving through the channel. Now, such trapped electrons do not participate in the current conduction mechanism and thus, I_D degrades in the presence of traps. Moreover, we know that the number of electrons in the linear region is low; thus trapping of charges decreases I_D to a great extent, which is visible in the plots (Fig. 11a, b). However, the scenario changes for higher V_G, where a large number of electrons accumulate in the channel and thereby trapping of carriers does not have much effect on the overall I_D. Therefore, device achieves almost ideal nature as considering no existence of traps. Figure 12a–c depict I_{off}, I_{on}, SS, plots versus V_G for different types of trap distribution. From those plots, it is clear that every parameter gets degraded with the introduction of UTD and GTD. However, degradation is more intense in the case of GTD. This happens due to the existence of more randomly distributed traps in the channel which significantly interfere with carrier dynamics.

4 Conclusion

The investigation of several attributes of *VSTB FET* is performed in order to develop a fundamental perspective of the same. The source and drain doping characteristics are optimized to improve the I_{off} value at a fixed channel doping concentration of 10^{15} cm^{-3}. The application of gate overlap over source/drain is observed to enhance I_{on}/I_{off} and SS. Additionally, I_{on}/I_{off} and SS show satisfactory achievement over a drain voltage range of 0.05–1 V. This demonstrates the operability of VSTB FET for ULP and HP applications. The roll-off of the threshold voltage with rising V_D is also small, which ensures sound gate control. For a wide collection of channel lengths from 15 to 100 nm, various electrical parameters such as DIBL, I_{on}/I_{off}, and *SS* show adequate values satisfying ITRS guidelines. The relative change in SS for different L_{ch} is also negligible. This nature depicts the good management of the electrostatic integrity of the device. Again, different *RF* of merit such as transconductance, gate/gate-drain capacitance, f_T, and *GBP* produce desired responses that ensure device operability for high-frequency applications. Furthermore, the device performance is robust to the stress/strain/velocity saturation effect too. The noise PSDs for different noises improve at high frequency ($f = 10$ GHz). Moreover, the influence of uniform and Gaussian trap on the transfer plot is also presented through I_{on}, I_{off}, and SS.

References

1. International Technology Roadmap for Semiconductors (2015). https://eps.ieee.org/images/files/Roadmap/ITRSFacInt2015.pdf
2. Skotnicki T, Hutchby JA, King T-J, Wong H-SP, Boeuf F (2005) The end of CMOS scaling: toward the introduction of new materials and structural changes to improve MOSFET performance. IEEE Circuits Devices Mag 21(1):16–26. https://doi.org/10.1109/MCD.2005.1388765
3. Balestra F (2014) Silicon-on-insulator devices. Wiley Encyclopedia of Electrical and Electronics Engineering
4. Jaju V, Dalal V (2004) Silicon-on-insulator technology. Adv MOSFETs EE 530:1–12
5. Cristoloveanu S (2001) Silicon on insulator technologies and devices: From present to future. Solid State Electron 45(8):1403–1411
6. Zhang X, Connelly D, Takeuchi H, Hytha M, Mears RJ, Liu TK (2017) Comparison of SOI versus bulk FinFET technologies for 6T-SRAM voltage scaling at the 7-/8-nm Node. IEEE Trans Electron Devices 64(1):329–332. https://doi.org/10.1109/TED.2016.2626397
7. Uchida K, Koga J, Takagi S-I (2003) Experimental study on carrier transport mechanisms in double- and single-gate ultrathin-body MOSFETs—Coulomb scattering, volume inversion, and δT_{SOI}—induced scattering. IEEE International Electron Devices Meeting 2003 Washington, DC, USA 33.5.1–33.5.4. https://doi.org/10.1109/IEDM.2003.1269402
8. Omura Y, Konishi H, Yoshimoto K (2008) Impact of fin aspect ratio on short-channel control and drivability of multiple-gate SOI MOSFETs. J Semicond Technol Sci 8(4):302–310
9. Liu Y, Ishii K, Masahara M, Tsutsumi T, Takashima H, Yamauchi H, Suzuki E (2004) Cross-sectional channel shape dependence of short channel effects in fin-type double-gate metal oxide semiconductor field effect transistors. Jpn J Appl Phys 43(4S):2151
10. Koldiaev V, Pirogova R (2014) Vertical super-thin body semiconductor on dielectric wall devices and methods of their fabrication. US Patent 8 796 085 B2, 5 August 2014
11. Veloso A, Keersgieter AD, Matagne P, Horiguchi N, Collaert N (2016) Advances on doping strategies for triple-gate finFETs and lateral gate-all-around nanowire FETs and their impact on device performance. Mater Sci Semicond Process 62:2–12. https://www.sciencedirect.com/science/article/pii/S1369800116304358
12. Sivasankaran K, Mallick PS, Chitroju TRKK (2013) Impact of device geometry and doping concentration variation on electrical characteristics of 22nm FinFET. 2013 IEEE International Conference ON Emerging Trends in Computing, Communication and Nanotechnology (ICECCN), Tirunelveli
13. Ferhati H, Douak F, Djeffal F (2017) Role of non-uniform channel doping in improving the nanoscale JL DG MOSFET reliability against the self-heating effects. Superlattices Microstruct 109:869–879
14. Vidhyadharan S, Yadav R, Hariprasad S, Dan SS (2019) A nanoscale gate overlap tunnel FET (GOTFET) based improved double tail dynamic comparator for ultra-low-power VLSI applications. Analog Integr Circuits Signal Process 101:109–117. https://doi.org/10.1007/s10470-019-01487-x
15. Abdi DB, Kumar MJ (2014) Controlling Ambipolar Current in Tunneling FETs Using Overlapping Gate-on-Drain. IEEE Journal of the Electron Devices Society 2(6):187–190. https://doi.org/10.1109/JEDS.2014.2327626

16. Abdi DB, Kumar MJ (2015) Dielectric modulated overlapping gate-on-drain tunnel-FET as a label-free biosensor. Superlattices Microstruct 86:198–202
17. Barman KR, Baishya S (2019) Performance analysis of vertical super-thin body (VSTB) FET and its characteristics in presence of noise. Appl Phys A 125(6):401. https://doi.org/10.1007/s00339-019-2682-x
18. Schrom G, Selberherr S (1996) Ultra-low-power CMOS technologies. 1996 International Semiconductor Conference. 19th Edition. CAS'96 Proceedings, Sinaia, Romania, 1:237–246 https://doi.org/10.1109/SMICND.1996.557352
19. Camacho-Galeano EM, Galup-Montoro C, Schneider MC (2004) An ultra-low-power self-biased current reference. Proceedings. SBCCI 2004, 17th Symposium on Integrated Circuits and Systems Design (IEEE Cat. No.04TH8784), Porto de Galinhas, Pernambuco, Brazil, pp. 147–150, https://doi.org/10.1109/SBCCI.2004.240769
20. Amerasekera A (2010) Ultra low power electronics in the next decade. 2010 ACM/IEEE international symposium on low-power electronics and design (ISLPED), Austin, TX, USA. https://doi.org/10.1145/1840845.1840892
21. Hu VP, Sachid AB, Lo C, Su P, Hu C (2015) Electrostatic integrity and performance enhancement for UTB InGaAs-OI MOSFET with high-k dielectric through spacer design. 2015 international symposium on VLSI technology, systems and applications, Hsinchu, 2015. https://doi.org/10.1109/VLSI-TSA.2015.7117568
22. Hu VP-H, Wu Y-S, Su P (2008) Investigation of electrostatic integrity for ultra-thin-body GeOI MOSFET using analytical solution of poisson's equation. 2008 IEEE international conference on electron devices and solid-state circuits, Hong Kong. https://doi.org/10.1109/EDSSC.2008.4760684
23. Gundapaneni S, Ganguly S, Kottantharayil A (2011) Enhanced electrostatic integrity of short-channel junctionless transistor with High-κ spacers. IEEE Electron Device Lett 32(10):1325–1327. https://doi.org/10.1109/LED.2011.2162309
24. Goswami R, Bhowmick B, Baishya S (2015) Electrical noise in circular gate tunnel FET presence of interface traps. Superlattices Microstruct 86:342–354
25. Andreev B, Titlebaum EL, Friedman EG (2006) Sizing CMOS inverters with miller effect and threshold voltage variations. J Circuits Syst Comput 15(3):437–454
26. Vijayvargiya V, Vishvakarma SK (2014) Effect of drain doping profile on double-gate tunnel field-effect transistor and its influence on device RF performance. IEEE Trans Nanotechnol 13(5):974–981. https://doi.org/10.1109/TNANO.2014.2336812
27. Cianca E, Rossi T, Yahalom A, Pinhasi Y, Farserotu J, Sacchi C (2011) EHF for satellite communications: the new broadband frontier. Proc IEEE 99(11):1858–1881. https://doi.org/10.1109/jproc.2011.2158765
28. Chen X, Tan CM (2014) Modeling and analysis of gate-all-around silicon nanowire FET. Microelectron Reliab 54:1103–1108
29. Iwai H, Pinto MR, Rafferty CS, Oristian JE, Dutton RW (1985) Velocity saturation effect on short-channel MOS transistor capacitance. IEEE Electron Device Lett 6(3):120–122. https://doi.org/10.1109/EDL.1985.26066
30. Baum G, Beneking H (1970) Drift velocity saturation in MOS transistors. IEEE Trans Electron Devices 17(6):481–482. https://doi.org/10.1109/T-ED.1970.17014
31. Jerdonek RT, Bandy WR (1978) Velocity saturation effects in n-channel deep-depletion SOS/MOSFET's. IEEE Trans Electron Devices 25(8):894–898. https://doi.org/10.1109/T-ED.1978.19198
32. Wu SY (1968) Theory of generation–recombination noise in MOS transistors. Solid State Electron 11:25–32
33. Anandan P, Nithya A, Mohankumar N (2014) Simulation of flicker noise in gate-all-around silicon nanowire MOSFETs including interface traps. Microelectron Reliab 54:2723–2727
34. Vandamme LKJ, Hooge FN (2008) What do we certainly know about 1/f noise in MOSTs? IEEE Trans Electron Devices 55(11):3070–3085. https://doi.org/10.1109/TED.2008.2005167
35. Huang XY et al (2010) Effect of interface traps and oxide charge on drain current degradation in tunneling field-effect transistors. IEEE Electron Device Lett 31(8):779–781. https://doi.org/10.1109/LED.2010.2050456

36. Chen I-C, Teng CW, Coleman DJ, Nishimura A (1989) Interface trap-enhanced gate-induced leakage current in MOSFET. IEEE Electron Device Lett 10(5):216–218. https://doi.org/10.1109/55.31725
37. Tsuchiya T (1987) Trapped-electron and generated interface-trap effects in hot-electron-induced MOSFET degradation. IEEE Trans Electron Devices 34(11):2291–2296. https://doi.org/10.1109/T-ED.1987.23234
38. Qiu Y, Wang R, Huang Q, Huang R (2014) A comparative study on the impacts of interface traps on tunneling FET and MOSFET. IEEE Trans Electron Devices 61(5):1284–1291. https://doi.org/10.1109/TED.2014.2312330
39. Saha R, Bhowmick B, Baishya S (2017) Statistical dependence of gate metal work function on various electrical parameters for an n-channel Si step-FinFET. IEEE Trans Electron Devices 64(3):969–976. https://doi.org/10.1109/TED.2017.2657233
40. Ma TP (1998) Making silicon nitride film a viable gate dielectric. IEEE Trans Electron Devices 45(3):680–690. https://doi.org/10.1109/16.661229
41. Sentaurus Device User Guide (Sep 2017) Version M-2017.09, Mountain View, CA, USA
42. Tsividis Y, McAndrew C (2012) The MOS transistor. Oxford University Press, Oxford
43. Eng Y et al (2018) Importance of $\Delta V_{DIBLSS}/(I_{on}/I_{off})$ in evaluating the performance of n-channel bulk FinFET devices. IEEE J Electron Devices Soc 6:207–213. https://doi.org/10.1109/JEDS.2018.2789922
44. Anandan P, Nithya A, Mohankumar N (2014) Simulation of flicker noise in gate-all-around silicon nanowire MOSFETs including interface traps. Microelectron Reliability 54:2723–2727
45. Barman KR, Baishya S (2019) An insight to the performance of vertical super-thin body (VSTB) FET in presence of interface traps and corresponding noise and RF characteristics. Appl Phys A 125:865. https://doi.org/10.1007/s00339-019-3165-9

Effect of Noise and Temperature on the Performance of Ferro-Tunnel FET

Basab Das and Brinda Bhowmick

Abstract This chapter discusses in detail the evolution of the Ferro-tunnel FET and presents an insight into its performance under the influence of noise and temperature. When a ferroelectric material is stacked with the gate oxide of a Tunnel FET, band-to-band tunneling mechanism of the Tunnel FET along with negative capacitance of the ferroelectric material ensure a subthreshold swing below the Boltzmann limit of 60 mV/dec. The analysis of noise is pivotal to understand the reliability of the nano-scale semiconductor device. The effect of the individual noise sources as well as the simultaneous presence of the components of noise: flicker, diffusion, and generation-recombination noise, is essential for predicting device efficiency in low and high-frequency applications. According to Landau- Ginzberg (LG) theory, the role of temperature is crucial to understand the changes in ferroelectric material properties from ferroelectric to paraelectric. This effect is studied on various electrical parameters and RF/Analog figures of merit.

Keywords Tunnel FET · Ferroelectrical material · Si:HfO$_2$ · Negative capacitance · Noise · Temperature

1 Introduction

In recent years, the advancement in microelectronics made human life more comfortable and easier. The advancement in microelectronics is guided by the landmark principle of Moore's law [1]. It states that the density of components in integrated circuits would double at regular intervals. The law saw over the years continued modification, and rapid progression in integrated circuit technology. The progress and applicability of Moore's law for the semiconductor industry since the 1960s are evidenced by IDRS [2]. Figure 1 shows the progress of Moore's law over time. From

B. Das (✉)
Department of ECE, GIMT Guwahati, Guwahati, Assam, India

B. Bhowmick
Department of ECE, NIT Silchar, Silchar, Assam, India

© The Author(s), under exclusive license to Springer Nature Singapore Pte Ltd. 2022   31
R. Goswami and R. Saha (eds.), *Contemporary Trends in Semiconductor Devices*,
Lecture Notes in Electrical Engineering 850,
https://doi.org/10.1007/978-981-16-9124-9_3

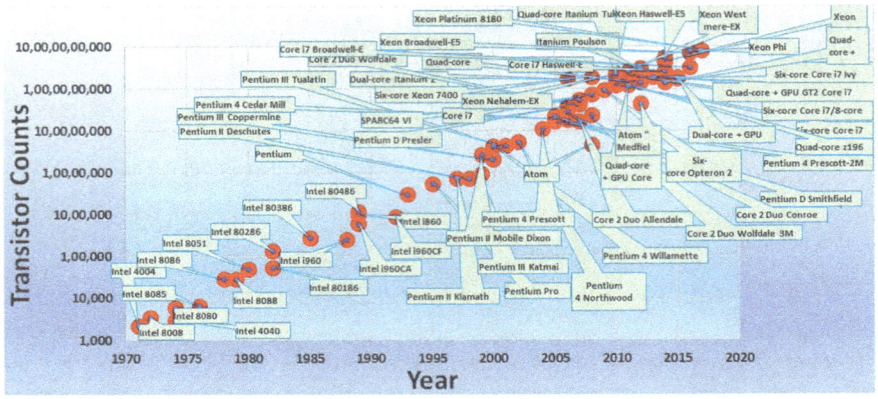

Fig. 1 Moore's Law progress over the year [1]

middle-scale integration (MSI) to ultra-large-scale integration (ULSI), it has moti-vated, since its introduction, numerous semiconductor researchers and chip manu-facturers to achieve efficient large-scale integrated circuits. The switching energy of a MOSFET is given as $0.5\varepsilon_d(V_{DD} \times L_{ch})^2/d_d$ where $\varepsilon_d, V_{DD}, L_{ch}$, and d_d are dielectric permittivity, supply voltage, channel length, and dielectric thickness, respectively. The significance of switching energy has become more important now than before because of the aggressive downscaling of devices and for continuous improvement in switching energy, it is supposed to go on without any obstacle [3]. Unfortunately, an increase in scaling density perturbs technology progress. As we increase the density of transistors, power consumption per chip also increases, which eventually tries to dissipate energy during the switching of transistors. The concern rises due to the increase in dissipation of heat generated by transistors per chip, which poses a serious threat to the life expectancy of a chip. Therefore, scaling of device dimensions became an important parameter to consider for the development of the semiconductor industry. In 1974, R.H. Dennard proposed the scaling rules [4], which put forth the idea of maintaining a constant electric field inside the gate dielectric in proportion to the scaling of the applied voltage. Moore's law along with Dennard's law paved the way for progress and advancement of the semiconductor industry. They made sure progression is in cohesion with improvement in the integration density, speed, and power consumption. The industry flourished for over years under the aegis of the two laws. Unfortunately, voltage scaling faces many challenges as one moves down the CMOS technology node of 65 nm. It is impossible to maintain good electro-static control of transistors [5]. Dennard assumed that we could maintain low power consumption requirements through reduction of supply voltage, although it became extremely difficult to scale it below 0.8 V [6]. The effect of reducing power supply was directly impacting the device threshold as it is directly proportional to the drive voltage shown in Fig. 2.

If the threshold voltage is reduced further, it increases the leakage current, thus degrading the device performance. If the supply voltage is further scaled

Fig. 2 Relation between
V_{DD} and V_{TH} [7]

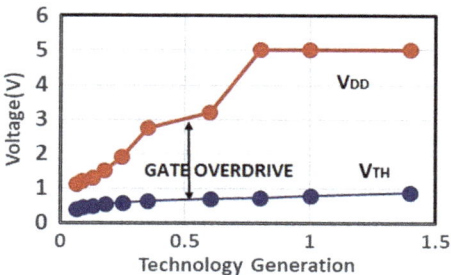

down without proper modeling of the threshold voltage, it reduces the overdrive voltage $V_{OV} = V_{DD} - V_{TH}$, thereby reducing on-current $I_{ON} = 0.5\mu C_{OX}(W/L)(V_{DD} - V_{TH})^2$ and speed of the device $C_g V_{DD}/I_{ON}$[7]. Hence, one can maintain acceptable performance by *slowing* down the supply voltage scaling. The solution was proving to be quite efficient but not effective for the future course of the advancement of semiconductor technology because if the supply voltage does not scale down in proportion to the device dimensions, power density increases. There is now an emerging need to discuss and thrive for a solution by analyzing what is causing the increase in power density. The total power dissipation consists of two components: dynamic power (power consumed during the switching of the transistor) and the other is static power (power consumed due to subthreshold leakage and gate-oxide leakage), which is expressed as shown in Eq. 1 [8].

$$P_{total} = P_{dynamic}\left(= f \times C_L \times V_{DD}^2\right) + P_{static} = I_{leakage} \times V_{DD} \qquad (1)$$

From Eq. 1, we can observe that the approach to scale down threshold voltage to maintain high drive voltage is the prominent reason for the increase in the static power consumption due to an increase in subthreshold leakage current. The only solution can be to find ways through research to reduce leakage current for the continued advancement of technology requiring low power consumption [9].

Subthreshold (SS) swing is a figure of merit for device capability for low-power applications. It is defined as the measure of minimum gate voltage to experience one order of magnitude increase in drain current. To have sharp switching characteristics between ON and OFF states, lower SS is required. It can be expressed as [10]:

$$SS = \left[\frac{\partial V_g}{\partial\left(\log_{10} I_d\right)}\right] = \frac{\partial V_g}{\partial \psi_s} \times \frac{\partial \psi_s}{\partial \log_{10} I_d} \qquad (2)$$

$$SS = \left(1 + \frac{C_s}{C_{ins}}\right)\frac{kT}{1}\ln 10 \qquad (3)$$

In Eq. 2, the first term is the body factor denoted as *m*-factor, expressed as for MOSFETs. It has always value greater than 1 and can come close to 1 only when oxide capacitance becomes greater than bulk depletion capacitance. The *m*-factor

Table 1 Parameters used in chapter

Symbol	Quantity	Symbol	Quantity
C_L	Total load capacitance	f	Frequency of switching
V_{DD}	Power supply	$I_{leakage}$	Leakage current
ε_0	Permittivity of free space	ε_{Fe}	Dielectric permittivity of ferroelectric material
C_{ox}	Oxide capacitance	C_{Fe}	Ferroelectric capacitance
C_s	Surface Capacitance	K_B	Boltzmann Constant
q	Charge of an electron	ϕ_F	Pseudo-Fermi level
ε_{si}	Silicon permittivity	V_{FB}	Flatband voltage
N_A	Doping concentration	E_g	Band gap energy
W	Gate Width	$E_g(0)$	1.1557 eV
L	Gate Length	ψ	Electrostatic potential
A, B	Kane's tunneling parameters	V_{onset}	Onset voltage of BTBT
α	7.021×10^{-4}	β	1108

has a limiting factor of $(kT/q) \ln 10$, which limits the minimum achievable SS of MOSFETs to 60 mV/decade at room temperature. The limit on SS is the result of thermally broadened Fermi distribution of carriers which is independent of dimension or the material. The investigation of SS can also be written as expressed in Eq. 3 and it can be interpreted to be reduced in two ways: We can lower SS by reducing m-factor, which we can if we employ a different method of injection of carriers into the channel rather than a thermal injection, which is used in MOSFET. Secondly, we can reduce SS by lowering the m-factor. This can be done by replacing the conventional gate stack with new material such that we can lower the m-factor below 1. Thereby, many attempts took place to modify the SS of the transistors by reducing the body factor (m) of transistors. We achieve desired results by moving toward a tunneling-based charge injection mechanism (Table 1).

Tunnel FET (TFET) incorporating the tunneling principle proved to be a promising candidate for lower power applications with SS lower than conventional MOSFET. Also, a new gate stack technique has been introduced by integrating ferroelectric material with conventional oxide to drop m-factor lower than 1. Transistor based on both methods of reducing SS will be discussed here. In this chapter, both methods are incorporated together to build ferroelectric gate stack tunnel FET (Ferro-TFET) and it is being thoroughly analyzed for the effect of different types of noise and temperature.

2 Tunnel FET

Tunnel FET has presented itself to be the most promising device for low-power applications. It has achieved SS below 60 mv/decade by implementing quantum mechanical band-to-band tunneling (BTBT) of carriers from source to the channel [11]. Leo Esaki [12] in 1958 presented the principle of tunnel effect to the world. Quantum mechanics states that an electron exhibits wave nature and can be represented in terms of the wave function. Hence, an electron can tunnel through a potential barrier when it is thin enough in comparison to the width of the region responsible for the potential barrier. The transfer characteristics of a germanium p–n junction are shown in Fig. 3; it shows non-conventional behavior in forward bias. In this device, when voltage is increased beyond V_P we see a decrease in current; thus we observe negative differential resistance (NDR) in tunnel diode which can't be explained in comparison to conventional P-N diode current expression comprising of only diffusion component of current [13, 14].

The P and N regions in the P-N tunnel diode are degenerately doped. Here, the Fermi level in the P-type region is below the maximum of the valence band, and in the N-type region above the minimum of the conduction band (Fig. 4). As there is heavy doping concentration on both sides, the depletion layer formed at the junction is thin enough to allow the electron from the valance band to tunnel through the forbidden energy gap to the conduction band. The energy of the electron is conserved during tunneling and thus moving on a horizontal line in the energy band diagram [15].

The potential barrier seen by the electrons can be approximated as a triangle and the probability of tunneling along with Wentzel-Kramers-Brillouin (WKB) can be expressed as [14].

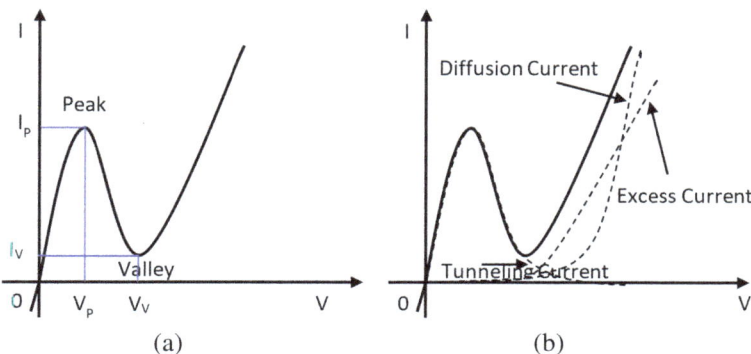

(a)   (b)

Fig. 3 **a** Transfer characteristic of a germanium p–n junction with negative differential resistance between the region (I_p, V_p) and (I_v, V_v). **b** Transfer characteristics showing total current in three components: tunneling current, excess current and diffusion current [14]

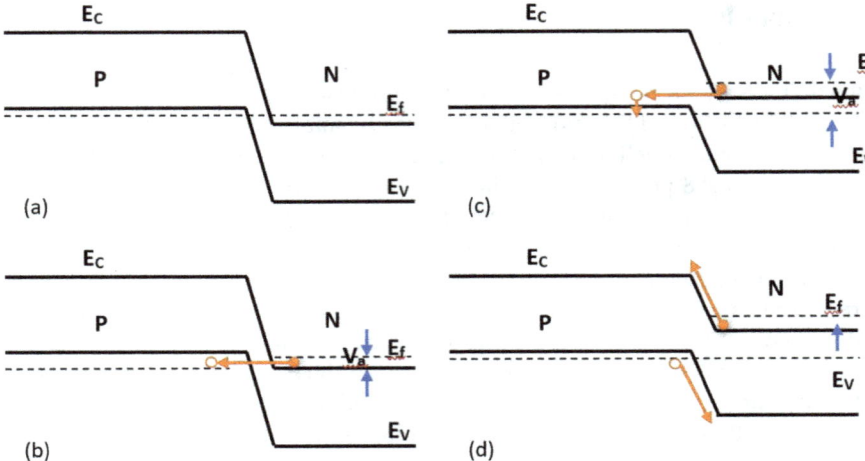

Fig. 4 Tunnel diode energy band diagrams of a tunnel diode in forward bias; **a** at zero bias; **b** peak tunneling current; **c** valley current; **d** diffusion current [15]

$$T_{\text{WKB}} \approx \exp\left(-\lambda \frac{4\sqrt{2m^* E_g^{3/2}}}{3qh\left(E_g + \Delta\Phi\right)} \right) \tag{4}$$

The Landauer equation describes the tunneling current of a TFET considering the tunneling probability T_{WKB}, source Fermi distribution function f_s, and drain distribution function f_d [16] and is expressed as

$$I_D = \frac{2q}{h} W \int_{E_V^{\text{ch}}}^{E_C^s} T_{\text{WKB}}(E)[f_s(E) - f_d(E)]\text{dE} \tag{5}$$

This principle of operation is used for charge injection mechanism to the channel of a device which gives us huge prospect in reducing leakage current and going down the Boltzmann limit for SS. Tunnel FETs (TFETs) are reversed biased gated p-i-n/n-i-p diodes [17–19]. BTBT generation takes place at p/i junction for N-TFET (p-i-n), and n/i junction for P-TFET (n-i-p) as shown in Fig. 5.

The gate bias modulates the channel potential of a TFET and controls the BTBT at the source-channel junction shown in Fig. 6a, b. At low gate voltage, TFET is switched off as bandgap cut-offs the Fermi tail of carrier concentrations (Fig. 6a). We get a low off-current as it is limited by the junction leakage. Off-current is limited by both the drift current and Shockley–Read–Hall generation. But the abrupt junction on the drain side proves to be detrimental to the TFET as it is responsible for the drain side BTBT contributing to the off-current of a TFET, and for high gate and drain voltage, source-side BTBT contributes on-current for TFET (Fig. 6b). Figure 6a, b.

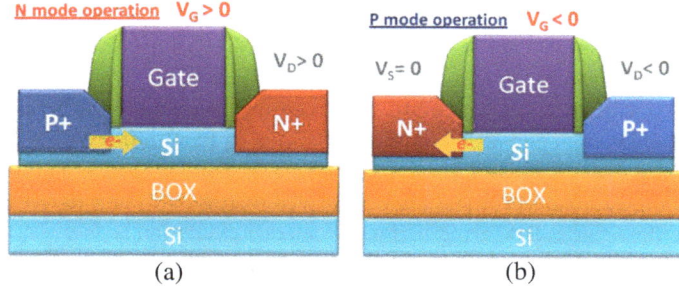

Fig. 5 a n-mode TFET operation with polarization for positive gate and the drain potential (cathode); **b** p-mode operation with polarization for negative gate and the drain potential (anode) [18]

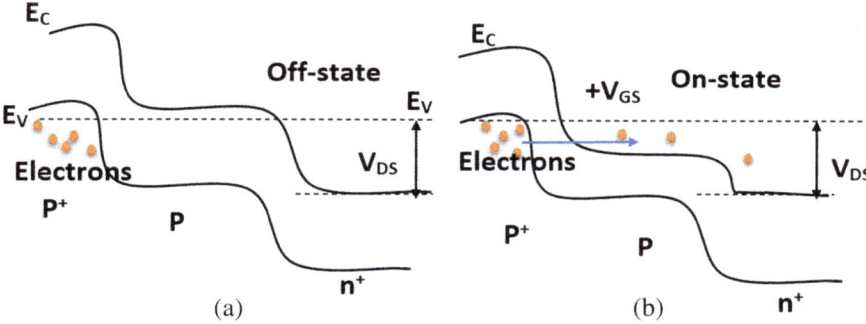

Fig. 6 Energy band diagrams extracted from the source to drain direction near the channel surface **a** OFF-state and **b** ON-state [17]

The tunneling-based transistor saw growth from its introduction in the 1980s. The prospects shown by gate-controlled tunnel transistors from an initial introduction by S. Banerjee [20] and T. Baba [21] have been immense and they directed researchers toward the inventing devices capable of low SS. W.M. Reddick in 1995[22] demonstrated Si surface tunnel transistor (STT). The progression depended on finding ways to find suitable tunnel FET and material to prove itself as an efficient replacement to MOSFET. The vertical tunneling path was then explored and introduced in 2000 by W. Hansh who fabricated a vertical Si TFET with an abrupt tunnel junction [23]. To get higher on the current using advantage of the low bandgap of germanium, TFET is fabricated with germanium [24]. Then the focus moved toward the improvement of electrostatic control for steeper subthreshold slope using Double-Gate (DG) [25] or Gate-all-around (GAA) architectures. Group III–V [26–28] material used in place of silicon provided flexibility in structural engineering as we can obtain different types of heterojunction, thereby enhancing the on-current with a lower gate voltage. The progress in BTBT materials and process fabrication based on strain engineering in Silicon [29], SiGe heterojunctions [30], strained Germanium [31, 32] demonstrated enhancement in tunnel currents. Heterojunction TFETs based on III–V materials provided higher on-currents but we did not have much improvement in SS [11]. There has always been a tradeoff between current and SS. SS degradation is due to the second-order effects which are not taken into account quite exhaustively in simulations. This happens at low gate voltage due to band tails of phonons, heavy doping, trap assisted tunneling, interface roughness, and density of interface states at the high-k dielectric [11]. The major concern of lower SS can be solved with a fresh concept of gate engineering using ferroelectric material.

3 Ferroelectric Materials

In 1920, Ferroelectrics were first discovered by Valasek when he was investigating the dielectric properties of Rochelle salt [33]. In early 1943, Barium Titanate (BaTiO3) discovered by A. Van Hippel was found to be a robust ferroelectric [34]. After that, it had been widely accepted as capacitors in industry. Though initially, it showed challenges in modeling the phase transition and difficulty in its integration, applicability got boost after 1960, when Cochran and Anderson [35] were successfully able to do the most significant description of ferroelectric transition. The applicability got boost with the introduction of thin films on silicon-integrated circuits [36]. Thin films paved the way for the electronic industry to have efficient integration of ferroelectric material with silicon ICs. The fundamental characteristic of ferroelectric materials is the hysteretic behavior characterized by polarization (P) and the electric field (E). The material changes polarization above the nominal electric field (E_c). Ferroelectric material typically has a critical coercive field $E = 50$ kV/cm in a 1 mm bulk device, which if applied in current-day technology will mark 5 kV switching voltage making it unsuitable. The integration issue of ferroelectric material in current form is solved with the introduction of the thin film with a thickness of micro/nano range through which we can get the coercive voltage of 5 V enabling its integration in ICs [37]. Today, they find a wide range of applications from ferroelectric memories to piezoelectric, nanotubes, ferroelectric transistors, etc.

Ferroelectric materials have the characteristics to show permanent electric dipole and it was observed at first by Rochelle in salt [33] where it is described as "analogous to the magnetic hysteresis in the case of iron". In ferroelectric crystals, the atomic arrangement of ions in the crystal structure produces spontaneous polarization (P_s). Spontaneous polarization is present in crystals with polar space groups. But polarization shall be switchable from one stable state to another for showing ferroelectric behavior. Ferroelectric materials have in general perovskite crystalline structure. It has general stoichiometry ABO_3 where "A" and "B" are cations and "O" is an anion [38].

In Fig. 7, perovskite barium titanate ($BaTiO_3$) ferroelectric material is shown with two stable states of the middle atom. Here, the middle atom moves with the direction of the applied electric field between two stable states, which are classified as up and down polarization according to the position of the atom. The polarization

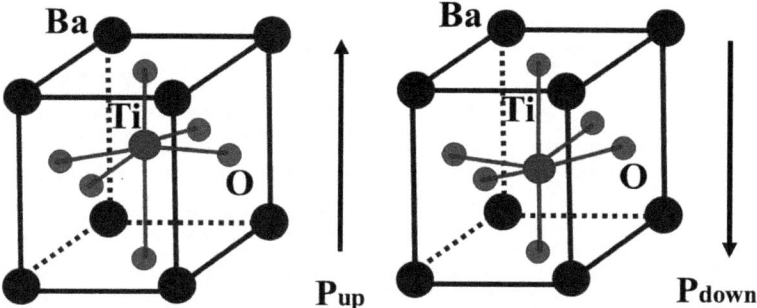

Fig. 7 $BaTiO_3$ middle atoms shows **a** up polarization **b** down polarization [38]

in ferroelectric materials can be explained by the presence of certain regions called domains, where dipoles point in a certain direction in absence of applied electric field thereby providing zero net polarization due to random orientations of dipoles. When the electric field is applied, the dipole moments realign themselves in the direction of the electric field to result in a net dipole moment, which exists even if we now remove an electric field. The effect of hysteresis results from these polarization effects is shown in Fig. 8 [38].

The polarization which is present when there is no applied electric field is known as remnant polarization (P_r). The coercive field (E_c) is the strength of the electric field which is needed to bring back the polarization to zero. A ferroelectric material switches state when the threshold electric field is greater than the coercive field (E_c) [38]. There is no switching if an electric field is applied in the same direction to the coercive field, and hence charge remains the same [38]. The two perovskite structure ferroelectric materials which have been widely used are $PbZn_{1-x}Ti_xO_3$ (PZT) and $SrBi_2Ta_2O_9$ (SBT). PZT has two stabilization points according to the zirconium and

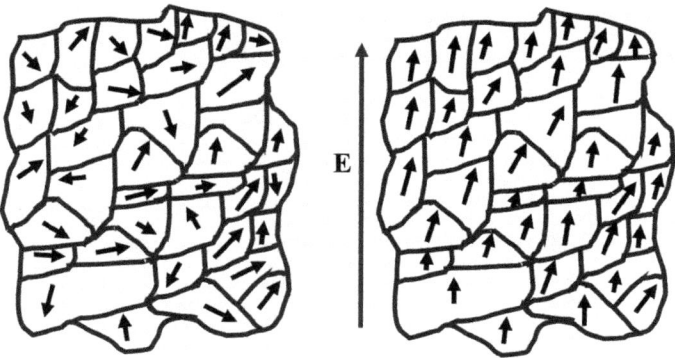

Fig. 8 The direction of dipole moment shown with and without the application of Electric field using the concept of domains [38]

titanium in the lattice. They move between the stabilization points according to the applied electric field resulting in polarization. The polarization remains in the crystal after removing the electric field. They show ferroelectric behavior below transition temperature known as Curie temperature, above which they lose their ferroelectric nature and show paraelectric behavior. The characteristic of PZT ferroelectric material due to its intermetallic inorganic compound results in remarkable piezoelectric effect, large polarization coefficient showing large spontaneous polarization and a transition temperature of about 370 °C. Their properties depend on alloy composition, whereas Ti/Zr ratios of 60/40 and 70/30 provide high polarization. On the other hand, SBT shows ferroelectric behavior in a few allotted directions of spontaneous polarization and has low remnant polarization concerning PZT. But its transition temperature is higher than PZT and is about 570 °C, which means it can withstand ferroelectric characteristics for a temperature range larger than PZT before losing it out to become paraelectric. Both SBT and PZT are widely used depending on their characteristics and have been majorly used in memory operations through integration with field-effect transistors (FETs) [38–40].

3.1 Landau's Theory

The Landau-Ginzburg-Devonshire theory (Landau's theory) provides a physical description of a system near a phase transition and explains its macroscopic properties [41]. In ferroelectric materials, temperature (T), polarization (P), electric field (E), stress (sigma), and strain (eta) are the variables used to determine the thermodynamic state of any system in equilibrium. Free energy (G) of a ferroelectric material is expressed as a function of 10 variables. The Landau theory expands G into powers of a dependent variable, with known coefficients that can be fitted with experiments. Landau's theory can be calculated from the first-order calculation or a macroscopic model and then validated experimentally [42]. The free energy can be expanded in terms of a single component of the polarization with other variables ignored. Hence, it can be expressed, assuming one-dimensional spatial variation as [43]

$$G = \frac{1}{2}aP^2 + \frac{1}{4}bP^4 + \frac{1}{6}cP^6 + \ldots - \text{EP} \tag{6}$$

where, "a", "b", and "c" are the material-dependent parameters. Free energy minima depict the equilibrium condition which is expressed as

$$\frac{\partial G}{\partial P} = 0 \tag{7}$$

Now, the free energy performs a minimum of origin when coefficients "a", "b", "c" are positive and Eq. 6 can be modified as

Fig. 9 a Paraelectric and **b** ferroelectric material is explained through graph for free energy as a function of polarization [44]

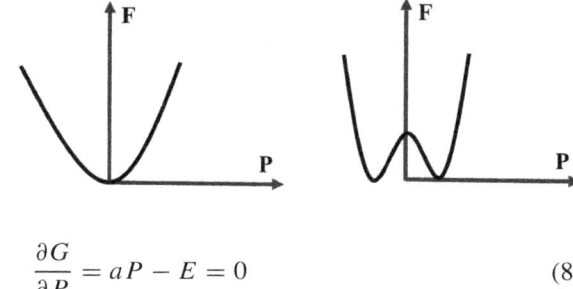

$$\frac{\partial G}{\partial P} = aP - E = 0 \tag{8}$$

It shows polarization is in a linear relationship to the electric field.

The free energy will be denoted as in Fig. 9b when coefficient "a" assumes negative values. The material behaves as ferroelectric in the above case because spontaneous polarization is in its ground state. The parameter "a" is a temperature-dependent parameter that changes signs at a specific temperature (T_0). Therefore, the ferroelectric property can be explained by approximating $a = a(T_0-T)$. It will help to predict the free energy, polarization, and susceptibility as shown in Fig. 10. Hence, ferroelectric material is modeled using a double-well energy landscape, wherein equilibrium the ferroelectric resides, providing spontaneous polarization. The capacitance of a ferroelectric material can be expressed as in

$$C_{FE} = \left[\frac{d^2 U_{FE}}{d Q_{FE}^2}\right]^{-1} \tag{9}$$

Where $Q_{FE} \propto P_{FE}$ and U_{FE} is the energy of the capacitor. As shown in Fig. 9, the capacitance of the ferroelectric capacitor is positive in its equilibrium states. Here, when a material is switching from one stable polarization state to the other, the curvature is negative around the origin. In experiments, negative capacitance has been elusive for ferroelectrics, but exhibits hysteric jumps in the polarization. The property of having negative capacitance is stabilized when it is integrated into series

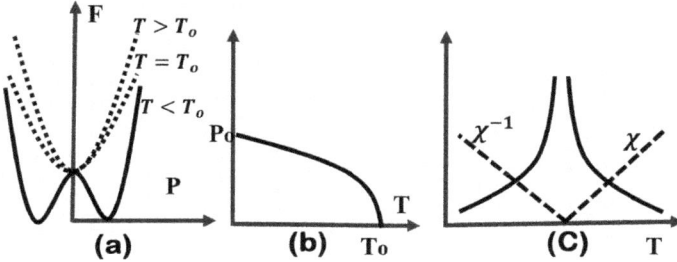

Fig. 10 a Free energy at $T > T_0$, $T = T_0$, and $T < T_0$ is shown with respect to the polarization. **b** Spontaneous polarization variation is shown with respect to the temperature and **c** Inverse of the susceptibility, χ is shown with respect to the temperature [44]

with a positive capacitor. This phenomenon has been explored for new innovative device architecture for having efficient performance in terms of current technology trends [41–44].

3.2 Negative Capacitance MOSFETs

The conventional insulator of a MOSFET when replaced with a stack including a ferroelectric capacitor can obtain values of m-factor below 1. Thereby, a subthreshold swing below the thermal limit of MOSFETs can be achieved. How one can reach a subthreshold swing below the thermal limit can be further explained by considering Eq. 10 along with the ferroelectric effect. The reduction of subthreshold swing doesn't disturb the principle of the MOS transistor and can be applied in parallel with any other performance boosters of CMOS technology [45]. The polarization–electric field relationship in a ferroelectric capacitor can be derived from the simplified form of Landau's theory [45].

$$E_{FE} = a'(T - T_0)P + b(T)P^3 + c(T)P^5 \qquad (10)$$

Now, the behavior of ferroelectric material to provide effective negative capacitance near the origin can be explained through the energy landscape and the dU/dQ vs Q curve of the ferroelectric capacitor. Though the negative slope part of the polarization is unstable and cannot be separately measured, the negative capacitance segment can be stabilized if a positive capacitor is attached in series with the ferroelectric thereby making the total capacitor of the structure positive. In addition to a reduction in subthreshold swing, the negative permittivity of ferroelectrics can provide a signal amplification that can be understood in terms of a positive feedback mechanism. A non-linear capacitor can be modeled using the following equation [46].

$$Q = C_0(V + \alpha Q) \qquad (11)$$

Eq. 11 shows the charge of the capacitor C_0 is a function of the applied voltage V along with voltage feedback αQ. Hence, capacitance can be written as

$$C_{ins} = \frac{C_0}{1 - \alpha C_0} \qquad (12)$$

Case 1: when $\alpha C_0 > 1$, it leads to instability.

Case 2: when negative capacitance part of the ferroelectric structure can be stabilized by the semiconductor capacitance C_s and hence the body factor can be defined as

$$\frac{\partial V_g}{\partial \psi_s} = 1 + \frac{C_s}{C_{ns}} = 1 - \frac{C_s}{C_0}(\alpha C_0 - 1) \qquad (13)$$

Therefore, it establishes that the negative capacitance of ferroelectrics can be used as a step-up transformer that amplifies the applied gate voltage and lowers the body factor below 1 [47].

4 Ferro-Tunnel FET

Low power and large-scale integration systems are the need of the hour for modern mobile computing including emerging IoT technologies [48, 49]. As mentioned previously, the active power can be reduced by reducing supply voltage and the threshold voltage accordingly to maintain the same performance. But the issue lies in an exponential increase in leakage current with the decrease in threshold voltage. The supply voltage can be reduced without performance loss and increasing the steepness of the off-to-on transition of the transistor [18]. The first term of the subthreshold swing is known as m-factor which is generally greater than 1 and is linked to the capacitance–voltage divider. The second term has a lower limit of 60 mV/decade at 300 K in a conventional MOSFET, where the limit is a consequence of the thermally broadened Fermi distribution of carriers irrespective of dimensionality or material in use [50–52]. However, to overcome it, steep switching devices such as Tunnel FETs [53, 54], impact ionization MOSFETs [55, 56], nanoelectromechanical FETs [57, 58], and negative capacitance MOSFETs [59–62] with 60 mV/decade are proposed. The most promising device which has attracted attention is the tunnel field-effect transistor (TFETs) which achieves steep subthreshold swing by employing quantum mechanical band-to-band tunneling (BTBT) of carriers from source to the channel [63]. Though we achieve low SS, the on-current of TFETs is unacceptably low for a technology that would most likely be thought of as a replacement for the MOSFET [63]. Also, achieving SS below 60 mV/decade is extremely challenging in most fabricated TFETs [63]. Hence, providing an acceptable I_{ON}/I_{OFF} ratio together with a sufficiently low SS has emerged as one of the most important technology issues involved in the fabrication of tunneling field-effect transistors. The SS can also be reduced by lowering the body factor by exploiting the negative capacitance region of ferroelectric materials as a replacement for conventional oxide in TFETs. This would result in lowering the body factor m below 1, hence reducing the SS and expanding the steep slope of the TFETs. Incorporation of negative capacitance will emerge as the most effective performance booster of TFETs ensuring a steep off-to-on transition together with an acceptable I_{ON}/I_{OFF}. There have been various works investigating Ferro-TFETs. The Ferro-TFETs are exploiting the benefit of the role of the band-to-band tunneling (BTBT) along with the introduction of negative capacitance into the gate stack of TFETs. The gate stack of conventional oxide and ferroelectric material will increase the energy band bending due to the increase of the internal voltage amplification thereby further enhancing the BTBT probability. In the next section, the reliability of Ferro-TFET will be explored and explained in detail for understanding its range of application in the electronics industry [64–71].

5 Device Structure and Simulation Setup

An SOI device structure is simulated with a doping concentration of 1×10^{20} and 5×10^{18} cm^{-3} for source and drain, respectively, for a gate length of 30 nm. Here, negative capacitance (using ferroelectric as oxide) is introduced to reduce subthreshold swing which makes TFET an efficient memory device. In the structure, the buffer layer is used to reduce lattice mismatch. The structure can have a stable state in total energy with negative curvature in Landau energy of ferroelectric as gate bias is applied. Relation between polarization charge density and electric field in ferroelectric oxide is analytically expressed and calculated by Landau equation [72]. The FTFET simulation has been performed on Sentaurus TCAD [73]. The models used for considering heavy doping Fermi Dirac Statistics and effect of bandgap narrowing are applied. The effect of doping concentration is considered by including the doping-dependent mobility model and for considering correctly the effect of polarization, electric field ENormal mobility model is considered. The interband tunneling effect is efficiently considered when the non-local band-to-band tunneling model along with Shockley–Read–Hall (SRH) Recombination Model is enabled and it uses Wetzel-Kramer-Brillouin (WKB) approximation. In the Sentaurus TCAD version earlier than 2016, we need to include polarization and coercive field value by manually calculating from P-E relation expressed in LG theory [72]. It is now included in versions 2016 and 2017. Though we can't still plot directly the P-E curve or can have ferroelectric material as other oxide material available, the features will be included in 2018 or future versions as announced [73].

6 Results and Discussions

In nano-scale geometries, the investigation of noise considering the effect of flicker, diffusion, and generation-recombination noise is one of the most crucial factors. Along with the effect in determining the reliability of the device, it has been explored concerning various scaling parameters. In addition to it, ferroelectric behavior changes at Curie temperature (T_C) hence it has become an important figure of merit for understanding ferroelectric device behaviors as stated by the Landau-Ginzberg (LG) theory resulting in a change in its properties from ferroelectric to paraelectric [42, 43, 72]. In this chapter, to understand the reliability of the device, the behavior of Ferro-TFET in presence of noise has been presented along with the effect of temperature on the threshold, memory window, off-current, transconductance, subthreshold swing, and RF parameters. The presented analysis is for the optimized structure shown in Fig. 11 where the conventional structure reliability analysis is presented in [74, 75]. The analysis parameter is shown in Table 2.

Fig. 11 Structure of the simulated device

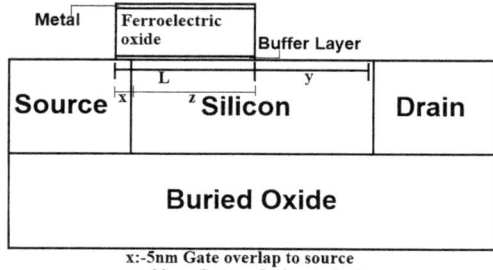

x:-5nm Gate overlap to source
y:- 20nm Gate underlap to drain
L=x+z=Effective Gate length=30nm

Table 2 Parameters for simulated structure

Parameter	Value	Parameter	Value
Concentration for source doping(p^+)	1×10^{20}/cm^3	Concentration for drain doping (n^+)	5×10^{18}/cm^3
Thickness of gate	30 nm	Channel doping	1×10^{16}/cm^3
Source and drain thickness	25 nm	Buried oxide thickness	20 nm
$P_s(\mu C/cm^2)$	9.25	$P_r(\mu C/cm^2)$	9.1
ε_r of Si:HfO$_2$	32.5	E_C(MV/cm)	1.1

6.1 Effect of Noise on Ferro-TFET

The electrical characterization of a device cannot fully describe the behavior of a device when second-order effects such as radiation, noise, temperature come into play. Hence, it is of utmost importance to carry out these analyses to know the reliability of the device. Here, detailed noise analysis of FTFET has been performed in terms of various noise performance parameters like drain current noise power spectral density (S_{id}) and gate voltage noise power spectral density (S_{vg}) along with other performance parameters. Figure 12 presents the effects of various ferroelectric film thicknesses under the presence of various noise components on different parameters of the device. The memory window depends on the coercive field and film thickness, from Fig. 12a, it can be observed that the memory window increases due to an increase in film thickness. The memory window is a significant criterion for memory devices: the larger the memory window, more is the data retention time and data detection [76]. The subthreshold swing (SS) value of the device considering various parameter variations is shown in Table 3. Figure 12b presents the S_{ID} variation of the device with ferroelectric thickness which increases with gate voltage and has the order in the range of 10^{-28} proving Ferro-TFET to have significant tolerances to noise in comparison to conventional TFETs [77, 78]. Hence, we can infer from the behavior that effect of noise is reduced due to the negative capacitance resulting in the reduction in trapping and de-trapping of charges. S_{VG} in Fig. 12c shows approximately constant value for the whole range of gate voltage dominantly due to band-to-band tunneling and thus

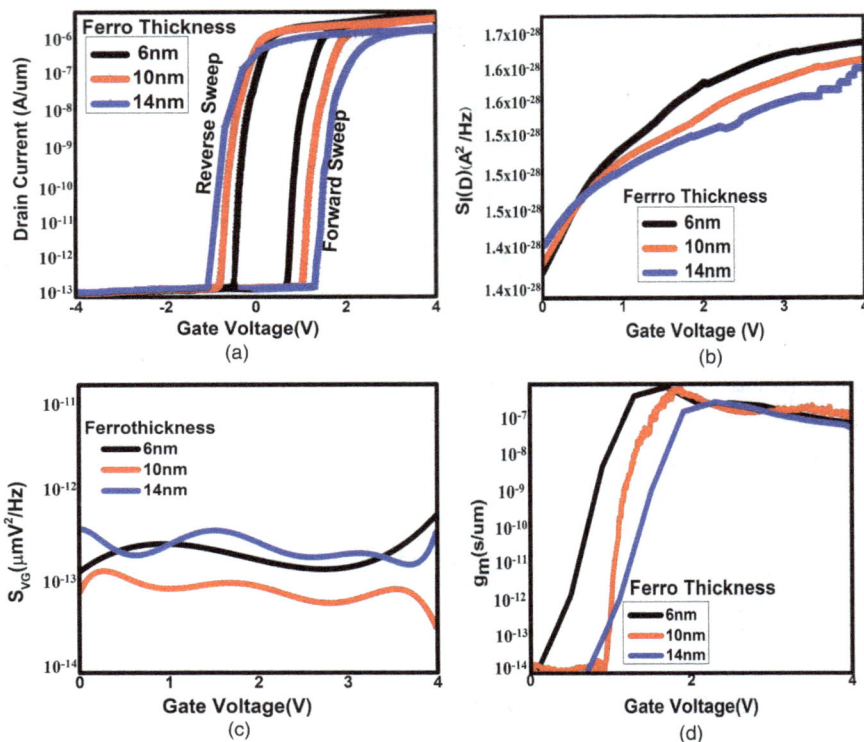

Fig. 12 Transfer characteristic, Noise current spectral density (S_{ID},) Noise voltage spectral density (S_{VG}) and transconductance plot with ferro-thickness at drain voltage of 0.5 V and gate voltage sweep from -4 to $+4$ V

Table 3 Extracted parameter value

Type of variation		SS (mV/Dec)	Memory window (V)	$I_{\mathrm{ON}}/I_{\mathrm{OFF}}$	V_{Th}(V)
Ferrothickness	6 nm	28.1	1.2	8.783E + 6	1.32
	10 nm	23.3	1.73	7.169E + 6	1.68
	14 nm	29.2	1.86	6.343E + 6	2.2
Buffer thickness	1 nm	23.3	1.73	7.169E + 6	1.68
	3 nm	28.9	1.25	6.986E + 6	1.99
	5 nm	45.2	1.14	2.678E + 6	3.16
Buffer type	SiO$_2$	53.7	0.51	5.865E + 6	2.91
	Si$_3$N$_4$	33.0.6	1.42	6.505E + 6	2.18
	HfO$_2$	23.3	1.73	7.169E + 6	1.68

we don't see any detrimental effect of noise in case of voltage spectral density. Figure 12d clearly explains the efficient characteristics of Ferro-TFET with a lower value of ferro-thickness as we can see maximum transconductance is achieved for 6 nm ferro-thickness and it is achieved for lower gate voltage in comparison to a higher value of ferro-thickness.

Figure 13 shows noise parameter variation for different types of buffers which reduces a large number of defects and lattice mismatch between ferroelectrics and silicon. The inclusion of the buffer layer results in detrimental effect on the memory window requirement as a voltage drop across it reduces the ferroelectric voltage. Therefore, high ferroelectric voltage is needed which can reduce the voltage drop across it for maintaining the working of a device in a saturation polarization loop [74]. We can observe a satisfactory memory window for HfO_2 in Fig. 13a. In Fig. 13b, noise current spectral density (S_{ID}) increases for the increase in dielectric constant as can be seen in the plot, because equivalent oxide thickness (EOT) for high-k dielectric constant is more which provides enhanced power spectral density. S_{VG} is shown in Fig. 13c and as it is relatively constant and invariant for different dielectric values, we

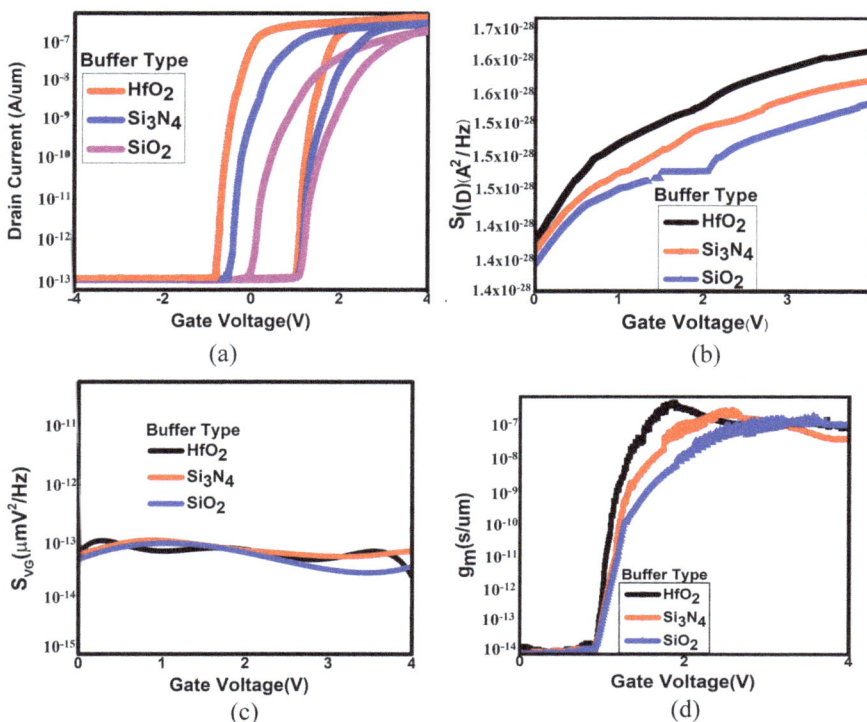

Fig. 13 Transfer characteristic, Noise current spectral density (S_{ID},) Noise voltage spectral density (S_{VG}) and transconductance plot with different types of buffer layer at drain voltage of 0.5 V and gate voltage sweep from -4 to $+4$ V

can conclude that the dominant tunneling mechanism in negative capacitance devices is the band-to-band tunneling mechanism. In Fig. 13d, the transconductance plot is shown for different k-values, and it can be observed that the transconductance efficiency of high-k dielectric is established by the fact that maximum transconductance is achieved for HfO_2 in comparison to SiO_2 and Si_3N_4.

In Fig. 14a, the effects of buffer thickness are shown. The increase in buffer thickness from 1 to 5 nm leads to a reduction in the memory window [74]. Noise current spectral density for different Ferro thicknesses is shown in Fig. 14b and noise voltage spectral density in Fig. 14c. It can be seen that S_{ID} is negligibly low in the order of 10^{-28} and S_{VG} is relatively constant for the whole range which establishes the ferro-thickness superiority in shielding the effect of noise on the device performance due to the dominance of BTBT over TAT as the tunneling mechanism. Through Fig. 14d, the transconductance plot establishes better performance in case of a thinner buffer layer to have superior control of gate over the channel.

Figure 15a, b show the S_{ID} and S_{VG} versus frequency plots, respectively, for $V_{DS} = 0.5$ V, $V_{GS} = 2$ V, and $I_D = 0.1 \mu A/\mu m$. Correlating Fig. 15a and Fig. 15b, it can

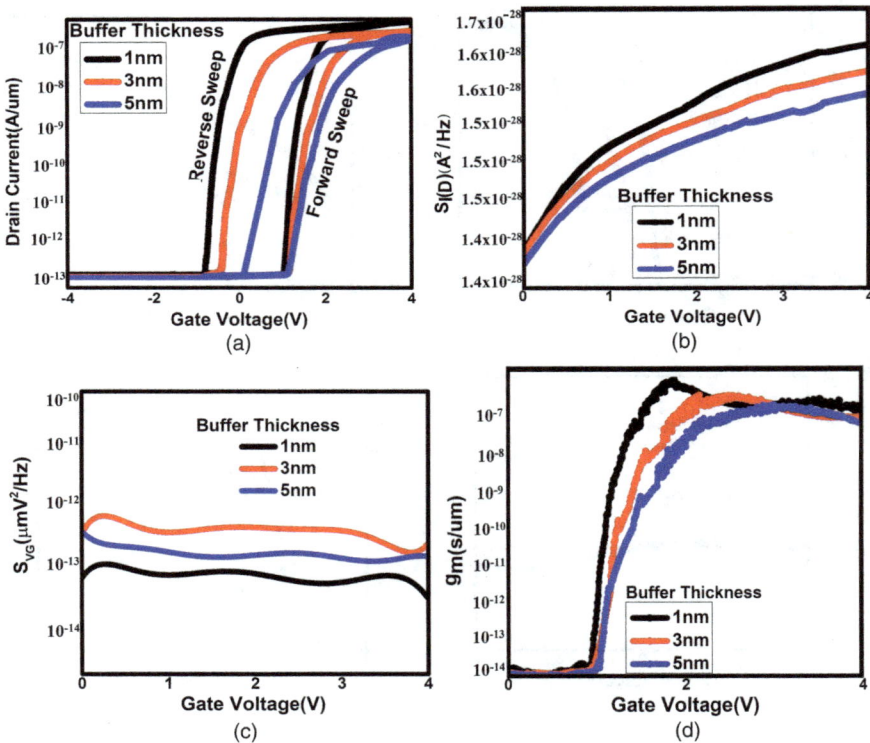

Fig. 14 **a** Transfer characteristic, **b** noise current spectral density (S_{ID},) **c** noise voltage spectral density (S_{VG}) and **d** transconductance plot with different types of buffer layer thickness at drain voltage of 0.5 V and gate voltage sweep from -4 to $+4$ V

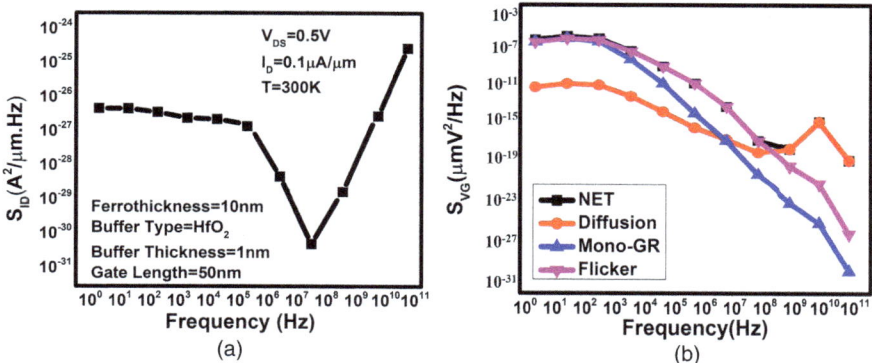

Fig. 15 Noise current and voltage spectral density with respect to variation in frequency form 1 Hz to 100 GHz

be observed that at low frequency for the range of 10 Hz to 10 kHz both generation-recombination and flicker noise are present. Above 10 kHz, flicker noise dominates over the others and the effect persists till 1 GHz after which diffusion noise has larger dominancy over other noise sources is represented through the rise in current and voltage spectral density above 1 GHz, this happens largely due to charge trapping and de-trapping at the oxide–semiconductor interface.

7 Effect of Temperature Variation on Ferro-TFET

The ferroelectric equation in LG theory [42, 43, 72] describes the thermodynamic properties of a ferroelectric material. In [75], they combined LG theory along with the capacitance model presented in [79, 80] and the drain current model of a tunnel FET in [81] to arrive at an analytical model for threshold voltage, transconductance, subthreshold swing, and drain current for considering the effect of temperature for Ferro-TFET.

The SS of the TFET with a gate stack of ferroelectric layer and conventional oxide is in parabolic dependence with temperature.

$$SS_{\text{Fe-stack}}(T) = \left[1 + \frac{C_S}{C_{\text{Ox}}} - \frac{dC_S T_C}{\varepsilon_0 \lambda C_{\text{CW}}}\right] \frac{K_B \ln 10}{q} T$$
$$+ \frac{dC_S}{\varepsilon_0 \lambda C_{\text{CW}}} \frac{K_B \ln 10}{q} T^2 \tag{14}$$

The threshold voltage (including the coercive field) dependence on temperature.

$$V_{\text{th}} = V_{\text{FB}} + 2\phi_F + \sqrt{4q\varepsilon_{\text{si}} N_a \phi_F} \frac{1}{C_{\text{Fe-stack}}(T)}$$

$$+ \begin{cases} dE_{vth}, & \text{for } T < T_C \\ 0, & \text{for } T > T_C \end{cases} \tag{15}$$

The expression of the current of a standard Ferro-tunnel FET [75] is

$$I \approx \frac{WLA\sqrt{q}}{BE_g^{1.5}} \left(\frac{E_g N_a}{2E_g}\right)^{D/2} e^{B\sqrt{2\varepsilon_s E_g q}/\sqrt{N_a}\left(\sqrt{\partial\psi}-\sqrt{E_g/q}\right)} \sqrt{\partial\psi} \tag{16}$$

$$\sqrt{\partial\psi} = (V_{GS} - V_{onset})/\gamma \tag{17}$$

$$\gamma = 1 + \frac{\varepsilon_s\varepsilon_0\lambda C_{CW} + (T - T_C)dC_{OX}}{\varepsilon_0\lambda C_{CW}C_{OX}} \sqrt{\frac{q^2 N_a}{2E_g\varepsilon_s}} \tag{18}$$

Therefore, the expression of the transconductance is given as

$$\begin{aligned} g_{m,\text{Fe-TFET}}(T) &= \frac{\partial I_D}{\partial V_{GS}} \\ &= \frac{WLA\sqrt{q}}{2BE_g^{1.5}} \left(\frac{E_q N_a}{2E_g}\right)^{D/2} e^{B\sqrt{2\varepsilon_s E_g q}/\sqrt{N_a}\left(\sqrt{\partial\psi}-\sqrt{E_g/q}\right)} \\ &\quad \left(\begin{array}{c} B\sqrt{2\varepsilon_s E_g q}/\sqrt{N_a} \\ + \dfrac{1}{\sqrt{\partial\psi}} \end{array}\right) \end{aligned} \tag{19}$$

In this section of the chapter, the effect of temperature variation to understand the reliability of the device is presented for the 30 nm Ferro-TFET structure. The temperature plays a crucial role in the phase determination of the ferroelectric material. The permittivity of ferroelectric material depends non-linearly on the temperature as it increases for an increase in temperature like a linear dielectric, and thereafter, it reduces again. This reduction in permittivity of the ferroelectric material happens due to a change of phase from ferroelectric to paraelectric [75]. The temperature at which the above phenomenon happens is known as the Curie temperature point. Here, we have varied temperature from lower to higher value for a drain-to-source voltage of 0.5 V, and gate-to-source voltage sweep is kept in the range: -4 V to 4 V. Figure 16a shows the transfer characteristics which increase with the temperature as bandgap reduces with the increase in temperature as shown in Fig. 16b. The drain current increases till 500 K and above it, drain current shows a reduction in current when it reaches V_{onset} due to a reduction in polarization loop. This phenomenon of reduction in drain current for temperature above 500 K can also be understood from the Eq. 18 as γ increases, which is the reason for reduction of $\delta\psi$ of (17), leading to the decrease in drain current of Eq. (16). For temperatures below 500 K, the increase in drain current stays as the dominant factor as bandgap is reduced (Fig. 16b). Hence, we can interpret that above 500 K, γ is the dominant

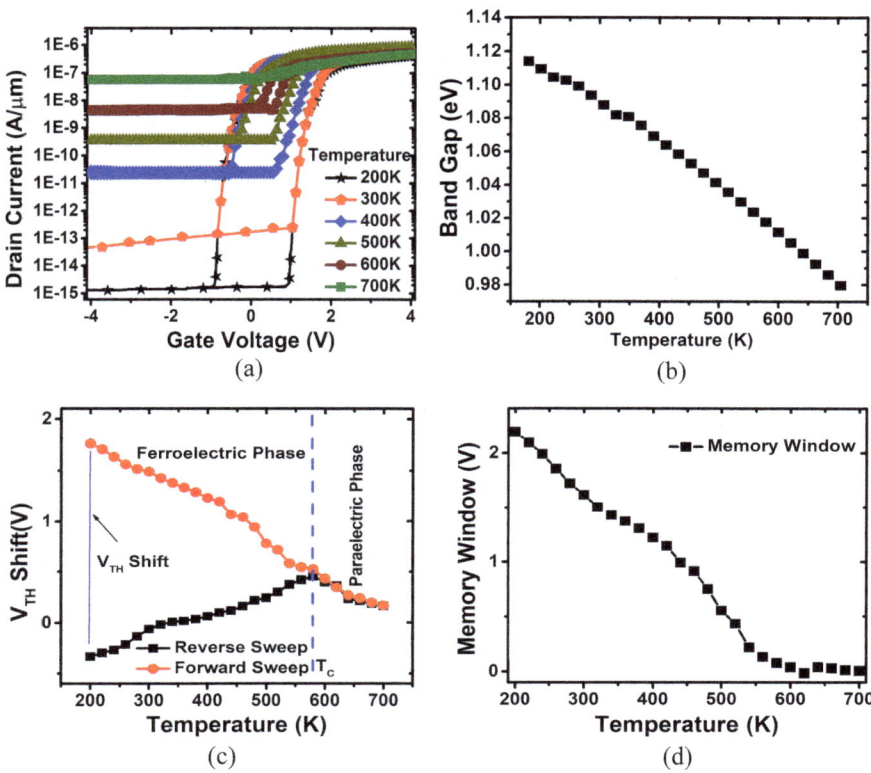

Fig. 16 **a** Transfer characteristic, **b** band Gap, **c** threshold voltage shift and **d** memory window plot for temperature variation from 200 to 700 K at drain voltage of 0.5 V and gate voltage sweep from −4 to + 4 V

parameter for variation in drain current behavior from the usual characteristics. The second effect we observe is a reduction in polarization loop when we approach T_c, due to the reduction in the coercive field, remnant polarization, and increase in its permittivity. This is the reason for which the device no longer exhibits ferroelectric properties once it reaches T_c, as shown in Fig. 16c, d. In the plot, we can observe a linear reduction in threshold voltage for forwarding path, whereas for reverse path, the threshold voltage increases till it reaches the transition point wherein it reduces again as supported by Eq. 13. Also, the memory window reduces with an increase in temperature. The device shows transition close to 500 ± 10 K, hence, it can be called Curie temperature for our device.

A large number of carrier generations in the reversed biased junction result in the increase in OFF-current due to the carrier generation due to excitation with the rise in temperature. Moreover, the rate of increase in ON-current is lower than the rate of increase in OFF-current, because the OFF-current is dependent on the two major temperature-dependent mechanisms: (a) trap assisted tunneling (TAT) and (b)

Shockley–Read–Hall (SRH) recombination current, whereas the ON-state current depends on the band-to-band tunneling (BTBT) current.

This is the reason for which we can observe a slow rate of increase in ON-current as it depends on less effective bandgap narrowing effect with increased temperature in (15) and it is shown in Fig. 17a. From Eq. 12, we can interpret that the subthreshold swing (SS) will not follow the linear relationship concerning the increase in temperature as in the case for MOSFET; rather it will show a parabolic relation for the ferroelectric-conventional oxide gate stack. In Fig. 17b, SS exactly corroborates the assumption. It reaches a minimum value near to 500 K ± 10 K as we know that the permittivity of ferroelectric material will reach maximum at Curie point which will result in exceptional improvement of gate coupling and electrostatic control of the device, thereafter it rises sharply again beyond the Curie point, thus satisfying the nature predicted by Eq. 12. The transconductance of a device describes the device's ability to amplify its input voltage and thus is a major factor in determining the gain of a device. In Fig. 17c, we can see that transconductance gradually increases till it

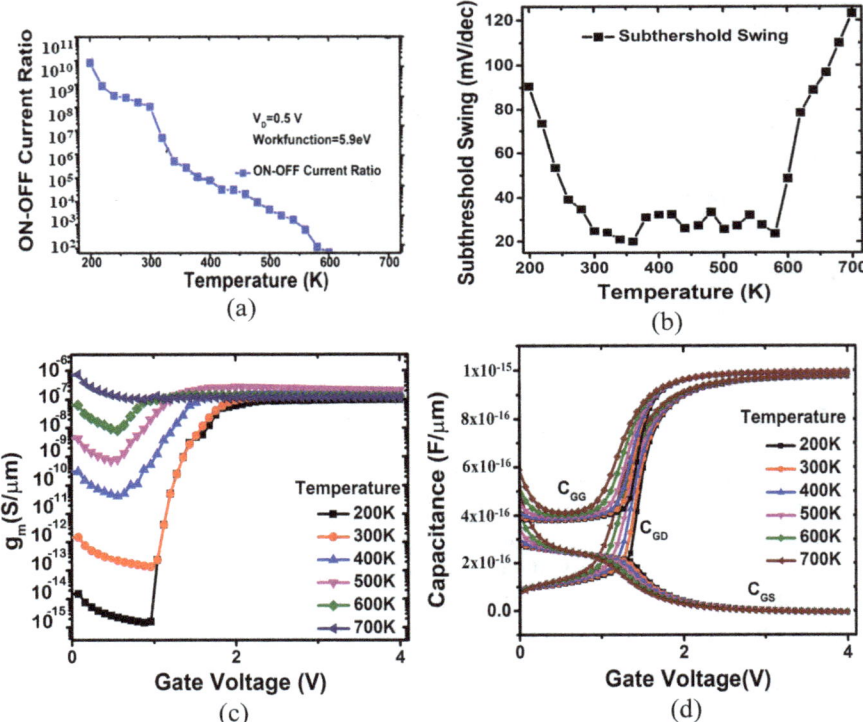

Fig. 17 **a** ON–OFF current ratio, **b** subthreshold Swing, **c** transconductance and **d** capacitance plot for temperature variation from 200 to 700 K at drain voltage of 0.5 V and gate voltage sweep from −4 to + 4 V

reaches the saturation region. This is due to the rise in the OFF-current which deteriorates the device characteristics in the subthreshold region. The transconductance shows an increasing trend with temperature till it reaches the Curie point, beyond which it reduces again in a strong inversion region. The conventional devices show a reduction in transconductance at higher gate voltage due to mobility degradation, which is also predicted by Eq. (17). In TFET, we know that gate-to-drain capacitance (C_{gd}) is largely due to the tunneling barrier on the drain side of the TFET and hence, C_{gd} dominates over C_{gs} in contribution to the total capacitance C_{gg}. The introduction of ferroelectric oxide also contributes to the better electrostatic coupling of the channel which aids the lowering of capacitance in comparison to the conventional TFET. In Fig. 17d, the capacitance increases with temperature as we move from subthreshold and then shows a reducing trend above 500 K due to the reduction in the ON-current for gate voltage beyond the onset voltage of band-to-band tunnelling.

8   Conclusion

This chapter is a thorough progression from the rise and progress toward finding a viable alternative of MOSFET for the current need of technology. The effect of noise and temperature on the device has been studied for exploring its viable application. Improved noise performance has been achieved in comparison to conventional MOSFET and TFET. The prime reason for the reduction in the detrimental effect of noise in the Ferro-TFET is the reduction of trapping and de-trapping of charges. Also, reduced noise has been observed in case of the inclusion of a high-k buffer layer. The Ferro-TFET in the high-frequency region will be most affected by diffusion noise, and in a low-frequency region by generation-recombination and flicker noise. The significant role of Curie temperature in designing Ferro-TFET for application in analog/RF and digital applications has been pointed out. For the Ferro-TFET under discussion, Curie temperature point is at 500 ± 10 K, and hence, it can work without losing ferroelectric properties for all applications having working regions of temperature below 500 ± 10 K. The interesting characteristics we got to observe that Ferro-TFET shows insulation for the detrimental effect of mobility degradation which is commonly observed for MOSFET and Conventional TFET at high gate voltage region of operation as reported in the literature.

References

1. Moore GE (1998) Cramming more components onto integrated circuits. Proc IEEE 86:82–85
2. Executive Summary (2017) International roadmap for device and systems (IRDS). https://www.itrsgroup.com/
3. Meindl JD, Chen Q, Davis JA (2001) Limits on silicon nanoelectronics for terascale integration. Science 293:2044–2049

4. Dennard RH, Gaensslen FH, Rideout VL, Bassous E, LeBlanc AR (1974) Design of ion-implanted MOSFET's with very small physical dimensions. IEEE J Solid-State Circuits 9:256–268

5. Omura Y, Mallik A, Matsuo N (2016) History of low-voltage and low-power devices. MOS devices for low-voltage and low-energy applications, Wiley

6. Ionescu AM (2017) Nano-devices with advanced junction engineering and improved energy efficiency. 17th international workshop on junction technology (IWJT). pp 1–6

7. Skotnicki T et al (2008) Innovative materials, devices, and CMOS technologies for low-power mobile multimedia. IEEE Trans Electron Devices 55:96–130

8. Nilsson P (2006) Arithmetic reduction of the static power consumption in nanoscale CMOS. Electronics, circuits and systems. 13th IEEE international conference on, 656–659

9. Zhang L, Huang J, Chan M (2016) Steep slope devices and TFETs. Tunneling field effect transistor technology. Springer, pp 1–31

10. Van Overstraeten RJ, Declerck GJ, Muls PA (1975) Theory of the MOS transistor in weak inversion-new method to determine the number of surface states. IEEE Trans Electron Devices 22:282–288

11. Lu H, Seabaugh A (2014) Tunnel field-effect transistors: state-of-the-art. IEEE J Electron Devices Soc 2:44–49

12. Esaki L (1958) New phenomenon in narrow germanium p-n junctions. Phys Rev 109:603–604

13. Esaki L (1974) Long journey into tunneling. Proc IEEE 62:825–831

14. Sze SM, Ng KK (2006) In: Physics of semiconductor devices. Tunnel Devices 3rd edn. Wiley

15. Cristoloveanu S, Wan J, Royer CL, Zaslavsky A (2013) Sharp switching SOI devices. ECS Trans 53:3–13

16. Knoch J, Mantl S, Appenzeller J (2007) Impact of the dimensionality on the performance of tunneling FETs: bulk versus one-dimensional devices. Solid-State Electron 51:572–578

17. Seabaugh AC, Zhang Q (2010) Low-voltage tunnel transistors for beyond CMOS logic. Proc IEEE 98:2095–2110

18. Ionescu AM, Riel H (2011) Tunnel field-effect transistors as energy-efficient electronic switches. Nature 479:329–337

19. Wan J, Le Royer C, Zaslavsky A, Cristoloveanu S (2011) Tunneling FETs on SOI: suppression of ambipolar leakage, low-frequency noise behavior, and modeling. Solid-State Electron 65:226–233

20. Low KL, Yang Y, Han G, Fan WJ, Yeo YC (2012) Electronic band structure and effective masses of Ge1-xSnx alloys. 222nd Electrochemical society meeting, Honolulu, HI USA, Oct 7–12

21. Guo P, Yang Y, Cheng Y, Han G, Chia CK, Yeo YC (2012) Tunneling field-effect transistor with novel Ge/In0.53Ga0.47As tunneling unction. 222nd electrochemical society meeting, Honolulu, HI USA, Oct 7–12

22. Liu B, Gong X, Han G, Lim PSY, Tong Y, Zhou Q, Yang Y, Daval N, Pulido M, Delprat D, Nguyen BY, Yeo YC (2012) High performance Ω-gate Ge FinFET featuring low temperature Si2H6 passivation and implantless Schottky-barrier NiGe metallic source/drain. 2012 Silicon Nanoelectronics Workshop (SNW), Honolulu HI, USA, June 10–11

23. Han G, Su S, Yang Y, Guo P, Gong X, Wang L, Wang W, Guo C, Zhang G, Xue C, Cheng B, Yeo YC (2012) High hole mobility in strained germanium-tin (GeSn) channel pMOSFET fabricated on (111) substrate. 222nd electrochemical society meeting, Honolulu, HI USA, Oct 7–12

24. Kotlyar R et al (2013) Bandgap engineering of group IV materials for complementary n and p tunneling field effect transistors". Appl Phys Lett 102:113106

25. Boucart K, Ionescu AM (2007) Length scaling of the double gate tunnel FET with a high-K gate dielectric. Solid-State Electron 51:1500–1507

26. Avci UE, Hasan S, Nikonov DE, Rios R, Kuhn K, Young IA (2012) Understanding the feasibility of scaled III-V TFET for logic by bridging atomistic simulations and experimental results. Symposium on VLSI Technol (VLSIT). 2012:183–184

27. Conzatti F, Pala MG, Esseni D, Bano E, Selmi L (2011) A simulation study of strain induced performance enhancements in InAs nanowire Tunnel-FETs. 2011 international electron devices meeting. 5.2.1–5.2.4
28. Sylvia SS, Khayer MA, Alam K, Lake RK (2012) Doping, tunnel barriers, and cold carriers in InAs and InSb nanowire tunnel transistors. IEEE Trans Electron Devices 59:2996–3001
29. Knoll L et al (2013) Inverters with strained si nanowire complementary tunnel field-effect transistors. IEEE Electron Device Lett 34:813–815
30. Kim SH, Kam H, Hu C, Liu TJK (2009) Germanium-source tunnel field effect transistors with record high ION/IOFF. 2009 Symposium on VLSI Technology
31. Krishnamohan T, Kim D, Raghunathan S, Saraswat K (2008) Double-gate strained-ge heterostructure tunnelling FET (TFET) with record high drive currents and <<60mV/dec subthreshold slope. 2008 IEEE Int Electron Devices Meeting 1–3
32. Yang Y et al (2012) Towards direct band-to-band tunneling in P-channel tunneling field effect transistor (TFET): technology enablement by Germanium-tin (GeSn) (2012) International Electron Devices Meeting 16.3.1–16.3.4
33. Valasek J (1921) Piezo-electric and allied phenomena in rochelle salt. Phys Rev 17:475
34. von Hippel A (1944) U.S. National defense research committee report 300, NDRC, Boston, MA
35. Kepler RG, Anderson R (1992) Ferroelectric polymers. Adv Phys 41:1–57
36. Scott JF, De Araujo CAP (1989) Ferroelectric memories. Science 246:1400–1405
37. Cross L, Newnham R (1987) History of ferroelectrics. Ceramics Civil 3:289–305
38. Dawber M, Rabe KM, Scott JF (2005) Physics of thin-film ferroelectric oxides. Rev Mod Phys 77:1083–1130
39. Matthias P (2014) The FRAM pentathlon. EP&Dee. Web http://www.epd-ee.eu/article/7686
40. Akira O, Fukunaga M, Takesada M (2012) Ferroelectric instability and dimensionality in bi-layered perovskites and thin films. Adv Condensed Matter Phys 2012:1–10
41. Lines ME, Glass AM (1977) Principles and applications of ferroelectrics and related materials. Oxford University Press
42. Tagantsev A, Sherman V, Astafiev K, Venkatesh J, Setter N (2003) Ferroelectric for microwave tunable applications. J Electroceram 11:5–66
43. Ginzburg V (2002) Phase transitions in ferroelectrics (some historical remarks). Ferroelectrics 267:23–32
44. Rabe KM, Ahn CH, Triscone JM (2007) Physics of ferroelectrics: a modern perspective. Springer Science and Business Media 105
45. Krishnamohan T, Kim D, Raghunathan S, Saraswat K (2008) Double-gate strained ge heterostructure tunneling FET (TFET) with record high drive currents and 60mV/dec subthreshold slope. Electron Devices Meeting 1–3
46. Salahuddin S, Datta S (2008) Can the subthreshold swing in a classical FET be lowered below 60 mV/decade?. Electron Devices Meeting, 1–4
47. Rusu A (2012) Negative capacitance transistor. Thesis EPFL
48. Ionescu AM (2017) Energy efficient computing and sensing in the Zettabyte era: from silicon to the cloud. IEEE Int Electron Devices Meeting (IEDM). https://doi.org/10.1109/IEDM.2017.8268307
49. Nikonov DE, Young IA (2015) Benchmarking of beyond-CMOS exploratory devices for logic integrated circuits. IEEE J Exploratory Solid-State Comput Devices Circuits 1:3–11
50. Takagi S et al (2008) Carrier-transport-enhanced channel CMOS for improved power consumption and performance. IEEE Trans Electron Devices 55:21–39
51. Boucart K, Ionescu AM (2006) Double gate tunnel FET with ultrathin silicon body and high-k gate dielectric. Solid-state device research conference, Proceeding of the 36th European, 383–386
52. Wang X et al (2017) Two-dimensional negative capacitance transistor with polyvinylidene fluoride-based ferroelectric polymer gating. NPJ 2D Materials and Applications 1. https://doi.org/10.1038/s41699-017-0040-4

53. Asra R et al (2011) A tunnel FET for VDD scaling below 0.6 V with a CMOS-comparable performance. IEEE Trans Electron Devices 58:1855–1863
54. Wu C et al (2014) An analytical surface potential model accounting for the dual modulation effects in tunnel FETs. IEEE Trans Electron Devices 61:2690–2696
55. Su P, Goto K-I, Sugii T, Hu C (2002) A thermal activation view of low voltage impact ionization in MOSFETs. IEEE Electron Device Lett 23:550–552
56. Hui K, Hu C, George P, Ko PK (1990) Impact ionization in GaAs MESFETS. IEEE Electron Device Lett 11:113–115
57. Sakurai T (2004) Perspectives of low-power VLSI's. IEICE Trans Electron 87:429–436
58. Bartsch ST, Rusu A, Ionescu A (2012) Phase-locked loop based on nanoelectromechanical resonant-body field effect transistor. Appl Phys Lett 101:153116
59. Appleby DJ et al (2014) Experimental observation of negative capacitance in ferroelectrics at room temperature. Nano Lett 14:3864–3868
60. Salvatore GA, Bouvet D, Ionescu AM (2008) Demonstration of subthreshold swing smaller than 60mV/decade in Fe-FET with P(VDF-TrFE)/SiO2 gate stack. Electron Devices Meeting, IEEE International, 1–4
61. Salahuddin S, Datta S (2008) Use of negative capacitance to provide voltage amplification for low power nanoscale devices. Nano Lett 8:405–410
62. Obradovic B, Rakshit T, Hatcher R, Kittl J, Rodder MS (2018) Modeling of negative capacitance of ferroelectric capacitors as a non-quasi static effect. arXiv preprint arXiv:1801.01842
63. Jiang C, Liang R, Xu J (2017) Investigation of negative capacitance gate-allaround tunnel FETs combining numerical simulation and analytical modeling. IEEE Trans Nanotechnol 16:58–67
64. Singh S, Kondekar P, Pal P (2016) Transient performance estimation of charge plasma based negative capacitance junctionless tunnel FET. J Semiconductors 37:024003
65. Liu C et al (2016) Simulation-based study of negative-capacitance double-gate tunnel field-effect transistor with ferroelectric gate stack. Japanese J Appl Phys 55:04EB08
66. Singh S, Kondekar P (2017) A novel electrostatically doped ferroelectric Schottky barrier tunnel FET: process resilient design. J Comput Electron 16:685–695
67. Singh S, Pal P, Kondekar P (2014) Charge-plasma-based super-steep negative capacitance junctionless tunnel field effect transistor: design and performance. Electron Lett 50:1963–1965
68. Tu L, Wang X, Wang J, Meng X, Chu J (2018) Ferroelectric negative capacitance field effect transistor. Advance Electron Mater 1800231:1–17
69. Saeidi A, Jazaeri F, Stolichnov I, Ionescu AM (2016) Double-gate negative-capacitance MOSFET with PZT gate-stack on ultra-thin body SOI: An experimentally calibrated simulation study of device performance. IEEE Trans Electron Devices 63:4678–4684
70. Saeidi A, Jazaeri F, Bellando F, Stolichnov I, Luong GV, Zhao QT, Mantl S (2017) Negative capacitance as performance booster for tunnel FETs and MOSFETs: an experimental study. IEEE Electron Device Lett 38:1485–1488
71. Saeidi A et al Effect of hysteretic and non-hysteretic negative capacitance on tunnel FETs DC performance Nanotechnology 29:1–8
72. Landau LD, Khalatnikov IM (1954) On the anomalous absorption of sound near a second order phase transition point. Dokl Akad Nauk SSSR 96:469–472
73. TCAD Sentaurus Device User's Manual (2017) Mountain view. USA, Synopsys Inc., CA
74. Das B, Bhowmick B (2019) Noise behavior of ferro electric tunnel FET. Microelectronic J 96:104677–104682
75. Das B, Bhowmick B (2020) Effect of curie temperature on ferro electric tunnel FET and its RF/analog performance. IEEE Trans Ultrasonics Ferroelectrics Frequency Control
76. Kobayashi M, Jang K, Ueyama N, Hiramoto T (2007) Negative capacitance for boosting tunnel FET performance. IEEE Trans Nanotechnol 16:253–258
77. Goswami R, Bhowmick B, Baishya S (2016) Effect of scaling on noise in Circular Gate TFET and its application as a digital inverter. Microelectronic J 53:16–24
78. Pandey R, Rajamohanan B, Liu H, Narayanan V, Datta S (2014) Electrical noise in heterojunction interband tunnel FETs. IEEE Trans Electron Devices 61:552–560

79. Salvatore GA, Lattanzio L, Bouvet D, Ionescu AM (2011) Modeling the temperature dependence of Fe-FET static characteristics based on Landau's theory. IEEE Trans Electron Devices 58:3162–3169. https://doi.org/10.1109/TED.2011.2160868
80. Salvatore GA, Lattanzio L, Bouvet D, Ionescu AM (2010) The Curie temperature as a key design parameter of ferroelectric Field Effect Transistors. 2010 Proceedings of the European Solid State Device Research Conference. https://doi.org/10.1109/ESSDERC.2010.5618386
81. Vandenberghe WG, Verhulst AS, Groeseneken G, Soree B, Magnus W (2008) Analytical model for a tunnel field-effect transistor. MELECON 2008—The 14th IEEE Mediterranean Electrotechnical Conference, Ajaccio. https://doi.org/10.1109/MELCON.2008.4618555

An Introduction to Photovoltaic Applications from Organic Material and Fabrication Perspective

Nidhi Sharma, Deeksha Kharkwal, Saral K. Gupta, and Chandra Mohan Singh Negi

Abstract The demand for the renewable energy sources is increasing in today's era due to limited supply of nuclear energy sources and their hazardous effects that is a prominent reason of rising greenhouse effect. Therefore, different solar cells are gaining importance among which, one specific the organic photovoltaic cells are gaining interest as they are environment friendly as well as economical for both developed and developing world. Although organic solar cells are achieving appreciable performance, there are challenges associated with organic photovoltaics such as low efficiency, less stability, and less strength as compared to silicon solar cells. However, utilization of conjugated polymers (CP) in the active layer of the photovoltaic device can lead to improvement in the organic photovoltaic efficiency.

Keywords Organic solar cells · Photodetectors · Fabrication · Characterization · Electronic structure · Charge transport

1 Introduction

The demand for low-cost and green energy production and utilization has increased over the years [1, 2]. The possibility of persistent growth in world's population is definitely expected to increase the demand of energy resources in the upcoming years. Currently, there are mainly two types of resources fulfilling the energy need of world population, one being fossil fuels and the other being nuclear energy. But, the limited supply of these resources accompanied by their hazardous long-term effects, suggest the necessity of developing renewable energy sources [3, 4]. The continuous burning of fossil fuels from the past one hundred and fifty years has resulted in huge

N. Sharma · D. Kharkwal · S. K. Gupta · C. M. S. Negi (✉)
Department of Physical Sciences, Banasthali Vidyapith, Rajasthan 304022, India
e-mail: nchandra@banasthali.in

S. K. Gupta
e-mail: gsaral@banasthali.in

© The Author(s), under exclusive license to Springer Nature Singapore Pte Ltd. 2022
R. Goswami and R. Saha (eds.), *Contemporary Trends in Semiconductor Devices*,
Lecture Notes in Electrical Engineering 850,
https://doi.org/10.1007/978-981-16-9124-9_4

Fig. 1 Sources of electricity in India by installed capacity

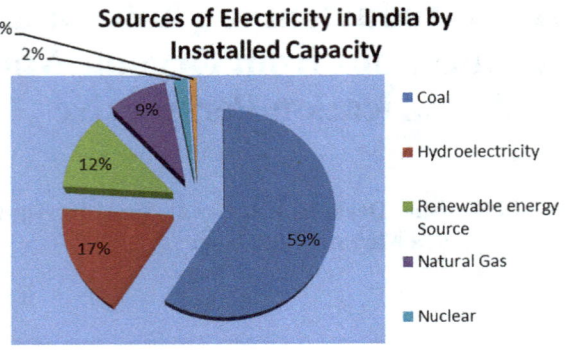

amount of carbon dioxide in the atmosphere causing increased greenhouse effect [5, 6] (Fig. 1).

The consecutive impacts of greenhouse effect are already being experienced in the form of frequently and severely occurring natural disasters [7]. These circumstances are emerging as a massive challenge for today's generation to bring requisite conversion in the conduct of energy being consumed and produced [8, 9]. Renewable resources like wind power, geothermal energy, biomass, hydroelectric, and solar energy due to their never-ending accessibility and absence of any detrimental effects to the environment can be utilized for sustainable energy production in the upcoming years [10, 11]. Compared to other renewable energy resources, green and clean solar energy has less geographical dependence, much lesser maintenance, good control over power generation, and vast access and can bring it forward as an efficient source of energy in future. The overall quantity of solar irradiation every year on the surface of earth is equal to around 10,000 times the energy needed by the world per year [12]. The sun has reservoir of approximately 86,000 TW renewable energy per year, while requirement of energy around the globe is nearly 15 TW per year, indicating that an ample amount of energy can be extracted from the sun than it has the capacity to power the complete world even after the twofold rise of energy consumption in imminent 50 years [13]. Yielding energy from direct sunlight by the use of photovoltaic (PV) technology is gaining worldwide recognition as a vital component of energy production in the recent years [14, 15]. It offers several advantages including being friendly to the environment, converting light straight away into a high-level energy, entailing lesser maintenance, and being modular, thus, serving both small and very large power demands. Therefore, solar energy can be utilized as a potential renewable and sustainable source of energy in future that will serve significant advantages to both environment as well as economy of developed as well as developing countries.

As seen in Fig. 2, renewable energy shares only 12% of the total electricity generation in India, in this also amount of electricity generation form solar cells is insignificant. Silicon technology based solar cells are in majority; however, the key issues, which limit widespread adoption of PV technology, include high manufacturing cost and extremely energy insensitive growth process of silicon ingots. The

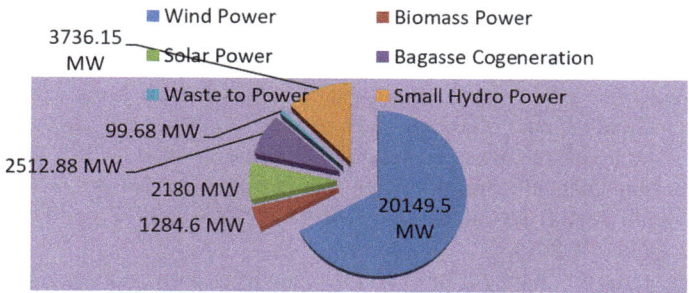

Fig. 2 Renewable energy in India by Installed Capacity

energy insensitive process is contributing to the greenhouse effect. Thus, to make PV technology more affordable, tenable technology pathways for PV require developing novel materials as well as device structures. Organic materials possess the potential of developing an enduring technology that is cost effective, eco-friendly, and have limitless accessibility. The extremely high optical absorption coefficients of organic semiconductors can lead to chances of developing very thin solar cells. The thin flexible devices can be produced through low temperature solution-processed printing approaches resulting in high throughput cost-effective manufacturing, is an important feature of organic PVs.

2 Types of Solar Cells

Depending on the technology and materials used, solar cells are commonly categorized into four generations. First-generation solar cells are mainly based on silicon wafers manufactured by diffusion process and have an efficiency of about 15–20%. The cell is competent to generate utilizable electrical energy through sunlight. These solar cells have demonstrated good performance and higher stability. Apart from energy intensive manufacturing process, these cells were very costly; despite this, they still remain dominated in the solar cell market. The year 2007 was marked with 89.6% of the commercial production of these first-generation cells [16]. Since crystal growth process is not required to manufacture organic solar cells, thus the process is less energy intensive as well as cost effective compared to first-generation cells. Further reduction in cost is possible by saving the amount of bulk material as the thin layer accumulates instead of the thick layer compared to the first-generation photovoltaic cells. Most importantly, these cells show poor performance compared to the first-generation cells. Second generation solar cells currently cover a small segment of the terrestrial photovoltaic market, while about 90% of the space market. Quantum well/dot structures and multilayer structure of III–V inorganic semiconductors are being experimented to develop the solar cells at lower cost and performance better than second generation solar cell. For global power generation, solar cells include photo electrochemical cells, nano-crystalline thin films, dye-sensitive solar cells, which are still under research [17]. The third-generation solar cells are not commercialized much due to their expensive production costs. Fourth-generation solar cells bring in the use of organic materials, such as small molecules or polymers making the polymer solar cells as the sub category of organic solar cells [18]. A novel category of thin film solar cells presently being investigated is perovskite solar cells that possess a high potential accompanied by recorded efficiencies of more than 20% on very small areas [19, 20].

Polymer or plastic solar cells have several benefits as they have flexible structures, easy accessibility and reasonable price, lower fabrication prices, easier processing, adaptable electrical properties leading to uncomplicated, speedy, and low-cost production on a large scale [21]. Use of safer materials and processing methods in terms of environment makes the solar cells eco-friendly and, thus would expand as a long-term technology.

2.1 Organic Solar Cells (OSCs)

Single crystal and thin film devices formed by the use of inorganic materials were brought into service as solar cells in both industrial as well as household applications since several years. In spite of the extensive research efforts, the cost of the cells was not reduced as expected. Therefore, recent trend in current years in the solar cell research has shifted to utilizing the materials of lower manufacturing expenses and obtaining decent performance from them. Lot of research from then was focused on the less expensive alternatives, such as organic materials. The organic solar cell technology is directed toward utilization of thin films of organic semiconductors for transduction of solar light into electrical energy. The advantages and disadvantages of this technology are presented in the Fig. 3. The benefits associated with OSCs as depicted in the figure make the OSC technology more appropriate in comparison with inorganic counterparts. However, the improvement in efficiency and operating life time is the biggest challenge to overcome to make these devices commercially viable.

2.1.1 Advantages

Easy Manufacturing Techniques

Numerous methods are available for the production of OSCs at the laboratory scale, among which the most importantly and commonly utilized technique, is spin coating. In spin coating technique, a liquid solution is deposited on the substrate and then

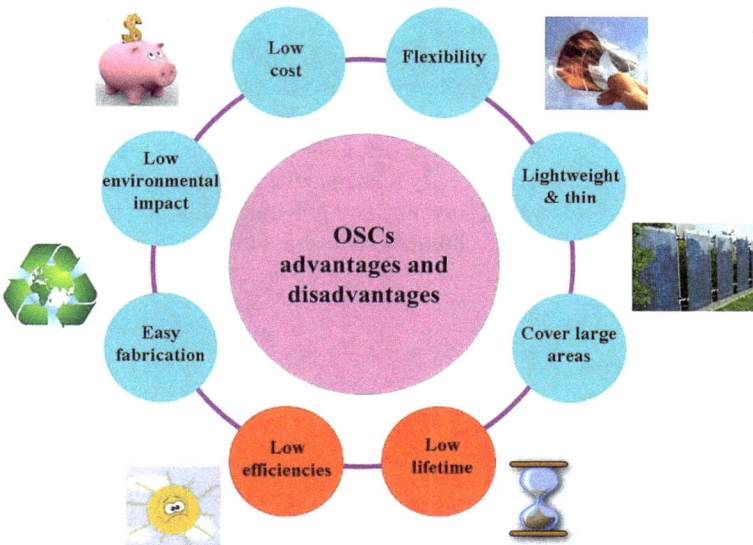

Fig. 3 Advantages and disadvantages of organic solar cells

substrate is subjected to spin at a particular angular or rotational speed [22]. One more currently utilized technique is the doctor's blade technique, which is beneficial as the solution amount required to cover the substrate is lesser than that spin coating. In this technique, a blade moving across the substrate is involved, that drags the previously placed solution in front of the blade and coats the surface.

For the production of OSCs at the large scale, a large area roll-to-roll technique (R2R) is usually preferred. The technique involves the deposition of thin films by inkjet printing or vacuum processing over flexible substrates in the form of large sheets, which can be bound into rolls [23]. These techniques offer the potential for cost-effective manufacturing of solar cells.

Flexibility and Lightweight

The vital role of technology of organic solar cells lies within the certainty that by the use of R2R method, the construction of devices can be made on flexible plastic substrates. Usage of metal foils as electrodes can result in light weight devices. The commonly utilized materials as flexible substrates are polyethylene terephthalate (PET) and polyethylene naphthalene (PEN) with indium tin oxide (ITO) generally utilized as transparent electrode for coating [24].

Low-Cost and Environmental Impact

The manufacture of organic solar cells on the basis of π-conjugated organic semiconductors can result in greater potential for technology development. The right choice of active materials, electrodes, and barrier foils can help in reducing the cost of OSCs. Environmental impact is directly related to the manufacturing process and the appropriate choice of substrates, solvents, and polymer materials. Low temperature solution-processed fabrication techniques reduce the environmental impact related to the manufacturing process [25].

2.1.2   Disadvantages

Low Power Conversion Efficiency

Till now, the power conversion efficiency (PCE) of organic solar cells has been reported in research up to 22% for small area devices. This relatively lower PCE has become a major disadvantage, which strongly depends on the choice of polymer materials, substrates, and active area, architecture and fabrication methods. There is possibility of resolving this issue by developing novel semiconductor materials, electrodes, and barrier films [26].

Poor Stability and Low Operating Lifetime

The major disadvantage comes out to be the poor stability of materials and lower lifetime observed in the devices upon testing in varying conditions. The successful commercialization can be achieved by necessarily unifying a higher efficiency in large area devices and by increasing the life time of devices. Better understanding regarding the stability of interface, materials, and mechanisms of degradation is necessary for enhancing the device lifetime. This drives the focus of investigations on factors affecting stability, like oxygen, humidity, temperature, light, loading conditions, electrodes, etc. The encapsulation materials and techniques are generally preferred for avoiding degradation [27].

3   Electronic Structure in Organic Semiconductors

The bond structure among the carbon atoms is the topic of attraction in conjugated polymers. Figure 4 shows the energy levels of Π-conjugated molecules. In most of the industrial plastics, their insulating properties arise through the forming of σ bonds among the adjacent carbons, whereas σ bonds in conjugated polymers acting as backbone are alternatively single or double. In other terms, every carbon atom can only form bond with its three out of four neighboring atoms in the backbone of

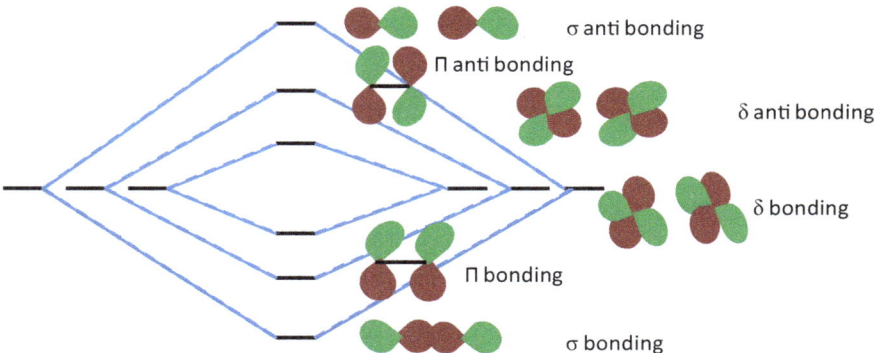

Fig. 4 Energy levels of Π-conjugated molecules

polymer, thus leaving one unattached electron per each carbon atom in the p_z orbital. As it continues to be analogous for each carbon atom in the backbone, all unbonded electrons mutually overlap among these p_z orbitals to form π bonds by the side of the backbone, in which π bonds in electrons can be delocalized along this conjugated path forming the conjugated polymer an intrinsic semiconductor [28]. The π bond, if filled with electrons, is known as the highest occupied molecular orbital (HOMO), or else, lowest unoccupied molecular orbital (LUMO). The electrons excited in this π bond leads to polymer chain staying together as a result of the σ bond formation among the neighboring carbon atoms. In addition, actual conjugated polymers exhibit energetic disorder. Molecular $\pi - \pi^*$ orbitals communicate to the highest HOMO and LUMO. LUMO and HOMO levels are analogues to conduction band (CB) and valence band (VB), respectively, in a crystalline based semiconductor [29, 30].

4 Operating Principle of Solar Cells

The procedure of the converting sun light into the electricity by the means of organic solar cell is carried out through following steps: (i) photon absorption resulting in the excited state development, i.e., the bound EHP (exciton) formation; (ii) diffusion of excitons to the interface zone, where excitons dissociate into free charge carriers; and (iii) charge transportation toward the appropriate electrodes.

4.1 Absorption of Light and Exciton Generation

A representative OSC structure includes an active layer, layers for extracting and transporting charge and electrodes. The active layer comprises of the composition of blend of donor and acceptor material. The active layer serves as the harvester of light, and at the same time ensures efficient separation of charges. The effectiveness of the absorption depends on the absorption coefficient (α) of the absorbing material and also the reflectance of the absorbing surface. Usually, organic materials show high absorption coefficient, as a result of which they provide efficient light absorption at certain wavelengths with merely relatively thin layer (100–200 nm). However, organic molecules cover only narrow absorption range of solar spectrum, with small absorption in nearby infrared region. Thus, to attain sufficient absorption, fine-tuning of the energy levels and expansion of the absorption range are required [31].

As stated earlier, the light absorption in OSCs is carried out by an active layer. One of the approaches to shift the absorption toward longer wavelengths is employing the donor materials with higher degree of delocalization resulting due to large-conjugated system. Once a light absorption by molecule takes place, an electron is excited from HOMO to LUMO as shown in Fig. 5. The excitons in organic semiconductors exhibit relatively large binding energy (0.3–1 eV). The binding energy needs to be surmounted by the built-in electric field to dissociate excitons into free electrons and holes for photocurrent production. The dissociation of excitons in OSCs occurs at the interface between the donor and acceptor semiconductors. Therefore, excitons produced anywhere inside the material need to migrate in the direction of the donor (D)/ acceptor (A) interfaces. The migration of excitons is done through process of diffusion from high exciton concentration regions to low excitons concentration regions. Approximately, only 10% of the photo-excitons result in free charge carriers within conjugated polymers [32]. The relation between diffusion length and exciton lifetime τ can be expressed as,

$$L_d = (D_\tau)^{1/2} \tag{1}$$

Here D represent diffusion coefficient.

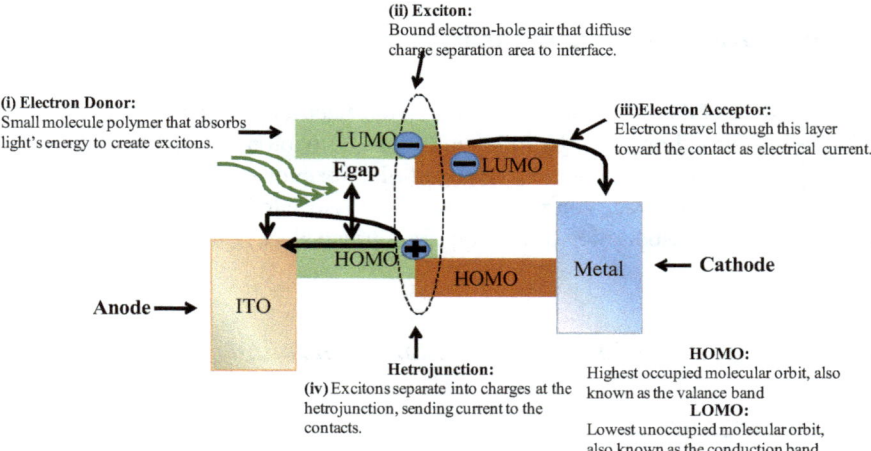

Fig. 5 Operating principle of BHJ solar cells

4.2   Exciton Dissociation

On arrival of the excitons at the D/A interface, its dissociation into free charges is required to produce photocurrent. The initial action is the transfer of electron from the exciton toward the acceptor's LUMO that results in D^+/A^- at the interface called charge transfer state (CT). The force that drives this event is supplied through the offset of the $LUMO_D$–$LUMO_A$, as illustrated in Fig. 5. During the state of CT, the location of the electron and the hole is at the neighboring molecules of the D/A interface, which are closely held through the Coulombic attraction [33].

4.3   Recombination

The bimolecular recombination depends on the mobility of charge carriers. Highly mobile charge carriers encounter frequently with one other, thus increasing the rate of recombination. Trap-assisted recombination is a two-step process, firstly the energetic trap states formed due to the material limitations captures a free charge carrier, and finally they recombine with the mobile charge carriers of opposite sign. Selective contacts are utilized for decreasing the OSC's surface recombination, thereby improving the efficiency of charge collection. Selectivity of merely single carrier type can be realized with interfacial layers of appropriate energy levels [34].

4.4   Charge Transport and Collection

The dissociated free charges are transported to the suitable electrodes during their lifespan to produce the photocurrent in the device. The force that drives the charge carriers to flow toward the electrodes arises from the built-in electric field. Additionally, utilization of uneven contacts that include one of lower work-function metal to collect electrons and the other of higher work-function metal to collect holes is expected to cause an external field during the short circuit in a metal–insulator-metal (MIM) configuration. The charge transport mechanism is significantly affected by the recombination process during the journey toward the electrodes. In the final stage, the extraction of charge carriers from the device is completed by two selective contacts. On the other hand, an evaporated aluminum metal contact having work function of about 4.3 eV, which matches with the acceptor LUMO, is commonly used as the electron selective contact [35].

5   Architecture of Organic Solar Cells

As an active layer of solar cells, organic semiconductors possess several distinct advantages relative to inorganic semiconductors. Organic semiconductors have an extremely high absorption coefficient which allows the fabrication of very thin devices (~100 nm) using minimum raw material. The lack of dangling bonds in organic materials can produce multilayer device structures. In addition, organic semiconductors are tunable, compatible with flexible substrate, highly abundant, and economical. The conversion of photon into photocurrent via organic PV devices is explained through four steps. Illumination with light causes absorption of photons which generate excitons, and then these excitons diffuse to the donor–acceptor interface, where they dissociate into the free electrons and holes owing to the built-in electric field at donor–acceptor interface. Finally, electrons and holes transport to the respective electrodes and generate photo current.

The thickness of the layer and the absorption coefficient of layer decide the amount of light absorption. The absorption efficiency is dependent on the number of photons absorbed with respect to the incident photons. This efficiency is also dependent on the absorption spectra of the active material, its matching with the incident light spectrum and the architecture of the device.

5.1   Bilayer Heterojunction Device

Bilayer heterojunction device architecture comprises of a sequential stack of donor and acceptor semiconductor one above the other and then sandwiched between two electrodes (see Fig. 6a). These bilayer devices utilizing organic semiconductor have been examined for several varying combinations of material systems. After dissociation of excitons, the electrons travel toward the n-type acceptor and hole toward the p-type donor. Thus, most of the photons absorbed away from the interface do not contribute to the photocurrent, due to which the bilayer devices exhibit lesser efficiency [36].

5.2   Bulk Heterojunction (BHJ) Device

When an active layer consisting of donor and acceptor material is sandwiched between two different work-function contacts, a BHJ organic photovoltaic (OPV) device is formed. The main potential active material combination is of conjugated polymer as a donor and fullerene derivative as an acceptor. The mixture of both of these materials leads to construction of a huge amount of interfacial area in between the donor and the acceptor that enables the bulk heterojunction to penetrate into the organic photovoltaic field. As a result of interfacial area for exciton dissociation being

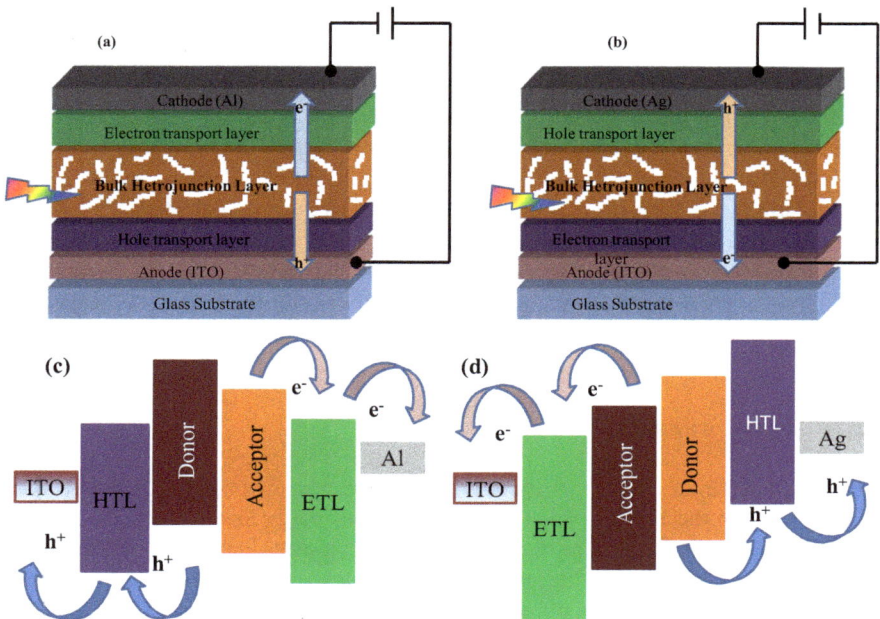

Fig. 6 Schematic of **a** standard structure of device, **b** inverted device structure, and **c, d** corresponding energy level diagrams with charge transportation process under illumination condition

preferably larger, none of the losses of carriers due to the diffusion of small excitons will occur and dissociation of all the excitons along with separation of charges will take place. Hence, suppression of recombination results, leading to improvement in efficiency [37].

Polymer solar cells are usually categorized in two groups on the basis of the solar cell stack geometry: one is the standard geometry and the other one is inverted geometry.

Both of these geometries are defined depending upon the direction of flow of charge. The standard geometry constructed on the substrate is fabricated with the transparent electrode acting as the solar cell anode terminal and the electrode mentioned before the active layer absorption, the uppermost electrode which is typically a cathode made of low work-function material [38]. The inverted geometry results through the switching of the two electrodes and the charge selective layers in such a way that the transparent electrode present on the substrate acts as cathode and the top electrode acts as anode, generally deposited using high work-function material.

5.2.1 Active Layer

The active layer present in the polymer solar cells comprises of two components, one being a donor that helps in light absorption and the other being an acceptor that facilitates the electron extraction from the excitons bound electron–hole leading to the travel of an electron in the active layer's acceptor phase and the travel of the hole in the donor phase. Moreover, as the traveling of holes as well as electrons is required from the active layer to the electrodes, there is a need for an interconnected network to be formed in between the donor and the acceptor that allows efficient dissociation of excitons along with efficient charge carrier transportation toward the respective electrodes [39].

5.2.2 Transport Layers

The materials being capable of transfer of either electrons or holes at the primary level because of the appropriate arrangement of the energy levels forms the basis of transport layers. The examples of the hole transport materials can be polymer-based PEDOT: PSS and a metal oxide such as molybdenum oxide (MoO_x). The examples of frequently utilized electron transport layers include lithium fluoride (LiF), calcium (Ca), and metal oxides like zinc oxide (ZnO) and titanium oxide (TiO_2) [40].

5.2.3 Electrodes

Indium tin oxide (ITO) is the mostly utilized electrode as a result of its higher optical transmission in combination with lower resistance. On the glass, it usually shows a transmission of >85% at <10 Ω/sq. The major challenge is to choose the electrodes that have an appropriate energy level in addition to transparency which can let entrance of adequate light into the solar cell. PEDOT: PSS is presently becoming a well-accepted material to be utilized as an electrode; through doping it can lead conductivities greater than 500 S/m along with >80% transmission. The doping of PEDOT: PSS solutions are suitable in allowing effortless fabrication on flexible substrates due to the polymer-based films having an enhanced tolerance for bending in comparison with an ITO electrode [41].

5.2.4 Substrates

The substrates utilized to support the layered solar cell stack for manufacturing of polymer solar cells can be categorized as two different groups namely, glass and plastics. The two usually utilized types are floated glass substrates with ITO transparent electrodes brought into use in the lab scale manufacture and flexible PET [42].

6   Organic Photodetectors

The contemporary/ fashionable electronics come out with detecting or sensing of a light as a key challenge due to many of the detectors being based on solid-state technology. The conversion of an optical signal into current is carried through photodetectors by majorly using them as optical receivers for the same. The basic principle forming the basis of the photodetectors is the photo electric effect where electron ejection from the metal surface after being struck by light takes place. Photodectectros are majorly utilized as safety devices and for the purpose of security through conjunction with other optical devices. Photodetectors can be categorized into photodiodes, PIN photodiodes, avalanche photodiodes, and phototransistors.

Photodiode is the commonly utilized photodetector that consists of a p-type and a n-type semiconductor placed together for the formation of a junction. A high impedance region, depleted from the charge carriers is formed through it. Its main mechanism is to work in reverse biased conditions. It can be used as a switch or relay because it also allows current to flow through the circuit in reverse bias and at its brightness by light, most of the transport is caused by the process of charge propagation. Thus, the PN photodiode reaction speed is usually low. Photodiodes are classified as pin photodiodes as an inner layer of high resistivity when introduced between p-type and n-type materials improves response speed. It contains a higher depletion region to make the generation of larger number of e–h pairs possible at a lower capacitance. An avalanche photodiode comprises of the construction of an additional thin layer for providing the multiplication of the charge carriers formed through optical generation. The photo generated charge carriers upon entrance into this high field region result in generation of additional carriers through the process of avalanche multiplication. Comparative improvement in sensitivity is seen as compared to PIN photodiode, but introduction of added noise component is seen through random avalanche multiplication process. The gain mechanism can also be attained through phototransistor. In the phototransistor, the highly absorptive base region is required and the illumination of the base region occurs through the light in place of applying voltage at the base region and internal gain is provided because of the transistor action. Conventional silicon (Si) photodetectors show higher responsivity for short wavelengths, while for longer wavelengths, less absorption coefficient ($<10^3$ cm^{-1}) leads to prevention of its usage in near infrared applications [43, 44]. However, organic material-based photodetectors are increasingly gaining attention as a result of them having noteworthy advantages in comparison with the inorganic counterparts including low-cost fabrication, large area sensing possibility, tunable response spectrum, lightweight, and flexibility. Many organic and organic/inorganic hybrid photodetectors have been observed in recent years of excellent linearity of production with incident light, quick response, wide spectral range, and low temperature sensitivity. The graded heterostructure offers extended volume for dissociation of excitons and bi-continuous pathways for carrier transportation [45]. Lim et al. fabricated a novel solution-processed high performance triisopropylsilyehhynyl anthracene derivative based photodetector [46].

7   Organic Light Emitting Diodes (OLEDs)

OLED's structure contains one or more organic layers or hybrid material (either small molecules or polymers) sandwiched in between the two electrodes [47]. Figure 7 presents the schematic of an OLED. OLEDs need lower power than the backlight-dependent display technology. Presently, they are gaining importance owing to their advantages, including wide viewing angle, flexibility, fast response, and low operating voltages. They are utilized in several applications like flat panel displays, solid-state lighting, and so on because of their outstanding features.

The significant breakthrough presenting the practical importance of OLED was acquired in 1987. The initiated use of a transparent hole injection electrode of ITO was intended toward the improvement of the extraction efficiency of electroluminescence, where the electron injection electrode of magnesium: silver (Mg:Ag) was employed. The OLED utilizing the conjugated polymer as the single active organic layer sandwiched in between the ITO and aluminum (Al) as bottom and top electrode, respectively, was first implemented in early 1990s [48]. They represented that large area organic light-emitting displays might be fabricated by utilizing the easily prepared solution-processed polymer. This work was instantly followed by development of OLED by the usage of a soluble derivative of PPV. On the basis of these initial investigations on polymer OLEDs, novel television display fabricated through polymer-printing techniques are completely available commercially these days.

There are several techniques used for the fabrication of discussed devices.

Fig. 7   Schematic diagram of OLED

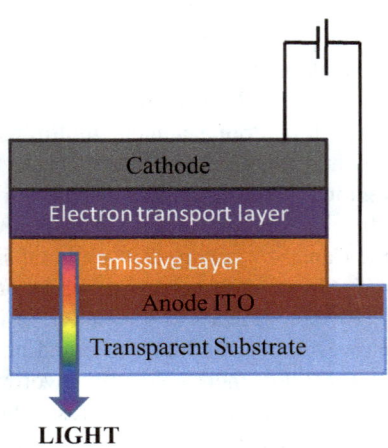

8 Experimental Techniques Used for Device Fabrication

The solution of polymer and fullerene solutions in 1,2-dichloro benzene can be prepared by magnetic stirrer for about 24 h separately and then blend of Polymer: Fullerene prepared by magnetic stirrer for about 3 h.

8.1 Magnetic Stirrer

The magnetic stirring technique is used to prepare the polymer solutions and polymer/fullerene composites. Magnetic stirrer is a commonly used device to mix liquids, low viscosity, or solid–liquid mixtures. It is a general and robust laboratory device which consists of a rotating magnet and the stationary electromagnets that create a rotation in magnetic field. This is used causing a stir bar immersed in the liquid to spin very rapidly, agitating/mixing the liquid. This device also often includes heater to heat the liquid during mixing. They are very much preferred over gear-driven motorized stirrers because of their mixing efficiency and quieter mode of operation [49].

The stirring bars are more easily cleaned, due to their small size, compared to other stirring devices. The magnetic stirrers avoid a few major issues associated with the motorized stirrers. The lubricants used in motorized stirrers can contaminate the reactions and the obtained products. The stirrer sealing between the rotating shaft and the vessel can be problematic, if a closed system is needed [50] (Fig. 8).

Fig. 8 Image of magnetic stirrer

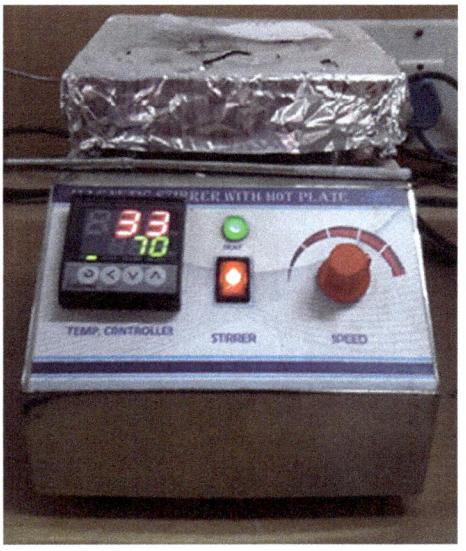

These devices also have some disadvantages; for example, it can only be used for relatively small (under 4 L) due to the limited size of the stirring bar. Additionally, viscous liquids and/or thick suspensions show difficulty in mixing by usage of these stirrers, though there are some stirrers having special magnets that can help in overcoming this problem.

9 Film Deposition Techniques

There are various existing techniques for solution-processed deposition of thin films of various materials.

9.1 Wet Chemical Deposition Techniques

There are various techniques to deposit thin films from the solutions. These techniques are;

(i) Spin coating, (ii) Chemical Bath, (iii) Dip coating, (iv) Doctor blade, (v) Metering rod, (vi) Spray-coating, (vii) Screen printing, (viii) Inkjet printing, and (ix) Aerosol Jet.

In all these techniques, the materials are already synthesized and particles are suspended as colloids in the solution, except only in the chemical bath deposition, in which there is in situ growth of material takes place. To obtain uniform/homogeneous coatings, the solution must fulfill the following requirements;

i.   The solubility of solute in the solution should be high.
ii.   For the higher wet ability, the contact angle between the solution and the substrate must be of a sufficiently small.
iii.   The solution must be of sufficient durability under certain conditions.

The spin coating technique, which is utilized in present research work for depositing the thin films from the solutions, are described below.

9.1.1 Spin Coating

The spin coating is the most versatile and efficient technique to deposit a very homogenous thin layer with very low interfacial roughness. This technique has very wide industrial applications, especially to prepare films of photo-resists for fabricating semiconducting devices, very large scale integrated (VLSI) circuits, anti-reflection coatings on solar cells/optical devices, sensors, detectors, magnetic disk for data storage. The image of the spin coating unit is shown in Fig. 9.

Fig. 9 Photograph of available spin coater technique

In the process of deposition of thin films, usually a small quantity of solution of the depositing material is dispensed over the substrate and the substrate spins the solution around an axis perpendicular to the depositing area.

Thereafter, the substrate containing the solution is rotated with high speed, so that excess solution can be removed by the centrifugal force and solution can uniformly spread over the substrate in order to form the uniform film. Figure 10 shows the scheme of spin coating technique.

The deposition process can be affected by various factors, including spin speed, spin time duration, acceleration, and exhaust among all, the spin speed mostly affects the overall process, the rotation speed of the substrate, the revolutions per minute (rpm) applied to the solution resin applied to the solution resin as well as the wind speed and specific turbulence. Directly above it. Particularly, the speed usually decides the resulting thin film thickness. Reasonably minor variations, for example, of $\pm$ 50 RPM during the process results in the thickness variation of 10%. Usually, high RPM and longer spin duration produce thinner films. The spin time, RPM, and viscosity of the liquid control the thickness of the films [51]. This can also be controlled with a good precision in the wide range from several microns to few nanometers [52].

10   Glove Box

Glove box is used to protect reactive samples against air and other mixture of gases. It is a very useful method for prevention of samples from any contamination through

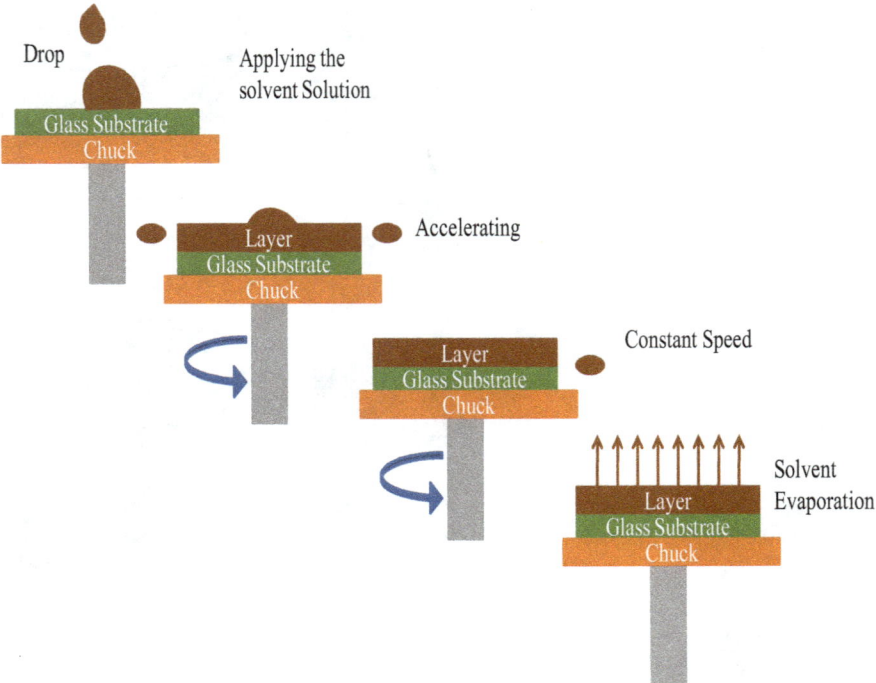

Fig. 10 Scheme of spin coating technique

air or surroundings. It is a sealed box and has a long attachment to the sides, allowing users to conduct materials processing, including sample preparation and other experiments inside the glove box. Glove box is the transparent box providing the transparency between the walls so that can clearly see the experiments. Argon/Nitrogen environment is maintained inside the glove box in order to provide an oxygen and moisture free environment [53]. Photograph of glove box is shown in Fig. 11. Two types of glove box are available, one of which is used in the processing of radioactive environment and the other for high passive environment.

11   Thermal Evaporation Technique for Thin Films Deposition

Thermal evaporation is basically a physical vapor deposition (PVD) technique for thin film deposition. In thermal evaporation, target material is heated in tungsten filament in order to evaporate the target material. These experiments have been done in vacuum better than 10^{-5} Torr. The vaporized atoms are further precipitated and they form a thin film. It is well-known that the melting temperature of materials

Fig. 11 Picture of the glove box used for the preparation of organic-based device

decreases when they are processed in high vacuum due to oxygen deficiency. The mentioned parts of the vacuum coating unit are discussed in details.

11.1 Vacuum Pumps

For creating vacuum in the chamber, the complete air from the chamber should be removed. It can be done in two different ways by providing sufficient speed transfer gas from the chamber by removing the current momentum. Mainly two types of pumps are used for making vacuum in the chamber i.e., Rotary pump and Diffusion pump. Rotary oil pump is used to remove gases in the atmosphere from the system with a rotary device. Initially the system is at atmospheric pressure, so the rotary pump removes the bulk air from the system. Further it provides support to diffusion pump as it cannot exhaust over atmospheric pressure. Generally, rotary pumps are also known as backing pumps and roughing pumps.

Second is diffusion pump—it requires a backing pressure of 10^{-2} Torr to start the operation, which is provided by the rotary mechanical pump. Its pumping action is a stream of the pumping fluid vapor emerging from an annular nozzle and expanding into the pump casing. The fast pumping is achieved using high-speed jets of oil vapor colliding with the gas molecules and then compressing in the direction of rotary pump. On heating the oil, the vapors are forced up by the jet stack. The vapors strike the umbrellas and move downward and outward through the nozzle. The oil vapors flow at a velocity that of the sound on passing through narrow jets. The oil molecules condense on the pump walls which are connected with coolers and this flow back to the bottom pool. Therefore, vaporization cycle, condensation, and evaporation continue to take place.

11.2 Gauges

The gauges are used to measure the pressure of the vacuum pumps. There are two gauges in our system: Pirani gauge and Penning gauge.

Pirani gauges are the simplest form of all gauges, wherein a thin wire of smaller diameter is placed along the axis of a tube with greater diameter, the open end of the tube is connected to the system in which the gas pressure is measured. The axial wire is heated by the passage of electric current. As the gas pressure changes around the wire, the heat loss from the wire takes place. This causes the change in temperature of wire, which increases with the decrease in pressure and vice versa. The variation of the temperature of the wire with pressure is reflected in change in the resistance of the wire with pressure. Consequently, the pressure may be measured by incorporating the gauge as a Wheatstone bridge.

In this gauge, a pair of identical cathode plates is placed equidistant on either side of a cylindrical anode within a glass envelope. Pressure is measured in terms of the discharge current. The discharge is initiated by applying a high voltage. A permanent magnet supplies the magnetic field to maintain the discharge at low pressure. The penning gauge provides a simple, rugged gauge for pressure measurement in the approximate range from 10^{-6} to 10^{-9} mbar. As the desired level of vacuum is achieved, the low-tension source is turned on, which causes evaporation of material and consequent deposition over the substrate.

This technique has its usage in depositing a range of materials, including organic, and metals. It is the most popular method for depositing metal electrodes in organic solar cells, LEDs, photo detectors, etc. Photograph of vacuum coating utilized for deposition of metal electrodes is shown in Fig. 12.

It also minimizes the contamination during deposition because the processing occurs at ultra-high vacuum conditions, which helps obtaining uniform coating of the depositing material. It can also be utilized to deposit multiple layers of different materials. Thickness of these layers can be controlled by loading specific quantities of the materials.

Fig. 12 Photograph of vacuum Coating utilized for deposition of metal electrodes

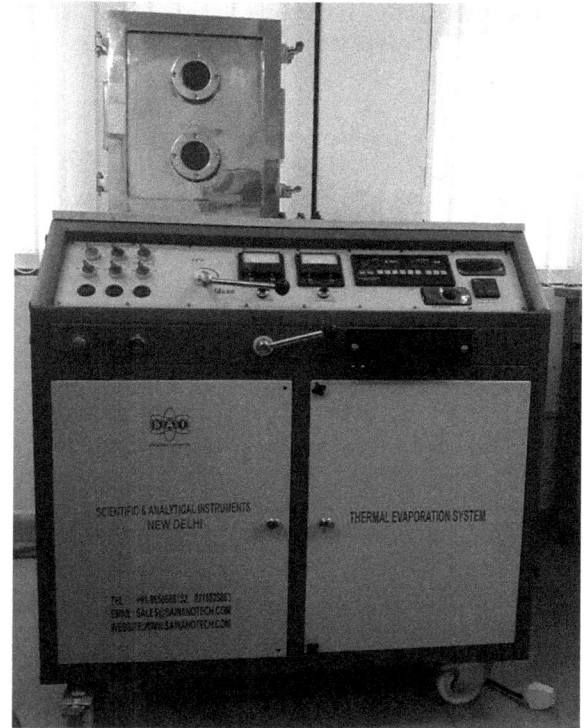

12 Materials

12.1 Conjugated Polymers

Conjugated polymers having continual delocalization all along the chain axis show several unusual properties. These properties have drawn lot of attention and interest due to having potential technological uses. Properties like electrical conduction approaching metals and nonlinear optical responses beyond most organics come out as the major focus of research. These conjugated polymers, can be utilized as active layer materials, hole transport materials and electron transport materials.

12.1.1 Poly(2-Methoxy-5(2-Ethylhexyloxy)-Phenylene Vinylene (MEH: PPV)

Poly (*p*-phenylene vinylene) (PPV, or polyphenylene vinylene) is a conducting polymer of the rigid-rod polymer family. When doped, PPV and its derivatives become electrically conducting. In spite of water insolubility, manipulation of its precursors can take place in an aqueous solution.

12.1.2 Poly(3-hexylthiophene) (P3HT)

The Π conjugated polymer P3HT has a crystalline structure and a broad optical absorption band. P3HT has emerged as an important material because of its solubility, high conductivity, and tunable photoluminescence property. It is well known that interesting physical, optical, and conducting characteristics are exhibited by The P3HT-CNT composites. The incorporation of Fullerene derivatives in the P3HT leads to the improvement in thermal, electrical, and optical properties of P3HT. While considering the PCE of the device, there is the need for using polymer with a high mobility as an active material. When compared among conjugated polymers Poly (3-hexylthiophene) (P3HT) is known to have the highest hole charge carrier mobility. P3HT possesses a hole mobility of 0.1 cm^2 V^{-1} s^{-1} and a band gap of 2 eV which matches with the strongest sunlight. The addition of fullerene to the P3HT increases photo current by creating charge separation. Gang Li et al. in 2005 produced P3HT solar cells having a higher efficiency of 4.4% [54].

12.1.3 Poly (3,4-Ethylenedioxythiophene) polystyrene Sulfonate (PEDOT: PSS)

PEDOT:PSS or poly(3,4-ethylenedioxythiophene) polystyrene sulfonate is a macromolecular salt which is formed together with the charged macromolecules. As a result of having higher transparency and conductivity, it is utilized as a hole transport material (HTM) in organic devices [55].

12.2 PCBM and the Likes

In spite of numerous organic compounds as electron acceptor materials having demonstrated possible properties, few of these electron acceptor materials have their usage in extremely efficient optoelectronic devices. The most successful electron acceptor materials among these are the fullerene and its derivatives.

Fullerene C60 having a well-symmetrical structure shows good electron mobility and is an identified fact that one C60 molecule can accept 4 electrons. It is mostly used in flexible electronics and plastic solar cells as an electron acceptor material in conjugation with P3HT or another polymer which is a donor material. It is soluble in chlorobenzene; so, it is the first choice as an electron acceptor in OPVs. The dominant usage of PC60BM and its corresponding C70 derivative (PC70BM) as acceptors in OPVs has been seen in the past years [56, 57]. The PC70BM resulting in stronger absorption within the visible range as compared to PC60BM is attracting more interest in recent time. However, the higher costs of C70 as a result of having tiresome procedure of purification, compared to that of C60 has resulted in limitation of its application [58].

13 Conclusion and Future Work

Over the past two decades, researchers have developed several next-generation solar cell technologies. Nowadays, inorganic-based semiconductors are dominated by organic-based optoelectronic and photovoltaic technology. The main drawbacks associated with organic solar cells are low efficiency, less stability, and less strength as compared to silicon solar cells. In a general perspective, the materials used for improvement of device performance are also slightly expensive. The low market penetration is one of the major issues of organic solar cell.

However, the fabrication process is relatively expensive and thus by using abundantly available material, the world needs the development of potentially cost-effective optoelectronics and photovoltaic technologies. Higher electrical conductivity, better mechanical flexibility, environmental stability, easier conductivity control, and easier solution process ability are the vital benefits offered by CPs. The impediment caused in providing economical, large area, and relatively efficient electronic devices is due to the issues of poor electrical conductivity and long-lasting stability. Incorporation of polymers with some other species can lead to radical improvement in the material properties of these polymers. Among several materials utilized like inorganic solar cell in the modern energy or optoelectronic technology, polymer: fullerene based organic solar cells have received attention due to their interesting optical, chemical, physical, electronic, and mechanical properties.

The fullerene derivatives have the potential to replace existing materials which are being used as electrodes, hole transportation layer, and electron transportation layer. So, our aim is to precisely control the film thickness, control and stabilize the efficiency of solar cell, increase the efficiency of organic-based solar cells, prepare an optimized blend of fullerene derivatives with polymers, and then using fullerenes and polymer: fullerene nanocomposite as an active layer with different doping (carbon nanotubes), hole transportation layer, and electron transportation layer.

References

1. Owusu PA, Asumadu-Sarkodie S (2016) A review of renewable energy sources, sustainability issues and climate change mitigation. Cogent Eng 3(1):1–14. https://doi.org/10.1080/23311916.2016.1167990
2. Zou C, Zhao Q, Zhang G, Xiong B (2016) Energy revolution: from a fossil energy era to a new energy era. Nat Gas Ind 36(1):1–10. https://doi.org/10.3787/j.issn.1000-0976.2016.01.001

3. Fischer M, López-Duarte I, Wienk M (2009) Functionalized dendritic oligothiophenes: ruthenium phthalocyanine complexes and their {…}. J Am Chem Soc 13:8669–8676. https://doi.org/10.1039/b904243a.VOL

4. Sharma N, Gupta SK, Singh Negi CM (2019) Influence of active layer thickness on photovoltaic performance of PTB7:PC70BM bulk heterojunction solar cell. Superlattices Microstruct 135(September):106278. https://doi.org/10.1016/j.spmi.2019.106278

5. Sorrell S (2015) Reducing energy demand: a review of issues, challenges and approaches. Renew Sustain Energy Rev 47:74–82. https://doi.org/10.1016/j.rser.2015.03.002

6. Zoombelt AP, Fonrodona M, Wienk MM, Sieval AB, Hummelen JC, Janssen RAJ (2009) Photovoltaic performance of an ultrasmall band gap polymer. Org Lett 11(4):903–906. https://doi.org/10.1021/ol802839z

7. Shao Y, Xiao Z, Bi C, Yuan Y, Huang J (2014) Origin and elimination of photocurrent hysteresis by fullerene passivation in CH3NH3PbI3 planar heterojunction solar cells. 1–7. https://doi.org/10.1038/ncomms6784

8. Karg S, Riess W, Meier M, Schwoerer M (1993) Electrical and optical characterization of light emitting poly-phenylene-vinylene diodes. Mol Cryst Liq Cryst Sci Technol Sect A Mol Cryst Liq Cryst 236(1):79–86. https://doi.org/10.1080/10587259308055212

9. Halls JJM, Friend RH, Road M (1997) oy ' I. 85:1307–1308

10. Abbasi SA, Abbasi N (2000) The likely adverse environmental impacts of renewable energy sources. Appl Energy 65(1–4):121–144. https://doi.org/10.1016/S0306-2619(99)00077-X

11. Zhang YHP (2013) Next generation biorefineries will solve the food, biofuels, and environmental trilemma in the energy-food-water nexus. Energy Sci Eng 1(1):27–41. https://doi.org/10.1002/ese3.2

12. Solangi KH, Islam MR, Saidur R, Rahim NA, Fayaz H (2011) A review on global solar energy policy. Renew Sustain Energy Rev 15(4):2149–2163. https://doi.org/10.1016/j.rser.2011.01.007

13. Grübler A, Nakićenović N, Victor DG (1999) Dynamics of energy technologies and global change. Energy Policy 27(5):247–280. https://doi.org/10.1016/S0301-4215(98)00067-6

14. Wilson GM, Al-Jassim M, Metzger WK, Glunz SW, Verlinden P, Xiong G, Mansfield LM, Stanbery BJ, Zhu K, Yan Y, Berry JJ, Ptak AJ, Dimroth F, Kayes BM, Tamboli AC, Peibst R, Catchpole K, Reese MO, Klinga CS, Denholm P, Morjaria M, Deceglie MG, Freeman JM, Mikofski MA, Jordan DC, Tamizhmani G, Sulas-Kern DB (2020) The 2020 photovoltaic technologies roadmap. J Phys D Appl Phys 53(49). https://doi.org/10.1088/1361-6463/ab9c6a

15. Uqaili MA, Harijan K (2012) Energy, environment and sustainable development. Energy, Environ Sustain Dev 12:1–349. https://doi.org/10.1007/978-3-7091-0109-4

16. Todorov TK, Tang J, Bag S, Gunawan O, Gokmen T, Zhu Y, Mitzi DB (2013) Beyond 11% effi ciency: characteristics of state-of-the-art Cu2ZnSn(S, Se)4 Solar Cells. Adv Energy Mater 3(1):34–38. https://doi.org/10.1002/aenm.201200348

17. Baxter JB, Aydil ES (2006) Dye-sensitized solar cells based on semiconductor morphologies with ZnO nanowires. Sol Energy Mater Sol Cells 90(5):607–622. https://doi.org/10.1016/j.solmat.2005.05.010

18. Docampo P, Ball JM, Darwich M, Eperon GE, Snaith HJ (2013) Efficient organometal trihalide perovskite planar-heterojunction solar cells on flexible polymer substrates. Nat Commun 4:1–6. https://doi.org/10.1038/ncomms3761

19. Liang PW, Chueh CC, Williams ST, Jen AKY (2015) Roles of fullerene-based interlayers in enhancing the performance of organometal perovskite thin-film solar cells. Adv Energy Mater 5(10):1–7. https://doi.org/10.1002/aenm.201402321

20. Yasuda T, Sakamoto K, Miki K (2017) Effects of neat C 60 doping on the performance of bulk-heterojunction solar cells based on P3HT:PCBM. Mol Cryst Liq Cryst 653(1):125–130. https://doi.org/10.1080/15421406.2017.1350043

21. Singh RP, Kushwaha OS (2013) Polymer solar cells: an overview. 128–149. https://doi.org/10.1002/masy.201350516

22. Krebs FC (2009) Fabrication and processing of polymer solar cells: a review of printing and coating techniques. Sol Energy Mater Sol Cells 93(4):394–412. https://doi.org/10.1016/j.solmat.2008.10.004

23. Machui F, Hösel M, Li N, Spyropoulos GD, Ameri T, Søndergaard RR, Jørgensen M, Scheel A, Gaiser D, Kreul K, Lenssen D, Legros M, Lemaitre N, Vilkman M, Välimäki M, Nordman S, Brabec CJ, Krebs FC (2014) Cost analysis of roll-to-roll fabricated ITO free single and tandem organic solar modules based on data from manufacture. Energy Environ

24. Jeong YJ, Woo S, Kim Y, Jeong SJ, Han YS, Lee DK, Ko JII, Jung SK, An BC (2011) Effects of solvents on ITO cracks in ultrasonic cleaning of ITO-coated flexible substrates for polymer solar cells. Mol Cryst Liq Cryst 551:212–220. https://doi.org/10.1080/15421406.2011.600655

25. Zhang Q, Ruan C, Xia G, Gong H, Wang S (2021) Low-temperature solution-processed InGaZnO thin film transistors by using lightwave-derived annealing. Thin Solid Films 723(September 2020):138594. https://doi.org/10.1016/j.tsf.2021.138594

26. Corzo D, Tostado-Blázquez G, Baran D (2020) Flexible electronics: status: challenges and opportunities. Front Electron 1(September):1–13. https://doi.org/10.3389/felec.2020.594003

27. Angmo D, Krebs FC (2015) Over 2years of outdoor operational and storage stability of ITO-free, fully roll-to-roll fabricated polymer Solar Cell Modules. Energy Technol 3(7):774–783. https://doi.org/10.1002/ente.201500086

28. Chen YC, Chang WH, Chang Y, Huang CM, Sung HW (2004) A natural compound (reuterin) produced by Lactobacillus reuteri for hemoglobin polymerization as a blood substitute. Biotechnol Bioeng 87(1):34–42. https://doi.org/10.1002/bit.20078

29. Coropceanu V, Li H, Winget P, Zhu L, Brédas J-L (2013) Electronic-structure theory of organic semiconductors: charge-transport parameters and metal/organic interfaces. Annu Rev Mater Res 43(1):63–87. https://doi.org/10.1146/annurev-matsci-071312-121630

30. Koehler A, Baessler H (2015) The electronic structure of organic semiconductors. Electron Process Org Semicond i:1–86. https://doi.org/10.1002/9783527685172.ch1

31. Stübinger T, Brütting W (2001) Exciton diffusion and optical interference in organic donor-acceptor photovoltaic cells. J Appl Phys 90(7):3632–3641. https://doi.org/10.1063/1.1394920

32. Günes S, Sariciftci NS (2017) Organic and inorganic hybrid solar cells

33. Frankevich EL, Lymarev AA, Sokolik I, Karasz FE, Blumstengel S, Baughman RH, Hörhold HH (1992) Polaron-pair generation in poly(phenylene vinylenes). Phys Rev B 46(15):9320–9324. https://doi.org/10.1103/PhysRevB.46.9320

34. Shoaee S (2012) Charge photogeneration in donor/acceptor organic solar cells. J Photonics Energy 2(1):021001. https://doi.org/10.1117/1.JPE.2.021001

35. He M, Chen Y, Liu H, Wang J, Fang X, Liang Z (2015) Chemical decoration of CH3NH3PbI3 perovskites with graphene oxides for photodetector applications. Chem Commun 51(47):9659–9661. https://doi.org/10.1039/c5cc02282g

36. Huang JH, Li KC, Kekuda D, Padhy HH, Lin HC, Ho KC, Chu CW (2010) Efficient bilayer polymer solar cells possessing planar mixed-heterojunction structures. J Mater Chem 20(16):3295–3300. https://doi.org/10.1039/b924147g

37. Sharma N, Mohan C, Negi S, Verma AS, Gupta SK (2018) C60 Concentration Influence on MEH-PPV: C60 Bulk Heterojunction-Based Schottky Devices. 47(12):7023–7033. https://doi.org/10.1007/s11664-018-6629-3

38. Sharma N, Singh Negi CM, Sharma M, SinghVerma A, Gupta SK (2019) A comparative analysis of the optoelectronic performance of conventional and inverted design organic photodetectors. Opt Mater (Amst) 95(July):109273. https://doi.org/10.1016/j.optmat.2019.109273

39. Facchetti A (2013) Polymer donor-polymer acceptor (all-polymer) solar cells. Mater Today 16(4):123–132. https://doi.org/10.1016/j.mattod.2013.04.005

40. Steim R, Kogler FR, Brabec CJ (2010) Interface materials for organic solar cells. J Mater Chem 20(13):2499–2512. https://doi.org/10.1039/b921624c

41. Alemu D, Wei HY, Ho KC, Chu CW (2012) Highly conductive PEDOT:PSS electrode by simple film treatment with methanol for ITO-free polymer solar cells. Energy Environ Sci 5(11):9662–9671. https://doi.org/10.1039/c2ee22595f

42. Tsai K-H, Huang J-S, Liu M-Y, Chao C-H, Lee C-Y, Hung S-C, Lin C-F (2009) High efficiency flexible polymer solar cells based on PET substrates with a nonannealing active layer. J Electrochem Soc 156(10):B1188. https://doi.org/10.1149/1.3184341

43. Jalali B, Paniccia M, Reed G, Fathpour S (2006) Silicon photonics. J Light Technol 24(June):58–68. https://doi.org/10.1109/JLT.2006.885782
44. Pospischil A, Humer M, Furchi MM, Bachmann D, Guider R, Fromherz T, Mueller T (2013) CMOS-compatible graphene photodetector covering all optical communication bands. Nat Photonics 7(11):892–896. https://doi.org/10.1038/nphoton.2013.240
45. Li C, Huang W, Gao L, Wang H, Hu L, Chen T, Zhang H (2020) Recent advances in solution-processed photodetectors based on inorganic and hybrid photo-active materials. Nanoscale 12(4):2201–2227. https://doi.org/10.1039/c9nr07799e
46. Lim BT, Kang I, Kim CM, Kim SY, Kwon SK, Kim YH, Chung DS (2014) Solution-processed high-performance photodetector based on a new triisopropylsilylethynyl anthracene derivative. Org Electron Phys Mater Appl 15(8):1856–1861. https://doi.org/10.1016/j.orgel.2014.04.017
47. Borchardt JK (2004) Developments in organic displays. Mater Today 7(9):42–46. https://doi.org/10.1016/S1369-7021(04)00401-8
48. Jung GY, Yates A, Samuel IDW, Petty MC (2001) Lifetime studies of light-emitting diode structures incorporating polymeric Langmuir—Blodgett films. Current 1–10. https://doi.org/10.1016/S0928-4931(01)00202-8
49. Phuse S (2017) Influence of extraction methods using different solvents on caesalpinia pulcherrima leaves. Influence of Extraction Methods Using Different Solvents. https://doi.org/10.22376/ijpbs.2017.8.2.b829-837
50. R HY, S JA, B ML, S KS, Prmod B (2017) Application of magnetic stirrer for influencing extraction method on tectona grandis as analgesic activity application of magnetic stirrer for influencing extraction method on tectona grandis as analgesic activity. https://doi.org/10.13140/RG.2.2.23134.74561
51. Yimsiri P, MacKley MR (2006) Spin and dip coating of light-emitting polymer solutions: matching experiment with modelling. Chem Eng Sci 61(11):3496–3505. https://doi.org/10.1016/j.ces.2005.12.018
52. Tyona MD (2013) A theoritical study on spin coating technique. Adv Mater Res 2(4):195–208. https://doi.org/10.12989/amr.2013.2.4.195
53. Cox ME, Mangels JI (1976) Improved chamber for the isolation of anaerobic microorganisms 4(1):40–45
54. Li G, Shrotriya V, Huang J, Yao Y, Moriarty T, Emery K, Yang Y (2005) High-efficiency solution processable polymer photovoltaic cells by self-organization of polymer blends. Nat Mater 4(11):864–868. https://doi.org/10.1038/nmat1500
55. Azeri Ö, Aktas E, Istanbulluoglu C, Hacioglu SO, Cevher SC, Toppare L, Cirpan A (2017) Efficient benzodithiophene and thienopyrroledione containing random polymers as components for organic solar cells. Polymer (Guildf) 133:60–67. https://doi.org/10.1016/j.polymer.2017.11.024
56. Ma Z, Wang E, Jarvid ME, Henriksson P, Inganäs O, Zhang F, Andersson MR (2012) Synthesis and characterization of benzodithiophene-isoindigo polymers for solar cells. J Mater Chem 22(5):2306–2314. https://doi.org/10.1039/c1jm14940g
57. Sharma N, Gupta SK, Negi CMS (2020) New insights into the impact of graphene oxide incorporation on molecular ordering and photophysical properties of PTB7:C70 blends. J Mater Sci Mater Electron 31(24):22274–22283. https://doi.org/10.1007/s10854-020-04728-2
58. Sariciftci NS, Smilowitz L, Heeger AJ, Wudi F (1988) Photoinduced electron transfer from a conducting polymer to buckminsterfullerene

Recent Development and Future Prospects of Rigid and Flexible Dye-Sensitized Solar Cell: A Review

Salam Surjit Singh and Biraj Shougaijam

Abstract The demand for solar-powered portable, wearable, lightweight and flexible electronic devices is increasing in the market. Hence, the development of flexible, lightweight and reliable solar cells is required to meet the market demand. Dye-sensitized solar cells (DSSCs) may be an alternative to fulfill this demand. The reported maximum power conversion efficiency (PCE) of DSSCs is ~14.1% only. Hence, there is a huge scope for increasing the PCE of DSSCs by using different nanostructure designs and materials in various layers of DSSCs. So, an extensive review of the available literature is done on recently developed fabrication and material synthesis techniques of various layers used in DSSCs for enhancing efficiency and durability. Again, the importance of using metal nanoparticles along with metal-oxide nanostructures as photoanode for enhancing light absorption and charge transport is also discussed in detail. Furthermore, the challenges currently faced by researchers in developing Flexible DSSCs (FDSSCs) are also addressed. Therefore, the main objective of this book chapter is to discuss the different materials and synthesis techniques for developing a novel photoanode layer. Another focus is to find out different synthesis techniques developed for the counter electrode (CE), electrolytes and dye layers for designing highly efficient rigid and FDSSCs.

Keywords DSSCs · Flexible · Metal nanoparticle · Metal-oxide · Photoanode

1 Introduction

People tend to use more portable devices like cellular mobile phones, flexible devices, electronic gadgets, etc., due to their lightweight, convenient and less energy consumption. Moreover, many researchers are trying to replace fossil energy with renewable energy sources. Hence, the development of flexible, lightweight and reliable solar cells will be exciting and may fulfill the market demand for the applications like e-vehicles, mobile phones, integration of flexible solar cells in housing design, charging

S. S. Singh · B. Shougaijam (✉)
Department of Electronics and Communication Engineering, Manipur Technical University, Takyetpat, Imphal 795004, India

portable devices and consumer appliances. Fossil fuels had been used for the past many decades as the main source of energy to supply electricity by burning it out. This leads to serious environmental issues such as pollutions, global warming, and the greenhouse effect. In this current era, fossil energy is continuously restoring by renewable energy sources like solar cell technology in e-vehicles, electronic devices, etc., to develop a pollution-free and better environment. The first solar cell was introduced in the early twentieth century, which was made up of Silicon and later on second-generation solar cell was developed with 20% more efficient as compared to Si-based solar cells [1]. The manufacturing cost of the second-generation solar cell was high due to the requirements of the high-temperature processing and a restricted number of fabrication techniques. Therefore, many researchers are trying to develop a new form of solar cell which is more cost-effective with better efficiency [2]. Later on, in the early 1990s, the third-generation solar cell was developed by introducing different materials like perovskite, polymer, quantum dot, organic solar cell and dye-sensitized solar cells (DSSCs). Among these third-generation solar cell technology, DSSCs has become research attention since its invention due to simple and comparatively low-cost fabrication as compared to the first and second-generation solar cells. DSSCs were first developed by O' Regan et al. in 1991 which is a third-generation type of solar cells [3, 4]. One of the most interesting things about the DSSCs is the usage of natural dye as a photochemical reagent to enhance the photo conversion efficiency (PCE) of the solar cells. DSSCs are one of the most trending next-generation solar cell technologies which have tremendous attraction worldwide because of their lightweight and economical [5, 6]. The highest reported PCE of DSSCs is ~14.1%, which is lower as compared to that of Si-based solar cells ~26.7% [7].

So, there is a huge potential to increase the efficiency of DSSCs having unique properties like flexibility and lightweight. To enhance the efficiency of DSSCs up to the level of Si-based solar cells, one needs to understand the function of different layers and their importance. DSSCs consists of four parallel structures, as shown in Fig. 1, a transparent electrode which passes the sunlight, mesoporous oxide (TiO_2, SnO_2, ZnO, etc.) layer call photoanode as a medium for photo-excited electrons, dye

Fig. 1 Shows the basic components of DSSCs and working principle

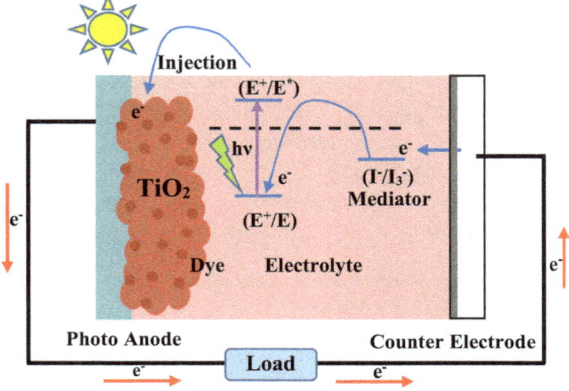

(organic dye, ruthenium dye, N3, natural dye, etc.) for the absorption of photons and transients of the electron, electrolytes (Co(II)/Co(III, I_3^-/I^- redox couples, etc.) as a transfer medium, a counter electrode (CE) (CNT, Graphene, Pt, etc.) for collecting the excited electrons from the external circuit. Among the four structural parts, photoanode plays an important role in increasing the efficiency of DSSCs. In this chapter, we focus on examining various synthesis and design techniques of photoanode and other layers. Researchers were using titanium dioxide (TiO_2), zinc oxide (ZnO), tin dioxide (SnO_2) nanostructures as a photoanode for DSSCs due to the wide bandgap and high electron mobility [8–10]. The maximum PCE of ZnO reported was 8.22%, whereas the maximum PCE of TiO_2 reported was 14.1% [11, 12]. Moreover, TiO_2 provides better performance as compared to ZnO in terms of chemical stability, recombination of dye molecule process, and injection of electrons. It can be synthesized in three forms namely rutile, brookite and anatase. Rutile and anatase are the most commonly used crystalline phases of the DSSCs application due to their high photocatalytic and stable nature. TiO_2 nanostructure is the most widely used photoanode in DSSCs due to its unique properties like thermal stability, high surface-to-volume ratio, excellent dye adsorbing capability and non-toxicity [13, 14].

It is well-known fact that the solar spectrum reaching the earth's surface has the highest intensity of the spectrum in the visible region. To increase the efficiency of a solar cell, different modification processes on the photoanode part were adopted to enhance the absorption of light in the visible region, like modification of nanostructure, decorating the nanowires with novel metal nanoparticles [15–17]. TiO_2 nanostructures can be synthesized using different techniques like screen printing [18, 19], spin coating [20], sol–gel [21], hydrothermal/solvothermal [22, 23], electrochemical anodization [24], electrospinning [25, 26], spray pyrolysis [27] and atomic layer deposition (ALD) [28]. Meidan Y. et al. reported that vertically oriented TiO_2 nanotube fabricated using a hydrothermal process as photoanode gives higher PCE as compared to pure TiO_2 thin film-based DSSCs [29]. Physical vapor deposition (PVD) provides an opportunity to deposit a wide range of materials and also gives an opportunity of depositing various nanostructures with large active surface areas and multiple scattering of light as compared to the thin film. However, to the best of the authors' knowledge, only a few reports are available in the literature for developing photoanode using the PVD process for the application of DSSCs. Therefore, a wide research avenue remains open for researchers to developed photoanode for DSSCs using this vacuum evaporation technique.

In this book chapter, every layer of DSSCs with the latest technology development is discussed. Further, we focus on various techniques for synthesizing photoanode to enhance the incident light to current efficiency (IPCE) of DSSCs. Moreover, recent advances in developing metal nanoparticles assisted metal oxide nanostructures photoanode for both rigid and FDSSCs are also discussed in detail. Also, issues related to the fabrication of FDSSCs are addressed. Finally, the review work is summarized and future prospects are highlighted for developing highly efficient rigid and FDSSCs.

2   Operational Principle

The basic working principle of DSSCs is manifested in Fig. 1. When a photon of light strikes the surface of the solar cell, the dye molecule gets excited to an energy level and releases excited electrons into the mesoporous TiO_2 nanostructures. This injected electron will diffuse to an external circuit through the electrode. Due to the transfer of an exciting free electrons from the TiO_2 photoanode to CE, some voltage potential is produced which can be stored in the battery. If more electrons transfer takes place between the anode and cathode of DSSCs then the rate of PCE will be increased. Herein, the whole process of electron generation inside the DSSCs is discussed to understand the working principle. Initially, dye molecules attached to the surface of TiO_2 nanoparticle adsorb the incident photons from the sunlight, the dye molecule is at a ground state position (E) when a photon of light falls on the cell and the electrons from the dye molecule jump from a ground state (E°) to a higher energy state (E^*) (Eq. 1) [30].

$$E^\circ + h\vartheta \rightarrow E^* \tag{1}$$

$$E^* \rightarrow E^+ + e^- (TiO_2) \tag{2}$$

$$2E^* + 3I^- \rightarrow 2E + I_3^- \tag{3}$$

$$I_3^- + 2e^- (Pt) \rightarrow 3I^- \tag{4}$$

The excited molecule E^* gets oxidized after injecting the excited electron into the conduction band (CB) of TiO_2 nanoparticles (Eq. 2). Finally, the excited electrons start transferring to the CE through an external load after diffusing the electrons from TiO_2 nanoparticles and also from the transparent conducting oxide layer (TCO). Further, the oxidized molecule (E^+) is regenerated due to the acceptance of electrons from the iodide ion in the redox electrolyte (Eq. 3). As a continuing process I^- is oxidizing to triiodide due to a concentration gradient. Hence, electrolyte forms an interface between the CE and photoanode which helps in increasing the recombination of ions (Eq. 4).

3   Modules of Dye-Sensitized Solar Cells

3.1   Recent Advances in Counter Electrodes (CE)

Counter electrode (CE) plays an important role in the regeneration of dye molecules in reducing I_3^- to iodide ion (I^-) from redox electrolytes after getting the electrons from the external circuit. In general, the CE should have robust stability, excellent transmittance, high catalytic activity, high conductivity and reflectivity, low charge transfer resistance, electrochemical and mechanical stability [31–33]. This electrode acts as a catalyst that increases the electron flow from the external circuit thereby reducing the oxidized redox couple at the electrolyte. The counter electrode is classified as composite-based, metal and carbon-based electrodes. Among them, Carbon (C) and Platinum (Pt) are commonly used CE in liquid and solid-electrolyte due to their high electro-catalyzing nature. Pt gives more PCE as compared to Carbon. However, the instability nature of the redox electrolytes and the high cost of Pt is a major challenge. So, the utilization of graphene film, carbon, and graphite sheet is an alternative way to overcome these issues. These electrodes can be prepared using different methods such as chemical vapor deposition, electrochemical deposition, thermal decomposition, etc. [34, 35]. The structure, morphology and particle size of the electrode materials are varied due to the implementation of different synthesis techniques for making the electrodes, which in turn affects the electrocatalytic activity of the solar cells. Kazuhiro S. et al. reported that by simply modifying the electrode designed of Pt CE through a one-step dipping process, the PCE of the solar cell can be enhanced [36]. Recently, materials like graphene, carbon nanostructures, carbon nanotubes (CNT), conducting polymer and transition metal compounds like sulfide and nitride are used as a CE because of their better catalytic properties, less expensive and simple fabrication process. In another study, the CE designed using graphene and CNT for DSSCs gives a promising efficiency of 7.19% and 7.75%, respectively [37]. Another type of promising CE material is transparent conducting oxide (TCO) materials like fluorine-doped tin oxide (FTO), indium tin oxide (ITO), etc. [38, 39]. Generally, TCO allows the incident photons to enter the cells which helps in charge transportation. Subrata S. et al. investigated four different TCOs specifically aluminum-doped zinc oxide (AZO), gallium doped zinc oxide (GZO), fluorine-doped tin oxide (FTO) and indium tin oxide (ITO) which are used as a substrate for DSSCs. GZO and AZO face a problem of Pt loading due to chemical reduction with the electrolytes. Hence, FTO is the most commonly used TCO substrates for photoanode and CEs due to good electrical conductivity, good adhesion, high chemical and good thermal stability up to 600 °C [40]. In recent years, researchers are also developing conducting polymer as CE such as PEN and PET which have flexible and lightweight properties [41, 42]. However, such materials have a low melting temperature of ~150 °C resulting in poor adhesion of the material which limits wider application. Although, the PET and PEN conducting polymer substrate can be used for the fabrication of flexible DSSCs by electrospray, screen printing, spin coating, electrodeposition and atomic layer deposition which are low-temperature techniques [43–46]. Pringle et al. reported the successfully deposition of PEDOT on ITO coated PEN substrate by electrodeposition technique. Moreover, the device achieved an efficiency of 8% which is compared to that of Pt coated FTO electrode-based device [47]. On the other hand, economical metals like Co, Al, Cu and Ni can also be used as metal substrates for making counter electrodes which reduce the internal resistance [48]. However, the efficiency of the device is less as compared to Pt coated CE due to corrosive nature of the metal after reacting with

Table 1 A comparative efficiency of DSSCs using different counter electrode

CE materials	Substrate	Redox couple	Dye	PCE/%	References
Carbon black	FTO	I_3^-/I^-	N719	9.10	[49]
PEDOT	–	I_3^-/I^-	N719	7.93	[50]
CoSe$_2$	FTO	–	N719	10.17	[51]
Pt/TiC	FTO	–	N719	7.68	[52]
Pt	FTO	I_3^-/I^-	N719	3.82	[53]
SWCNTs	FTO	I_3^-/I^-	N719	7.81	[54]

the redox electrolyte medium. A comparative efficiency of DSSCs using different counter electrodes is shown in Table 1.

3.2 Recent Advances in Electrolytes

Electrolytes work as a medium for regenerating the dye molecule in DSSCs bypassing the charge from counter electrodes. This excited electron is transferred from the photoanode via dye molecule through an external circuit. It can be of three types: liquid electrolyte (organic solvent and ionic liquid), quasi-solid gel, and solid-state electrolyte. Among the electrolytes, the liquid electrolyte is most commonly used due to its stable nature, a good solvent for redox couple and more charge generation between the electrolytes and electrode, making the DSSCs more efficient. It is made by mixing either acetonitrile or ethylene carbonate. But, such DSSCs fabricated using liquid electrolyte required to seal it properly because of the higher vapor pressures and volatilely of acetonitrile (solvent) [55]. An alternative is the use of 3-methoxypropionitrile (MNP) which has long-term durability with a less volatile solvent. Therefore, MNP is commonly used solvent electrolytes which gives high efficiency [56]. It may be mention that the liquid electrolyte may be classified as an organic and ionic solvent. So, the ionic solvent is also an alternative electrolyte. This ionic solvent has high chemical and thermally stability, non-flammability and low vapor pressure [57]. The stability and the ability to redevelop the oxidized molecule is enhanced by introducing the iodide-triiodide (I_3^-/I^-) between the CE and photoanode which enhances the overall performance of the cell. Hence, the typical electrolyte must have low viscosity and good solvent properties for the redox couple. An ideal redox electrolyte should have less negative redox potential than that of the oxidized dye molecules, slow electron recombination, high rate of electron transfers and good photochemical stability [58]. But due to the complex chemistry and corrosive nature of the I_3^-/I^- many researchers are looking for an alternative way to replace this redox electrolyte with iodine-free redox couple. Another issue of liquid electrolytes is the problem of evaporation, leakage, photo-degradation and instability that reduce the efficiency of the DSSCs after long-term operation. To overcome these issues, the volatile solvent in the liquid electrolyte is replaced with low viscosity ionic electrolytes. Another alternative is the employment of solid-electrolyte which can be used as an electrolyte. This alternative partially solves the problem of volatilization. But due to high resistance and low charge mobility, the solid electrolyte is hardly preferred for DSSCs. On the other hand, quasi-solid gel and liquid electrolytes are often used to enhance PCE for both DSSCs and FDSSCs. Even though this semisolid state gives less efficiency as compared to a liquid electrolyte, it can be used as an alternative way to overcome the stability and sealing problem. Depending on the electrolyte used there is an impact on the open-circuit voltage (V_{oc}), short circuit current (J_{sc}), and fill factor (FF) which affect the efficiency of DSSCs [59]. Qingjiang Y. et al. reported a remarkable efficiency of DSSCs ~11% using I_3^-/I^- as an electrolyte [59]. In another case study, Yella A. et al. confirmed that the DSSCs incorporated with Co (II/III) in the redox electrolyte can produce an interesting efficiency of ~12.3% [60]. It is also reported that bromine (Br^-/Br^2), iodine (I_3^-/I^-), and hydroquinone can be utilized as an electrolyte for DSSCs application [61, 62]. But at room temperature iodide triiodide electrolyte (I_3^-/I^-) gives better results as a liquid electrolyte in terms of efficiency because of its fast-oxidizing nature of I^- and slow reduction of I_3^- at the interface between photoanode and counter electrode, respectively.

3.3 Recent Advances in Sensitizers

The major role of the dye-sensitizer is to inject the photo-excited electrons into the mesoporous oxide film which is generated after the dye molecule absorbs the incident photon. A typical dye-sensitizer must have deep absorbance in the visible regions, excessive absorption or adsorption of dye molecules onto the photoanode surface. A dye can be classified as natural, metal complexes and organic dyes which are used for DSSCs application [63–65]. Natural dyes can easily prepare using minimal chemical procedures from naturally available plants which are eco-friendly, economical, easy of availability in abundance and extremely cheaper [63]. It can also be extracted through the process of a squeeze, grind filtration, and purification. The extracted pigment from raspberries, purple cabbage, pomegranate, purple grape skin, dragon fruit and mulberry can be employed as a sensitizer. Most of the commonly used dye has an effective charge injection of electrons at the visible and near-infrared region.

However, N3 (Ruthenium 535) and N719 (Ruthenium 535- bis TBA), which are known as an organic sensitizer, extends the absorption of photons in the ultraviolet (UV) region [66]. These chemical dyes are more preferred in DSSCs due to their excellent absorption of photons which provides high conversion efficiency as compared to other organic sensitizers. Nazeeruddin M. K. et al. reported that the DSSCs made up of Ru (II) dye gives better efficiency due to their higher absorption rate in the UV and visible range [68]. Also, Jacqueline M. C. et al. reported an interesting DSSCs efficiency (η) of 11.2% and 10.0% using N719 and N3 under ~1 sun intensity, respectively [66]. The comparative efficiency of DSSCs using various organic dyes is illustrated in Table 2. Ruthenium (II) dyes become a more preferred dye compared to commonly used dyes like porphyrin, natural dye, organic dye etc.,

Table 2 A comparative efficiency of DSSCs using different dyes

Dye	V_{oc} (V)	J_{sc} (mA cm^{-2})	FF	Efficiency (η) (%)	References
N719	0.846	17.73	0.75	11.2	[66]
N3	0.720	18.20	0.73	10.0	[66]
Porphyrin based	0.745	1.5	0.82	9.5	[14]
Natural dye	0.507	0.491	0.60	0.15	[67]

despite the cost factor. One of the most important advantages of Ru (II) based dyes is the metal–ligand charge transfer mechanism which allows injection of charge into the photoanode making the device more efficient [69]. However, the used of organic sensitizer have an impact on the environment due to its toxic nature because of the presence of heavy metal in Ruthenium complexes. Moreover, Ruthenium complexes can under goes degradation in presence of water [70]. Therefore, researchers are looking toward the natural dyes as an alternative, which is more environmentally friendly and cost-effective.

3.4 Recent Advances in Photoanode

Proper designing of the photoanode helps in enhancing the overall efficiency of DSSCs. Photoanode generates the free electrons from the incident photons after reacting with dye molecules. Ideal photoanode should have high dye loading capability with high resistance to photo corrosion, large surface areas, and fast transfer of photo-excited electrons from the dye to the mesoporous oxide layer. Among the metal oxides, TiO$_2$ and ZnO are the most commonly used photoanode material by researchers for the last two decades. TiO$_2$ has a wide bandgap of ~3.2 eV which restricts the absorption of light in the visible spectrum. ZnO also has a similar bandgap of ~3.3 eV with higher electron mobility and more excitation binding energy against photo corrosion [71]. However, ZnO gives less efficiency as compared to TiO$_2$ due to its poor corrosion resistance from dye and electrolytes. Therefore, TiO$_2$ is superior in terms of stability and photocatalytic nature as a photoanode for DSSCs application.

3.4.1 Zinc Oxide (ZnO)

ZnO has been used as photoanode material for DSSCs application due to large excitation binding energy against photo corrosion and higher electron diffusivity. It can be synthesized as nanotubes, hollow spheres, nanowire and nanoflower. Few research papers based on ZnO as a photoanode for DSSCs application are discussed in this section. Sanjay P. et al. reported the analysis of DSSCs using a natural dye with ZnO nanorods synthesized using hydrothermal method, which gives maximum efficiency of 4.62% [72]. Mian En Yeoh et al. reported a synthesis technique for

Table 3 Performance of ZnO photoanode versus photovoltaic properties of DSSCs [75–77]

Photoanode	η (%)	V_{oc} (mV)	I_{sc} (mA cm^{-2})	FF (%)	References
ZnO-NWs nanoforest	2.63	680	8.78	53	[75]
ZnO nanofibers	1.34	600	3.58	62	[76]
ZnO-NRs	0.66	580	2.8	40.4	[77]

ZnO as blocking layers using a sol–gel spin-coating technique, which decreases the electron recombination at electrolytes giving out an interesting PCE of 4.34% [73]. Again, Kumara G. R. A. et al. achieved a PCE of 5.2% by spray pyrolysis technique forming a dense layer of ZnO followed by a mesoporous layer of ZnO nanoparticle on FTO substrate [74]. The performance of ZnO photoanode versus photovoltaic properties of DSSCs is shown in Table 3.

The demand for low-temperature growth process for developing FDSSCs is increasing day-by-day. An attempt was made by Ahmad U. et al. for the synthesis of ZnO nanoflowers (NF) on FTO substrate using a low-temperature process which gives an interesting efficiency of ~1.40% [78]. In another study, Periyayya U. et al. presented the development of different morphologies of ZnO nanostructures synthesized by a hydrothermal process using Zn (OH)$_2$ solution to enhance the performance of DSSCs [79]. Therefore, further research may be carried out to enhance the PCE of DSSCs using modified ZnO nanostructures.

3.4.2 Titanium Dioxide (TiO$_2$)

TiO$_2$ is known for its unique properties like chemical stability, non-toxic to the environment, low-cost, and good photo-catalyst. Because of these unique properties, TiO$_2$ nanostructures are being widely employed as photoanode since the invention of DSSCs [12]. However, continuous research is going on to enhance the photo conversion activity by developing different geometrical nanostructures of TiO$_2$. In recent studies, nanotubes [80], nanorods (NRs) [81, 82], nanoparticles (NPs) [83], nanowires (NWs) [84], nanoflowers (NFs) [85], and hollow hemispheres [86] were fabricated on FTO/ITO substrates for DSSCs application. These TiO$_2$ nanostructures were synthesized using various techniques to slow down the recombination of electrons. Further, the dye molecules absorption is also improved by modifying the surface morphology of the TiO$_2$ nanostructures. The comparisons of different photoanode nanostructures with their corresponding parameter for DSSCs are shown in Table 4 [87–90]. It is noteworthy to mention that vertically aligned nanostructures like NRs and NWs, offer faster and more direct electron pathways to the substrate [91–94]. Moreover, a large surface-to-volume ratio enhances photon absorption by reducing scatterings of light [95, 96]. These modifications lead to an improvement in the overall performance of the DSSCs.

In recent years, many researchers are also trying to develop DSSCs on flexible substrate. However, photoanode deposited on the ITO film layer develops cracks and

Table 4 Performance of photoanode versus photovoltaic properties of DSSCs [87–90]

Photoanode	η (%)	V_{oc} (mV)	I_{sc} (mA cm^{-2})	FF (%)	References
TiO$_2$-NPs	11.28	846	17.73	75	[87]
TiO$_2$-NRs	4.35	740	8.41	69	[88]
TiO$_2$-NRs/TiO$_2$-NPs	7.1	756	4.45	65	[88]
SnO$_2$ hollow spheres	6.02	765	14.59	54	[89]
TiO$_2$-NTs	4.32	670	13.48	48	[90]

spalling off under repeated bending of the substrate which is a major issue of developing FDSSCs [97]. It was also observed that the efficiency of FDSSCs decreases with repeated bending of ITO coated flexible substrates due to morphological changes on the surface of TiO$_2$ nanostructures [97]. The change in surface morphology of TiO$_2$ photoanode after bending several times under different angles is illustrated in Fig. 2a before bending (b–d) after bending multiple times. Hence, a less stable nature is observed in FDSSCs as compared to normal DSSCs. On the other hand, recently, Xing G. et al. reported the designing of highly efficient FDSSCs by using CuS mesh film as a CE on the ITO coated substrate which gives out PCE to be ~4.4% [98]. More challenges and opportunities are also discussed in last section of this book

Fig. 2 Morphologies of TiO$_2$ photoanode at a different bending angle for FDSSCs **a** original state before bending, **b** cracking R = 12 mm at 10,000 cycles, **c** cracking R = 9 mm at 10,000 cycles, **d** cracking R = 8 mm at 5800 cycles, **e** spalling R = 8 mm at 5800 cycles, **f** spalling R = 7 mm at 4400 cycles. Reprinted by permission from Elsevier: Ref. [97], copyright 2015

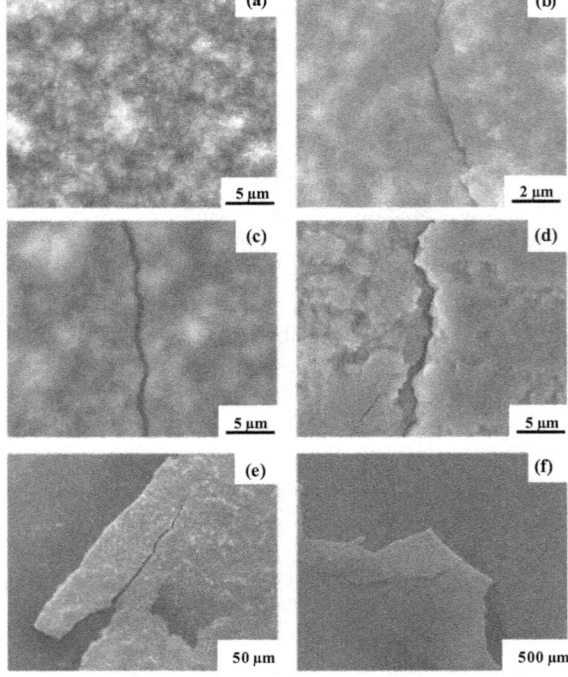

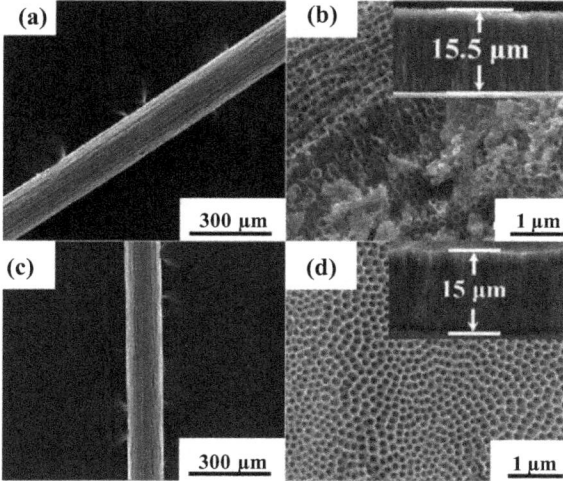

Fig. 3 **a** Shows cross-sectional SEM image of TiO_2 nanotube arrays of single nanotube array after 10 h of first anodization process, **b** SEM image (top view) of nanotubes at different lengths and surface contaminants by TiO_2 due to chemical dissolution, **c** SEM image of single nanotube array after '10 h of second anodization process, and **d** shows the top view image of TiO_2 nanotube arrays with uniform alignment and clean surface. Reprinted by permission from Elsevier: Ref. [99], copyright 2013

chapter. Further, the most commonly used photoanode based on various TiO_2 nanostructures and uniquely modification of TiO_2 nanostructure by incorporating metal nanoparticles are also discussed in the subsequent section.

TiO_2 Nanostructures

Nanorods (NR) provide a direct conductive pathway for the transfer of electrons which increases the charge collection efficiency of DSSCs. In a recent study, Jia Liang et al. reported that hierarchical TiO_2 nanotubes arrays were prepared by anodization process, thereby increasing the surface area, which intern enhances the PCE as compared to traditional photoanode [99]. The hierarchical structure of TiO_2 nanotube arrays is shown in Fig. 3.

Again, Chang K. et al. achieved a PCE of ~6.1% using the composite structure of NRs and NPs photoanode using electro-spinning technique which enhances the transport of electrons [100]. Scanning Electron Microscope (SEM) image of TiO_2 nanorod at different sintering temperatures is illustrated in Fig. 4. Table 5 summarized different TiO_2 photoanode structures developed by various deposition techniques like PVD, spin coating and sputtering, etc. [101–109]. Hsiao Y. et al. reported the deposition of TiO_2 nano-columnar structure on the ITO substrate by glancing angle deposition (GLAD) techniques. Here, the thin film TiO_2 photoanode was deposited at

Fig. 4 **a** Shows SEM images of TiO$_2$ nanorods (NR) sintering at 450 °C for 30 min composed of single crystallites with size less than 15 nm, **b** SEM image of the TiO$_2$ NP/NR composite electrode sintered at 450 °C reveal randomly oriented interconnecting NRs, **c** shows cross-sectional SEM image of a TiO$_2$ NP/NR composite electrode with 7 μm thickness dense TiO$_2$ NP, **d** TiO$_2$ NR sintering at 850 °C composed of the nanocrystalline size of 30–50 nm, **e** S SEM images NP/NR composite film with small particles randomly oriented in NR arrays and **f** SEM image reveal the thickness of the TiO$_2$ NP/NR composite electrode is 5.5–6 μm. Reprinted by permission from Elsevier: Ref. [100], copyright 2014

a different incidence angle ranging from 53° to 86°. The porous TiO$_2$ film deposited at 73° gives out the highest PCE of 2.78% due to more dye absorption at this particular glancing angle [105]. On the other hand, Bo-Lei et al. reported that TiO$_2$ NWs-NPs were synthesized using an inverted fabrication process as photoanode for FDSSC, which gives a PCE of 2.7%. This TiO$_2$ nanostructure was deposited by D.C magnetron sputtering and further annealing at 500 °C to enhance the efficiency of the FDSSCs [107].

On similar research reported by Lijian M. et al., TiO$_2$ NR arrays were deposited by DC reactive magnetron sputtering on the ITO substrate. These NRs are highly order perpendicular structures to the substrates and can be employed as a photoanode for DSSCs application. Due to this well-aligned TiO$_2$ nanostructure PCE of ~4.78% is achieved. And, the diameter of the TiO$_2$ NRs-arrays can be adjusted by changing the

Table 5 Summary of physical processes employed for development of TiO_2 photoanode for DSSCs application

Sr. No.	Photoanode	Technique	Remarks	References
1	TiO_2	Electron beam PVD	TiO_2 layer on Ti electrode	[101]
2	TiO_2	Electron beam PVD	Compact TiO_2 layer between FTO glass and TiO_2 porous layer	[102]
3	TiO_2	Electron beam PVD	Porous film of TiO_2	[103]
4	TiO_2	Electron beam PVD	NanoporousTiO_2 films (6 μm thick)	[104]
5	TiO_2	Electron beam PVD	Nano columnar TiO_2 films	[105]
6	TiO_2	Spin coating	TiO_2 nanoparticle thin film	[106]
7	TiO_2	Sputtering	TiO_2 nanowires/nanoparticles hybrid DSSCs	[107]
8	TiO_2	DC magnetron sputtering	TiO_2 nanotube array on conducting substrate	[108]
9	TiO_2	DC magnetron sputtering	TiO_2 nanorods array	[109]

distance between the target and the substrate [109]. W. H. Jung et al. prepared a highly porous TiO_2 nanofiber by using the electrospinning process. They have pointed out that different spinning voltage affects the thickness of the TiO_2 nanowire. Figure 5 manifests the SEM images of TiO_2 nanofiber manufactured at applied voltages of 12 kV, 21 kV, 20 kV, and 18 kV (a–d), respectively. These SEM images reveal that the surface area of the TiO_2 nanofibers form by electrospun are varied by an applied voltage. In another report, the quantity of glycerin added during the electro-spinning affects the porosity of the TiO_2 nanofiber surface. If more amount of glycerin is added during the process, the porosity of TiO_2 nanofiber will increase as illustrated in Fig. 5e–f [110]. Wenwu L. et al. reported the preparation of TiO_2 NWs arrays having a diameter of 80 nm for FDSSCs applications by hydrothermal reaction. Figure 6a–d shows cross-sectional FESEM images of the TiO_2-NW as-synthesized at different reaction duration of 12, 16, 20, and 24 h. Here, TiO_2-NW synthesized were employed as a photoanode for FDSSCs which achieved an overall efficiency of 3.42% [111]. The thickness of TiO_2 film can be mainly controlled through the reaction time of the hydrothermal process. Therefore, the shape and sizes of TiO_2 nanostructure can be adjusted by twinning various parameters like reaction time, catalyst used and spinning voltage, etc. It is also observed that most of the TiO_2 nanostructure developed for DSSCs are not well-aligned to the substrate.

TiO_2 Nanostructure with Nanoparticles Synthesis

One of the most interesting nanostructure modifications of TiO_2 in DSSCs application is the combination of nanostructure with NPs. This unique structure provides

Fig. 5 SEM photographs of TiO$_2$ nanofiber prepared at **a** 12 kV spinning voltage, **b** 21 kV spinning voltage **c** 20 kV spinning voltage **d** 18 kV spinning voltage, and **e** 0.1 g of glycerin and **f** increase porosity at 0.4 g of glycerin. Reprinted by permission from Elsevier: Ref. [110], copyright 2012

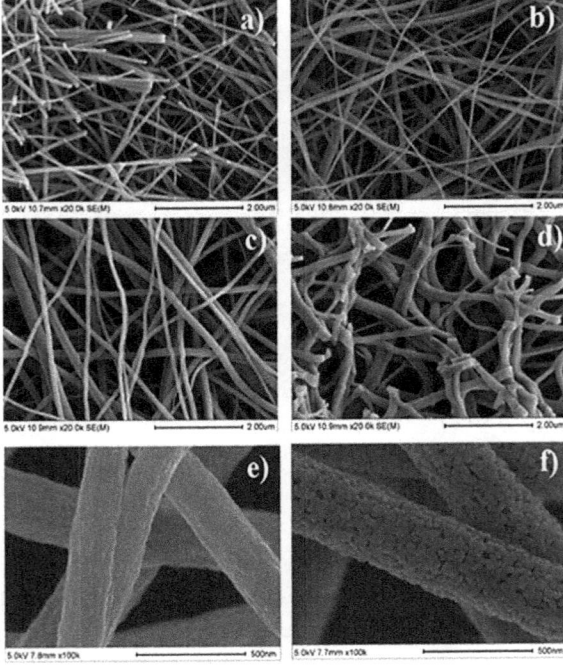

Fig. 6 Cross-sectional FESEM images of the TiO$_2$ nanowires synthesized for different hydrothermal reaction times **a** 12 h, **b** 16 h, **c** 20 h, and **d** 24 h. Reprinted by permission from Elsevier: Ref. [111], copyright 2015

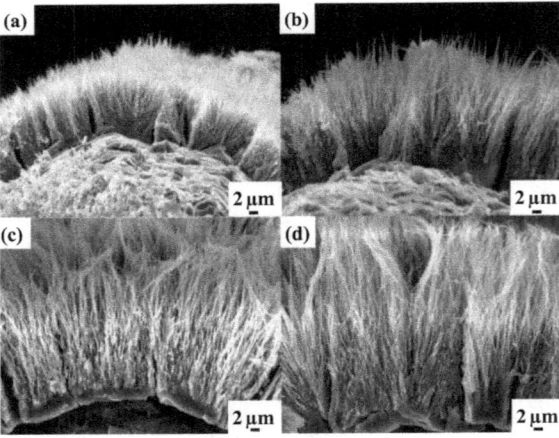

better charge separation, large surface area, and high charge recombination. Further, it can absorb a greater number of dye molecules on the surface of nanoparticles which can increase dye loading. Recently, Liqiang et al. has reported that DSSCs are being designed by fabricating a double-layer photoanode of a nanotube (NT) and nanoparticle (NP) which gives PCE of 6.43% by electrochemical anodization [112].

In another report, Klein et al. worked on a modified TiO_2 photoanode of nanotube and nanoparticle for FDSSCs. They found out that TiO_2 nanoparticles enhanced the photocurrent generation as compared to the normal TiO_2 nanotube. Further, they have reported that the use of PEDOT: PSS film as a CE enhanced the efficiency [113]. Similarly, Yan et al. designed DSSCs which able to achieved 10.1% efficiency by inlaid gold nanoparticles of thickness 2 nm on the surface of TiO_2 photoanode. It also concludes that putting metal nanoparticles can enhance the overall performance of DSSCs as compared to pure TiO_2 photoanode due to plasmon resonance taking place at the interface of TiO_2–Au nanoparticles thereby increasing the dye adsorption [114]. In another study, Prasenjit et al. reported that TiO_2 mesoporous structure embedded with noble metal nanoparticles photoanode gives an interesting PCE of ~7.1% compared to a normal TiO_2 photoanode which gives a PCE of ~5.63%. This enhanced efficiency is due to the plasmonic coupling which enhances the electron injection [115]. Again, Cao et al. reported that the bilayer growth of photoanode (Nanorods-Nanoparticle) significantly impacts the PCE by achieving 7.39%, which enhances the performance of DSSCs [91]. Further, Wan-Yu et al. reported that DSSCs are made up of composite photoanode by sintering TiO_2 (P25) photoelectrode along with Ag nanoparticle which enhances the efficiency from 2.75 to 5.66% [116]. Similarly, Seung et al. reported that Au-NPs were incorporated into the open-ended free-standing TiO_2 nanotube arrays using the electrodeposition technique. The reported efficiency of DSSCs using this technique is 7.12%. This method of incorporating Au-NPs into the vertically aligned TiO_2 nanotube arrays significantly enhanced the photon absorption thereby increasing the flow of electrons which intern enhances the PCE from 5.12 to 7.12% [117]. Yu-H. et al. reported the fabrication of DSSCs using TiO_2 nanofiber with Ag-NPs as photoanode which enhances the efficiency from ~3.99 to ~6.23% by using the sol–gel process technique follow by electrospinning. The increase in the PCE is due to the suppression of electron recombination and an increase in light absorption because of the good conductive nature of Ag-NPs present in the TiO_2 nanofiber. Also, this worked shows good performance of DSSCs even at low intensity of light which may be applicable in various indoor devices [118]. Similarly, Ho-S. et al. demonstrate the fabrication of Ag-NPs and carbon-coated highly ordered TiO_2 nanotube arrays (TNTAs) using the electrochemical deposition technique. The fabricated DSSCs device shows an increase in PCE from 5.23 to 6.14% due to the plasmonic effect and multiple scattering of light inside the highly ordered TNTAs [119]. In another report, TiO_2 nanoparticles were synthesized using both sol–gel and hydrothermal techniques and loaded Ag-NPs on the TiO_2 film using a chemical reduction process. The Ag-NPs loaded TiO_2 film-based device shows enhance efficiency from 7.30 to ~8.82% compare to pure TiO_2 film due to the plasmonic resonance effect thereby increasing the photon absorption at the anode [120]. Even though metal NPs embedded in TiO_2 nanostructures play an important role in enhancing the PCE of DSSCs. It is observed that most of the above work used double fabrication techniques to embed metal NPs on the surface of TiO_2 nanostructures and also employed chemical processes. Therefore, if the metal NPs embedded TiO_2 nanostructures are developed using only a single fabrication technique which may be helpful in designing cheaper photoanode for DSSCs application.

4 Flexible Dye-Sensitized Solar Cells (FDSSCs)

In the modern electronic industry, the demand for portable, eco-friendly, and flexible, renewable energy source devices is exponentially increasing [121, 122]. Further, in the solar cell design industry, the demand for flexible cells is rapidly increasing due to their lightweight and capable of integration in different curve structures. However, the performance, stability, and efficiency of FDSSCs need to be improved for commercial production. The flexible substrate like PET and PEN became popular due to an engineering design problem of a rigid substrate of solar cells. This flexible substrate gives design ease to modern architects while designing an e-car, wearable solar cell, and smart house integrated with solar photovoltaic cells on the rooftop compared to the rough, heavy, and breakable substrate. However, FDSSCs have low PCE and poor stability in nature. Many researchers are trying to implement FDSSCs with different fabrication techniques for enhancing the overall efficiency of the solar cells. But the main challenge is to make the efficiency of the FDSSCs stable even after repeated bending of the substrate. In order to achieve robust FDSSCs, different fabrication techniques have been implemented like dry-spray deposition, electrophoretic deposition and electrospray deposition for designing photoanode [123, 124]. But, most of the fabricated nanomaterials are in the form of a thin film or mesoporous structure. Also, the choice of types of substrates is very important because it will decide the nature of the DSSCs whether it will be rigid or flexible. If the metal-oxide coated glass is used as a substrate, then the DSSCs will be rigid and when the plastic substrates like PET or PEN are used, then the DSSCs will be flexible DSSCs. Substrates like ITO coated PET or PEN, or FTO coated PET or PEN substrates are mainly used for FDSSCs due to high conductivity and flexibility.

Recently, Qiuchen H. et al. reported that the fabrication of flexible counter electrode CE by single-layer and double-layer using silver (Ag) and copper (Cu) NPs. The DSSCs made up of this composite CE achieved an efficiency of 4.39%, which was deposited by drop-casting of Ag NWs on the polymer substrate [125]. On another report, Hsiu-P. J. et al. reported that FDSSCs made of a single-cell module and parallel modules achieve an efficiency of 5.40% and 4.77%, respectively. Here, TiO_2 nanotube (TNT) arrays were used as a photoanode to fabricate FDSSCs on ITO coated PET substrates. The measured length of TNT was ~35 μm deposited on the flexible substrate by anodization process [126]. Again, Hyun-G. H. et al. have reported a technique that can be used for the fabrication of large-scale production of FDSSCs particularly for flexible indoor devices. These FDSSCs achieved a PCE of ~4.1% using pre-dye coating ultrasonic spraying techniques. Using this method, the amount of dye adsorption on TiO_2 can be easily controlled without applying a high-temperature sintering process [127]. Similarly, Daibing et al. reported that designing flexible DSSSs using vertical TiO_2 nanotubes TNT on the Ti mesh substrate gives an efficiency of ~2.66%. The structures consist of a Pt CE of 100 nm thickness, which was separated by the polytetrafluoroethylene membrane and also a layer of Teflon is used as heat sealed. However, the PCE of DSSCs depends on the thickness of the Teflon and Ti mesh along with the incident angle of the incoming light. The efficiency of FDSSCs retains at an incident angle ranging from 30° to 90°, beyond that the performance of FDSSCs gets reduced [128].

Again, Haijun S. et al. reported a unique fabrication technique of FDSSCs on Ni-PET substrate by electrodeposition and hot-press transferred method [42]. The maximum PCE measured for these Ni-PET-based FDSSCs is ~7.89%, which is 10.4% more as compared to the ITO coated PET substrate-based FDSSCs. Furthermore, the preparation of Ni-PET film gives uniformity and low resistance of ~0.18 Ω of the substrate, which can withstand the I^-/I_3^- electrolytes. This method of fabrication for FDSSCs can also be applied for designing flexible optoelectronic devices [42]. Zhaosheng X. et al. reported the deposition of mesoporous TiO_2 film at a low-temperature environment using atomic layer deposition (ALD) and electrophoretic deposition (EPD) method. The fabricated FDSSCs show an overall efficiency of 1.93%. This low-temperature dense TiO_2 film deposition technique gives a big milestone for designing a FDSSCs device with roll-to-roll fabrication on polymer substrates [46]. In another report, Jia L. et al. developed FDSSCs with a multi-working (MW) electrode using a flexible fiber-type that can harvest light from all directions. They used a special plastic capillary tube to assemble the highly ordered TiO_2 nanotube arrays with Ti microwires and Pt microwires to work as an electrode as shown in Fig. 7a. The performance of a highly ordered multi-working electrode is better than that of single working (SW) electrode, thereby enhancing the PCE

Fig. 7 **a** Multi working electrode inside capillary tube, **b** flexible fiber which contain SW electrode, **c** flexible fiber which contain MW electrode, and **d, e** SW-MW at particular bending angle. Reprinted by permission from Elsevier: Ref. [129], copyright 2015

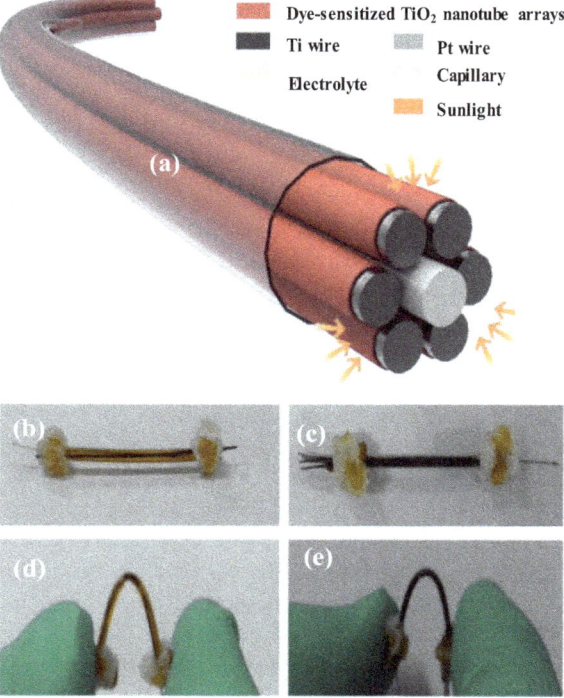

(~6.6%) of DSSCs. Figure 7b, c shows SW and MW flexible fiber types DSSCs at a normal bending angle. Figure 7d, e shows SW and MW flexible fiber types DSSCs at a particular bending angle [129]. In another approach, Yue et al. reported the fabrication of flexible fibrous DSSCs using a multilayer structure of TiO_2 and double-layer Pt. This double-layer CE enhances the electrocatalytic nature thereby reducing the triiodide ion. The FDSSCs shows a remarkable enhance efficiency of 6.35%, which is higher than that of monolayer fiber TiO_2 photoanode [130]. It is observed that no report has been made on designing FDSSCs using vertical metal oxide nanowires as photoanode which may provide advantages like more bending capacity and robustness with higher efficiency. To resolve the issue of reduction of efficiency after repeated bending of the substrate vertically oriented nanowires can be developed on flexible substrates as a possible alternative for FDSSCs applications.

5   Conclusion and Future Scope

In this book chapter, we have thoroughly explained the importance of each layer's presence in DSSCs and the latest technology development for designing photoanode for efficient DSSCs and FDSSCs. Even though DSSCs gives low efficiency compare to Si solar cell, they have unique properties in terms of lightweight and nature friendly. Various advanced fabrication techniques and emerging materials for designing photoanode, organic dye, electrolytes, and the CE to improve the efficiency as well as device stability for DSSCs and FDSSCs are discussed. Further, recent advancements in designing a new type of photoanode by different fabrication techniques such as screen printing, sputtering, doctor-blade, and hydrothermal processes were also explained. It is observed that unique nanostructured photoanode like the CNT, modified TiO_2 nanostructure, TiO_2 nanostructure with the metal nanoparticle, ZnO nanostructure, graphene, etc., gives better performance in terms of efficiency as compared to normal photoanode. In our investigation, it is also observed that limited research papers were available as literature on vertical photoanode for DSSCs application which has unique properties like large surface area and fast electron transport. Therefore, an enormous research scope is still left for designing photoanode for highly efficient DSSCs or FDSSCs. In another observation, most of the literature on DSSCs were designed using FTO and ITO substrate; however, the demand for flexible and lightweight photovoltaic is increasing in the market day-by-dye. Therefore, it is important to develop a highly efficient and flexible photovoltaic solar cell to fulfill the market demand. On the other hand, research is going to developed new solar cells to achieve more than 50% efficiency using advanced fabrication techniques for space and industrial applications. However, developing a solar cell design comparable with conventional Si-based solar cells is underway using various techniques. One of the most promising low-cost and environmentally friendly devices is DSSCs. Even though the efficiency of DSSCs is less compared to Si-based solar cells, the efficiency can be improved since the maximum efficiency reported till now is far less than the maximum theoretical efficiency of DSSCs of ~33%. Therefore, enormous research scope is there to improve the efficiency in the area of designing novel photoanode and developing more environmentally friendly dye materials. Finally, it is very important to understand the overall efficiency, degradation mechanism, and lifetime of each layer to developed stable DSSCs or FDSSCs.

Acknowledgements The authors would like to acknowledge the Department of Electronics and Communication Engineering, Manipur Technical University (MTU), Imphal for providing research facilities.

Funding The authors would like to thank Science and Engineering Research Board (SERB), Department of Science and Technology (DST), Government of India for funding this work under File no: ECR/2018/000834.

References

1. Green MA, Emery K, Hishikawa Y, Warta W, Dunlop ED (2014) Solar cell efficiency tables (version 44). Prog Photovoltaics Res Appl 22(7):701–710
2. Bokalic M, Topic M (2015) Spatially resolved characterization in thin-film photovoltaics. Springer, US, New York
3. O'Regan B, Gratzel M (1991) A low-cost, high-efficiency solar cell based on dye-sensitized colloidal TiO_2 films. Nature 353:737–740
4. Gratzel M (2005) Solar energy conversion by dye-sensitized photovoltaic cells. Inorg Chem 44(20):6841–6851
5. Blakersa A, Zina N, McIntosh KR, Fong K (2013) High-efficiency silicon solar cells. Energy Procedia 33:1–10
6. Green MA (2002) Third-generation photovoltaics: solar cells for 2020 and beyond. Phys E 14(1–2):65–70
7. Green MA, Hishikawa Y, Dunlop ED, Levi DH, Hohl-Ebinger J, Ho-Baillie AW (2018) Solar cell efficiency tables (version 51). Progr Photovolt Res Appl 26:3–12
8. Ullattil SG, Thelappurath AV, Tadka SN, Kavil J, Vijayan BK, Periyat P (2017) A sol-solvothermal processed Black TiO_2 as photoanode material in dye-sensitized solar cells. Sol Energy 155:490–495
9. Wang D, Zhu X, Fang Y, Sun J, Zhang C, Zhang X (2017) Simultaneously composition and interface control for ZnO-based dye-sensitized solar cells with highly enhanced efficiency. Nano-Struct Nano-Objects 10:1–8
10. Li KN, Wang YF, Xu YF, Chen HY, Su CY, Kuang DB (2013) Macroporous SnO_2 synthesized via a template-assisted reflux process for efficient dye-sensitized solar cells. ACS Appl Mater Interfaces 5(11):5105–5111
11. Xie Y, Zhou X, Mi H, Ma J, Yang J, Cheng J (2018) High-efficiency ZnO-based dye-sensitized solar cells with a 1H, 1H, 2H, 2Hperfluorodecyltriethoxysilane chain barrier for cutting on interfacial recombination. Appl Surf Sci 434:1144–1152
12. Kakiage K, Aoyama Y, Yano T, Oya K, Fujisawa J, Hanaya M (2015) Highly-efficient dye-sensitized solar cells with collaborative sensitization by silyl-anchor and carboxy-anchor dyes. Chem Commun 51:15894–15897
13. Ahmad MS, Pandey AK, Rahim NA (2017) Advancements in the development of TiO_2 photoanodes and its fabrication methods for dye-sensitized solar cell (DSSC) applications. Renew Sustain Energy 77:89–108

14. Ye M, Wen X, Wang M, Iocozzia J, Zhang N, Lin C, Lin Z (2015) Recent advances in dye-sensitized solar cells: from photoanodes, sensitizers, and electrolytes to counter electrodes. Elsevier Ltd 18(3):155–162
15. Ni S, Guo S, Wang D, Jiao S, Wang J, Zhang Y, Wang B, Feng P, Zhao L (2019) Modification of TiO_2 nanowire arrays with Sn doping as photoanode for highly efficient dye-sensitized solar cells. Curr Comput-Aided Drug Des 9(2):113
16. Sim YH, Yun MJ, Cha SI, Seo SH, Lee DY (2018) Improvement in energy conversion efficiency by modification of photon distribution within the photoanode of dye-sensitized solar cells. ACS Omega 3(1):698–705
17. Liu C, Lia T, Zhanga Y, Konga T, Zhuang T, Cui Y, Fang M, Zhu W, Wu Z, Li C (2019) Silver nanoparticle modified TiO_2 nanotubes with enhanced the efficiency of dye-sensitized solar cells. Microporous Mesoporous Mater 287:228–233
18. Mariani P, Vesce L, Di Carlo A (2015) The role of printing techniques for large-area dye-sensitized solar cells. Semicond Sci Technol 30:104003
19. Lee H, Hwang D, Jo SM, Kim D, Seo Y, Kim DY (2012) Low-temperature fabrication of TiO_2 electrodes for flexible dye-sensitized solar cells using an electrospray process. ACS Appl Mater Interfaces 4(6):3308–3315
20. Song L, Du P, Shao X, Cao H, Hui Q, Xiong J (2013) Effects of hydrochloric acid treatment of TiO_2 nanoparticles/nanofibers bilayer film on the photovoltaic properties of dye-sensitized solar cells. Mater Res Bull 48(3):978–982
21. Bahramian A (2013) High conversion efficiency of dye-sensitized solar cells based on coral-like TiO_2 nanostructured films: synthesis and physical characterization. Ind Eng Chem 52(42):14837–14846
22. Wu HP, Lan CM, Hu JY, Huang WK, Shiu JW, Lan ZJ, Tsai CM, Su CH, Guang Diau EW (2013) Hybrid Titania photoanodes with a nanostructured multi-layer configuration for highly efficient dye-sensitized solar cells. J Phys Chem Lett 4(9):1570–1577
23. Bao ZQ, Xie H, Zhu Q, Qian J, Ruana P, Zhou X (2013) Microsphere assembly of TiO_2 with tube-in-tube nanostructures: anisotropic etching and photovoltaic enhancement. Cryst Eng Comm 15:8972–8978
24. Mir N, Lee K, Paramasivam I, Schmuki P (2012) Optimizing TiO_2 nanotube top geometry for use in dye-sensitized solar cells. Chem Eur J 18(38):11862–11866
25. Chen HY, Zhang TL, Fan J, Kuang DB, Su CY (2013) Electrospun hierarchical TiO_2 nanorods with high porosity for efficient dye-sensitized solar cells. ACS Appl Mater Interfaces 5(18):9205–9211
26. Kumar EN, Jose R, Archana PS, Vijila C, Yusoffb MM, Ramakrishna S (2012) High performance dye-sensitized solar cells with record open-circuit voltage using tin oxide nanoflowers developed by electrospinning. Energy Environ Sci 5:5401–5407
27. Huo J, Hu Y, Jiang H, Huang W, Li Y, Shao W, Li C (2013) Mixed solvents assisted flame spray pyrolysis synthesis of TiO_2 hierarchically porous hollow spheres for dye-sensitized solar cells. Ind Eng Chem Res 52(32):11029–11035
28. Son HJ, Prasittichai C, Mondloch JE, Luo L, Wu J, Kim DW, Farha OK, Hupp JT (2013) Dye Stabilization and enhanced photoelectrode wettability in water based dye-sensitized solar cells through post-assembly atomic layer deposition of TiO_2. J Am Chem Soc 135(31):11529–11532
29. Ye M, Xin X, Lin C, Lin Z (2011) High-efficiency dye-sensitized solar cells based on hierarchically structured nanotubes. Nano Lett 11(8):3214–3220
30. Sharma K, Sharma V, Sharma (2018) Dye-Sensitized solar cells: fundamentals and current status. NRL 13:381
31. Wu J, Xiao Y, Tang Q, Yue G, Lin J, Huang M, Huang Y, Fan L, Lan Z, Yin S (2012) A large-area light-weight dye-sensitized solar cell based on all titanium substrates with an efficiency of 6.69% outdoors. Adv Mater 24:1884–1888
32. Wu J, Li Y, Tang Q, Yue G, Lin J, Huang M, Meng L (2014) Bifacial dye-sensitized solar cells: a strategy to enhance overall efficiency based on transparent polyaniline electrode. Sci Rep 4:4028

33. Liu D, Zhao M, Li Y, Bian Z, Zhang L, Shang Y, Xia X, Zhang S, Yun D, Liu Z (2012) Solid-state, polymer-based fiber solar cells with carbon nanotube electrodes. ACS Nano 6(12):11027–11034

34. Tang Z, Wun J, Zheng M, Huo J, Lan Z (2013) A microporous platinum counter electrode used in dye-sensitized solar cells. NANO 2(5):622–627

35. Kakroo S, Suran K, Bhattachary B (2019) Counter electrode in polymer-electrolyte-based DSSC: platinum versus electrodeposited MnO_2. Macromol Symp 388(1):1900011

36. Shimada K, Toyoda T Shahiduzzaman Md, Taima T (2019) Platinum counter electrodes for dye-sensitized solar cells prepared by a one-step dipping process. Jpn J Appl Phys 58(12):124001–124004

37. Ouyang J (2019) Applications of carbon nanotubes and graphene for third-generation solar cells and fuel cells. Nano Mater Sci 1(2):77–90

38. Zhang S, Jin J, Li D, Fu Z, Gao S, Cheng S, Yu X, Xiong Y (2019) Increased power conversion efficiency of dye-sensitized solar cells with counter electrodes based on carbon materials. RSC Adv 9:22092–22100

39. Zatirostami A (2020) Electro-deposited SnSe on ITO: a low-cost and high-performance counter electrode for DSSCs. J Alloys Compd 844(5):156151

40. Sarkera S, Seoa HW, Jina YK, Azizb MdA, Kima DM (2019) Transparent conducting oxides and their performance as substrates for counter electrodes of dye-sensitized solar cells. Mater Sci Semicond Process 93:28–35

41. Qing FuN, Xiao XR, Zhou XW, Zhang JB, Lin Y (2012) Electrodeposition of platinum on plastic substrates as counter electrodes for flexible dye-sensitized solar cells. J Phys Chem C 116(4):2850–2857

42. Su H, Zhang M, Chang YH, Zhai P, Hau NY, Huang YT, Liu C, Soh AK, Feng SP (2014) Highly conductive and low cost Ni-PET flexible substrate for efficient dye-sensitized solar cells. ACS Appl Mater Interfaces 6(8):5577–5584

43. Kim SS, Nah YC, Noh YY, Jo J, Kim DY (2005) Electrodeposited Pt for cost-efficient and flexible dye-sensitized solar cell. Electrochim Acta 51(18):3814–3819

44. Popoola IK, Gondal MA, Ghamdi JM, Qahtan TF (2018) Photofabrication of highly transparent platinum counter electrodes at ambient temperature for bifacial dye sensitized solar cells. Sci Rep 8:12864

45. Liu J, Yi L, Yong S, Arumugam S, Beeby S (2019) Flexible printed monolithic structured solid-state dye sensitized solar cells on woven glass fibre textile for wearable energy harvesting applications Sci Rep 9:1362

46. Xue Z, Jiang C, Wang L, Liu W, Liu B (2013) Fabrication of flexible plastic solid-state dye-sensitized solar cells using low temperature techniques. J Phys Chem C 118(30):16352–16357

47. Pringle JM, Armel V, Mac Farlane DR (2010) Electrodeposited PEDOT-on-plastic cathodes for dye-sensitized solar cells. Chem Commun 46:5367–5369

48. Vyas N, Charbonneau C, Carnie M, Worsley D, Watson T (2013) An inorganic/organic hybrid coating for low cost metal mounted dye sensitized solar cells. ECS Trans 53(24):29–37

49. Murakami TN, Ito S, Wang Q, Nazeeruddin MK, Bessho T, Cesar I, Liska P, Humphry-Baker R, Comte P, Pechy P, Gratzel M (2006) Highly efficient dye-sensitized solar cells based on carbon black counter electrodes. J Electrochem Soc 153(12):A2255–A2261

50. Ahmad S, Yum JH, Xianxi Z, Gratzel M, Butt HJ, Nazeeruddin MK (2010) Dye-sensitized solar cells based on poly (3,4-ethylenedioxythiophene) counter electrode derived from ionic liquids. J Mater Chem 20:1654–1658

51. Jia J, Wu J, Dong J, Bao Q, Fan L, Lin J, Hu L, Daib S (2017) Influence of deposition voltage of cobalt diselenide preparation on the film quality and the performance of dye-sensitized solar cells. Sol Energy 151(15):61–67

52. Wu M, Wang Y, Lin X, Guo W, Wu K, Lin Y, Guo H, Ma T (2013) TiC/Pt composite catalyst as counter electrode for dye-sensitized solar cells with long-term stability and high efficiency. J Mater Chem A 1:9672–9679

53. Zhou R, Guo W, Yu R, Pan C (2015) Highly flexible, conductive and catalytic Pt networks as transparent counter electrode for wearable dye-sensitized solar cells. J Mater Chem A 3:23028–23034

54. Mei X, Cho SJ, Fan B, Ouyang J (2010) High-performance dye-sensitized solar cells with gel-coated binder-free carbon nanotube films as counter electrode. Nanotechnology 21(39):395202

55. Patil DS, Sonigara KK, Jadhav MM, Avhad KC, Sharma S, Soni SS, Sekar N (2018) Effect of structural manipulation in hetero-tri-aryl amine donor-based D-A' -p-A sensitizer in dye-sensitized solar cells. New J Chem 42:4361–4371

56. Zakeeruddin SM, Gratzel M (2009) Solvent-free ionic liquid electrolytes for mesoscopic dye-sensitized solar cells. Adv Funct Mater 19(14):2187–2202

57. Hilmy NIMF, Yahya WZN, Kurnia KA (2020) Eutectic ionic liquids as potential electrolytes in dye-sensitized solar cells: physicochemical and conductivity studies. J Mol Liq 320:114381

58. Bidikoudi M, Zubeirb LF, Falaras P (2014) Low viscosity highly conductive ionic liquid blends for redox-active electrolytes in efficient dye-sensitized solar cells. J Mater Chem A 2:15326–15336

59. Yu Q, Wang Y, Yi Z, Zu N, Zhang J, Zhang M, Wang P (2010) High-efficiency dye-sensitized solar cells: the influence of lithium ions on exciton dissociation, charge recombination, and the surface states. ACS Nano 4(10):6032–6038

60. Yella A, Lee HW, Tsao HN, Yi C, Chandiran AK, Nazeeruddin MdK, Diau EWG, Yeh CY, Zakeeruddin SM, Gratzel M (2011) Porphyrin-sensitized solar cells with Cobalt (II/III)- based redox electrolyte exceed 12 percent efficiency. Science 334(6056):629–634

61. Wang ZS, Sayama K, Sugihara H (2005) Efficient Eosin Y dye-sensitized solar cell containing Br-/Br 3-electrolyte. J Phys Chem B 109(47):22449–22455

62. Kloo L (2014) Iodine in dye-sensitized solar cells. In: Kaiho T (ed) Iodine chemistry and applications, 1st edn. Wiley, New York, pp 501–502

63. Garcia-Salinas MJ, Ariza MJ (2019) Optimizing a simple natural dye production method for dye-sensitized solar cells: examples for Betalain (Bougainvillea and beetroot extracts) and anthocyanin dyes. Appl Sci 9(12):2515

64. Takashi F, Hiromi F, Ki O, Nobuko OK, Kazuyuki K, Kazuhiro S, Hideki S (2012) Cyclometalated ruthenium (II) complexes as near-IR sensitizers for high efficiency dye-sensitized solar cells. Angew Chem Int Ed 51(30):7628–7531

65. Richhariya G, Kumar A (2021) Performance evaluation of mixed synthetic organic dye as sensitizer-based dye sensitized solar cell. Opt Mater 111:110658

66. Cole JM, Gong Y, Cree-Grey JM, Evans PJ, Holt SA (2018) Modulation of N3 and N719 dye TiO_2 interfacial structures in dye-sensitized solar cells as influenced by dye counter ions, dye deprotonation levels and sensitizing solvent. ACS Appl Energy Mater 1(6):2821–2831

67. Ayalew WA, Ayele DW (2016) Dye-sensitized solar cells using natural dye as light-harvesting materials extracted from *Acanthus sennii chiovenda* flower and *Euphorbia cotinifolia* leaf. J Sci Adv Mater 1(4):488–494

68. Nazeeruddin MK, Kay A, Rodicio I, Baker RH, Miiller E, Liska P, Vlachopoulos N, Gratzel M (1993) Conversion of light to electricity by cis - XzBis (2,2'-bi-pyridyl-4,4'-dicarboxylate), ruthenium (11) charge-transfer sensitizers (X = Cl-, Br-, I-, CN-, and SCN-) on nanocrystalline TiO_2 electrodes. J Am Chem Soc 115(14):6382–6390

69. Huang Y, Chen W, Zhang XX, Ghadari R, Fang XQ, Yu T, Kong FT (2018) Ruthenium complexes sensitizers with phenyl-based bipyridine anchoring ligands for efficiently dye-sensitized solar cells. J Mater Chem C 6:9445–9452

70. Dongshe Z, Suzanne ML, Jonathan AD, Jason LA, June L, Jeanne LMH (2008) Betalain pigments for dye-sensitized solar cells. J Photochem Photobiol A Chem 195(1):72–80

71. Kuo SY, Yang JF, Lai FI (2014) Improved dye-sensitized solar cell with a ZnO nano tree photoanode by hydrothermal method. Nanoscale Res Lett 9:206

72. Sanjay AP, Isaivani I, Deepa K, Madhavan J, Senthil S (2019) The preparation of dye-sensitized solar cells using natural dyes extracted from *Phytolacca icosandra* and *Phyllanthus reticulatus* with ZnO as Photoanode. Mater Lett 244:142–146

73. Yeoh ME, Chan KY (2019) Efficiency enhancement in dye-sensitized solar cells with ZnO and TiO_2 blocking layers. J Electron Mater 48:4342–4350

74. Kumara GRA, Deshapriya U, Ranasinghe CSK, Jayaweera EN, Rajapakse RMG (2018) Efficient dye-sensitized solar cells from mesoporous zinc oxide nanostructures sensitized by N719 dye. J Semicond 39(3):033005

75. Ko SW, Lee D, Kang HW, Nam KH, Yeo JY, Hong SJ, Grigoropoulos CP, Sung HJ (2011) Nanoforest of hydrothermally grown hierarchical ZnO nanowires for a high efficiency dye-sensitized solar cell. Nano Lett 11(2):666–671

76. Zhang Z, Li X, Wang C, Liu LWY, Shao C (2009) ZnO hollow nanofibers: fabrication from facile single capillary electrospinning and applications in gas sensors. J Phys Chem 113(45):19397–19403

77. Kim D, Hong J M, Lee B H, Kim D Y (2007) Dye-sensitized solar cells using network structure of electrospun ZnO nanofiber mats. Appl Phys Lett 91:163109

78. Umar A, Akhtar MS, Almas T, Ibrahim AA, Al AMS, Masuda Y, Rahman QI, Baskoutas S (2019) Direct growth of flower-shaped ZnO nanostructures on FTO substrate for dye-sensitized solar cells. Curr Comput-Aided Drug Des 9(8):405

79. Uthirakumar AP (2011) Fabrication of ZnO based dye-sensitized solar cells, solar cells—dye-sensitized devices. Prof. Leonid A. Kosyachenko (ed). InTech, 435–456

80. Hu J, Cheng J, Tong S, Zhao L, Duan J, Yang Y (2016) Dye-sensitized solar cells based on P25 nanoparticles/ TiO_2 nanotube arrays/hollow TiO_2 boxes three-layer composite film. J Mater Sci Mater Electron 27:5362–5370

81. Suriani AB, Mohamed A, Mamat MH, Hashim N, Isa IM, Malek MF, Kairi MI, Mohamed AR, Ahmad MK (2018) Improving the photovoltaic perfor-mance of DSSCs using a combination of mixed-phase TiO_2 nanostructure photoanode and agglomerated free reduced graphene oxide counter electrode assisted with a hyperbranched surfactant. Optik 158:522–534

82. Ahmad MK, Soon CF, Nafarizal N, Suriani AB, Mohamed A, Mamat MH, Malek MF, Shimo-mura M, Murakami K (2016) Effect of heat treatment to the rutile based dye-sensitized solar cell. Optik 127(8):4076–4079

83. Zhang D, Yoshida T, Oekermann T, Furuta K, Minoura H (2006) Room-temperature synthesis of porous nanoparticulate TiO_2 films for flexible dye-sensitized solar cells. Adv Funct Mater 16(9):1228–1234

84. Faisal A (2014) Synthesis and characteristics study of TiO_2 nanowires and nanoflowers on FTO/glass and glass substrates via hydrothermal technique. J Mater Sci Mater Electron 26(1):317–321

85. Ahmad MK, Murakami K (2015) Rutile-phased TiO_2 nanorods/nanoflowers based dye-sensitized solar cell. Appl Mech Mater 773–774:725–728

86. Wang J, Qu S, Zhong Z, Wang S, Liu K, Hu A (2014) Fabrication of TiO_2 nanoparti-cles/nanorod composite arrays via a two-step method for efficient dye-sensitized solar cells. Prog Nat Sci Mater Int 24(6):588–592

87. Hafez H, Lan Z, Li Q, Wu J (2010) High efficiency dye-sensitized solar cell based on novel TiO_2 nanorods/nanoparticle bilayer electrode. Nanotechnol Sci Appl 3:45–51

88. Shao F, Sun J, Gao L, Chen J, Yang S (2014) Electrophoretic deposition of TiO_2 nanorods for low-temperature dye-sensitized solar cells. RSC Adv 4:7805–7810

89. Chen J, Li C, Xu F, Zhou Y, Lei W, Sunb L, Chen J, YZ (2012) Hollow SnO_2 microspheres for high-efficiency bilayered dye-sensitized solar cell. RSC Adv 2(19):7384–7387

90. Roy P, Albu SP, Schmuki P (2010) TiO_2 nanotubes in dye-sensitized solar cells: higher efficiencies by well-defined tube tops. Electrochem Commun 12(7):949–951

91. Cao Y, Li Z, Wang Y, Zhang T, Li Y, Liu X, Li F (2016) Influence of TiO_2 nanorod arrays on the bilayered photoanode for dye-sensitized solar cells. J Electron Mat 45(10):4989–4998

92. Biraj S, Ngangbam C, Lenka TR (2018) Enhancement of broad light detection based on annealed Al-NPs assisted TiO_2-NWs deposited on p-Si by GLAD technique. IEEE Trans Nanotechnol 17(2):285–292

93. Wu W, Liao J, Chen H, Yu X, Su C, Kuang D (2012) Dye-sensitized solar cells based on a double-layered TiO_2 photoanode consisting of hierarchical nanowire arrays and nanoparticles with greatly improved photovoltaic performance. J Mater Chem 22(34):18057–18062

94. Biraj S, Ngangbam C, Lenka TR (2017) Plasmon-sensitized optoelectronic properties of Au nanoparticle-assisted vertically aligned TiO_2 nanowires by GLAD technique. IEEE Trans Electron Dev 64(3):1127–1133

95. Hua B, Lin Q, Zhang Q, Fan Z (2013) Efficient photon management with nanostructures for photovoltaics. Nanoscale 5(1):6627–6640

96. Zhu J, Yu Z, Fan S, Cu Y (2010) Nanostructured photon management for high-performance solar cells. Mater Sci Eng R Rep 70:330–340

97. He XL, Yang GJ, Li CJ, Liu M, Fan SQ (2015) Failure mechanism for flexible dye-sensitized solar cells under repeated outward bending: Cracking and spalling off of nano-porous titanium dioxide film. J Power Sourc 280:182–189

98. Guo X, Xu Z, Huang J, Zhang Y, Liu X, Guo W (2019) Photoelectrochromic smart windows powered by flexible dye-sensitized solar cell using CuS mesh as a counter electrode. Mater Lett 244(1):92–95

99. Liang J, Yang J, Zhang G, Sun W (2013) Flexible fiber-type dye-sensitized solar cells based on highly ordered TiO_2 nanotube arrays. Electrochem commun 37:80–83

100. Hong CK, Jung YH, Kim HJ, Park KH (2014) Electrochemical properties of TiO_2 nanoparticle/nanorod composite photoanode for dye-sensitized solar cells. Curr Appl Phys 14(3):294–299

101. Kim YG, Shim CH, Kim DH, Lee HJ (2012) Fabrication of transparent conductive oxide-less dye-sensitized solar cells consisting of Ti electrodes by an electron-beam evaporation process. Thin Solid Films 520(6):2257–2260

102. Manca M, Malara F, Martiradonna L, Marco LD, Giannuzzi R, Cingolani R, Gigli G (2010) Charge recombination reduction in dye-sensitized solar cells by means of an electron beam-deposited TiO_2 buffer layer between conductive glass and photo-electrode. Thin Solid Films 518(23):7147–7151

103. Kiema G, Colgan M, Brett M (2005) Dye-sensitized solar cells incorporating obliquely deposited titanium oxide layers. Sol Energy Mater Sol Cells 85(3):321–331

104. Wong MS, Lee MF, Chen CL, Huang CH (2010) Vapor deposited sculptured nanoporous titania films by glancing angle deposition for efficiency enhancement in dye-sensitized solar cells. Thin Solid Films 519(5):1717–1722

105. Yang HY, Lee MF, Huang CH, Lo YS, Chen YJ, Wong MS (2009) Glancing angle deposited titania films for dye-sensitized solar cells. Thin Solid Films 518(5):1590–1594

106. Seo YG, Kim MA, Lee H, Lee W (2011) Solution-processed thin films of non-aggregated TiO_2 nanoparticles prepared by mild solvothermal treatment. Sol Energy Mater Sol Cells 95(1):332–335

107. Chena BL, Hua H, Tai QD, Zhang NG, Guo F, Sebo B, Liu W, Yuan JK, Wang JB, Zhao XZ (2012) An inverted fabrication method towards a flexible dye-sensitized solar cell based on a free-standing TiO_2 nanowires membrane. Electrochim Acta 59(1):581–586

108. Wang H, Li H, Wang J, Wu J (2012) High aspect-ratio transparent highly ordered titanium dioxide nanotube arrays and their performance in dye sensitized solar cells. Mater Lett 80(1):99–102

109. Meng L, Ren T, Li C (2010) The control of the diameter of the nanorods prepared by dc reactive magnetron sputtering and the applications for DSSC. Appl Surf Sci 256(11):3676–3682

110. Junga WH, Kwaka NS, Hwanga TS, YiKB, (2012) Preparation of highly porous TiO_2 nanofibers for dye-sensitized solar cells (DSSCs) by electro-spinning. Appl Surf Sci 261:343–352

111. Liu W, Lu H, Zhang M, Guo M (2015) Controllable preparation of TiO_2 nanowire arrays on titanium mesh for flexible dye-sensitized solar cells. Appl Surf Sci 347:214–223

112. Cao L, Wu C, Hu Q, Jin T, Chi B, Pu J, Jian L (2013) Double-layer structure photoanode with TiO_2 nanotubes and nanoparticles for dye-sensitized solar cells. J Am Ceram Soc 96(2):549–554

113. Klein M, Szkoda M, Sawczak M, Cenian A, Lisowska-Oleksiak A, Siuzdak K (2017) Flexible dye-sensitized solar cells based on Ti/TiO_2 nanotubes photoanode and Pt-free and TCO-free counter electrode system. Solid State Ion 302:192–196

114. Li Y, Wang H, Feng Q, Zhou G, Wang ZS (2013) Gold nanoparticles inlaid TiO_2 photoanode: a superior candidate for high efficiency dye-sensitized solar cells. Energy Environ Sci 6(7):2156–2165

115. Kar P, Maji TK, Sarkar PK, Sardar S, Pal SK (2016) Direct observation of electronic transition-plasmon coupling for enhanced electron injection in dye-sensitized solar cells. RSC Adv 6(101):98753–98760

116. Wu WY, Hsu CF, Wu MJ, Chen CN, Huang JJ (2017) Ag-TiO_2 composite photoelectrode for dye-sensitized solar cell. Appl Phys A 123(357):1–8

117. Han SH, Rho WY, Jun BH (2019) Au-nanoparticle-embedded open-ended free-standing TiO_2 nanotube arrays in dye-sensitized solar cells for better electron generation and electron transport. ACS Omega 4(23):20346–20352

118. Nien YH, Chen HH, Hsu HH, Rangasamy M, Hu GM, Yong ZR, Kuo PY, Chou JC, Lai CH, Ko CC, Chang JX (2020) Study of how photoelectrodes modified by TiO_2/Ag nanofibers in various structures enhance the efficiency of dye-sensitized solar cells under low illumination. Energies 13:2248

119. Kim HS, Chun MH, Suh JS, Jun BH, Rho WY (2017) Dual functionalized free-standing TiO_2 nanotube arrays coated with Ag nanoparticles and carbon materials for dye-sensitized solar cells. Appl Sci 7(6):576

120. Garmaroudi ZA, Mohammadi MR (2016) Plasmonic effects of infiltrated silver nanoparticles inside TiO_2 film: enhanced photovoltaic performance in DSSCs. J Am Ceram Soc 99(1):167–173

121. Gupta S, Navaraj, WT, Lorenzelli L, Dahiya R (2018) Ultra-thin chips for high-performance flexible electronics. npj Flex Electron 2(8):1–17

122. Wu C, Chen B, Zheng X, Priya S (2016) Scaling of the flexible dye-sensitized solar cell module. Sol Energy Mater Sol Cells 157:438–446

123. Kim MS, Chun DM, Choi JO, Lee JC, Kim YH, Kim KS, Lee CS, Ahn SH (2012) Dry-spray deposition of TiO_2 for a flexible dye-sensitized solar cell (DSSC) using a nanoparticle deposition system (NPDS). J Nanosci Nanotechnol 12(4):3384–3388

124. Chen LC, Ke CR, Hon MH, Ting JM (2015) Electrophoretic deposition of TiO_2 coatings for use in all-plastic flexible dye-sensitized solar cells. Surf Coat Technol 284(25):51–56

125. Han Q, Liu S, Liu Y, Jin LD, Cheng S, Xiong Y (2020) Flexible counter electrodes with a composite carbon/metal nanowire/polymer structure for use in dye-sensitized solar cells. Sol Energy 208(18):469–479

126. Jen HP, Lin MH, Li LL, Wu HP, Huang WK, Cheng PJ, Diau EWG (2013) High-performance large-scale flexible dye-sensitized solar cells based on anodic TiO_2 nanotube arrays. ACS Appl Mater Interfaces 5(20):10098–10104

127. Han HG, Weerasinghe HC, Kim KM, Kim JS, Cheng YB, Jones DJ, Holmes AB, Kwon TH (2015) Ultrafast fabrication of flexible dye-sensitized solar cells by ultrasonic spray-coating technology. Sci Rep 5(14645):1–9

128. Luo D, Liu B, Fujishima A, Nakata K (2019) TiO_2 nanotube arrays formed on Ti meshes with periodically arranged holes for flexible dye-sensitized solar cells ACS appl. Nano Mater 2(6):3943–3950

129. Liang J, Zhang G, Sun W, Dong (2015) High efficiency flexible fiber-type dye-sensitized solar cells with multi-working electrodes. Nano Energy 12:501–509

130. Yue G, Liu X, Chen Y, Huo J, Zheng, (2018) H Improvement in the photoelectric conversion efficiency for the flexible fibrous dye-sensitized solar cells. Nanoscale Res Lett 13(188):1–10

Theory of Nanostructured Kesterite Solar Cell

Soumyaranjan Routray and K. P. Pradhan

Abstract Kesterite crystals such as copper-zinc-tin-sulphide (CZTS) are greatly explored as bulk and nanostructured absorber materials for future generation high efficiency and low-cost solar cell. Kesterite materials have attractive optical parameters, high absorption coefficients, and symmetrical structure with adjustable bandgap of 1.0–1.5 eV. In practice, any kesterite material possesses high density of different defects, traps, and interface defects which degrade the performance metrics of the solar cells. The current chapter elaborates a basic mathematical model, considering all the defects, and traps as centres for recombination. The performance metrics of the solar cell are well explored and analysed with different types of recombination centres. Finally, the current–voltage characteristics and efficiency of the solar cell are demonstrated, which may give a clear visualization of low-cost next generation thin film solar cells.

Keywords Kesterite · Defects · Traps · Efficiency · CZTS quantum well · CZTSe

1 Introduction

Humans rely on different forms of energy, ranging from traditional source to renewable energy for their civilized existence. In other words, it can be said that energy is the only currency for global civilization. Now-a-days, harvesting of renewable sources of energy has drawn the remarkable attention of developed and developing provinces. The instability in the future for fossil fuels, demand and supply challenges, adverse effects of nuclear power, and climate change are the driving force behind the transition towards sustainable sources of energy [1]. Photovoltaic (PV) cells are the

S. Routray (✉)
SRM Institute of Science and Technology, Chennai, Tamil Nadu 603203, India
e-mail: s.r.routray@ieee.org

K. P. Pradhan
Indian Institute of Information Technology Design and Manufacturing Kancheepuram, Chennai, Tamil Nadu 600127, India
e-mail: k.p.pradhan@ieee.org

© The Author(s), under exclusive license to Springer Nature Singapore Pte Ltd. 2022
R. Goswami and R. Saha (eds.), *Contemporary Trends in Semiconductor Devices*,
Lecture Notes in Electrical Engineering 850,
https://doi.org/10.1007/978-981-16-9124-9_6

only stable solution for decarbonized and unlimited source of energy. Furthermore, high efficiency solar cell is a viable solution against fast increasing price of traditional energy sources. Currently, the world faces severe threats of climate change due to emission of toxic gases and a number of countries stands as front leaders of introducing a new sustainable aspect in space stations. In 1958, the 'Vanguard 1' satellite was first introduced as a fully solar powered space station [2].

In general, solar cells are categorized generation-wise depending on their use of technology such as (i) 1st generation (1G), (ii) 2nd generation (2G), (iii) 3rd generation (3G), and (iv) 4th generation/Future generation (4G) as shown in Fig. 1. Although 80% of the current PV market emphasises on first generation silicon-based technology, however, the Si solar cell performance is saturated and further improvement of efficiency and reduction of cost is a myth. The wastage of photon energy as heat energy due to bandgap of Si is also another cause towards the search for alternative materials. To address the above limitations, there is a need of thin film solar cells (TFSCs), which can absorb the irradiation with a thickness of few nanometres [3]. However, due to the presence of heavy and rare-earth elements such as Te, In, Cd, and Ga, high defects and poor stability limit the commercial production of such solar cells. In addition, third generation solar cells use technology to add multiple layers of compound semiconductors, which further increase the fabrication cost. It is noteworthy to mention that cost-effective solar cells are unlocked with implementation of thin film technology in early 70s. The advantages of thin film solar cells are low-cost fabrication process and flexibility. Mostly, I-III-VI or II-VI compound semiconductors are used as absorber layer with micrometre or few hundreds of nanometres of thickness in thin film solar cells.

Additionally, these absorber layers can be coated on low-cost metal films, glass plates, or plastic material, enabling a new scope for building integrated technology. The limitations of the material are its stability and lifetime as compared to existing technology. In the late 1980s, copper-zinc-tin-sulphide-based (Cu_2ZnSnS_4) kesterite

Fig. 1 Pictorial presentation of different generation of solar cell

quaternary compound semiconductors based thin films are introduced as a PV material. In this chapter, fundamental principle, physical properties, and chemical properties for the thin film solar cells are emphasized considering Cu_2ZnSnS_4 kesterite compound semiconductor as an absorber layer.

2 Kesterite Material

Cupper-indium-gallium-selenide (CIGS) material belongs to chalcopyrite crystal group and can achieve record-breaking efficiency as an absorber layer. CIGS solar cells could achieve laboratory efficiencies of 22.6% which, along with other advantages, can challenge the current Si solar cell technology [4, 5]. CIGS solar cells have the potential for low-cost and high-volume production as well as roll-to-roll manufacturing. However, one of the major challenges in CIGS thin film solar cells is the presence of indium content, having estimated abundance of 0.049 ppm in the earth's crust. This makes it a rare-earth material, even rarer than silver [6]. In addition to this, the use of gallium increases the risk of high cost. Hence, the stability of CIGS is a big question mark to supply terawatt energy for global consumption.

Kesterite (Cu_2ZnSnS_4) solar cells can be projected as a potential alternative candidate to overcome the issues of CIGS solar cell. The CZTS absorber layer uses non-toxic materials such as copper, zinc, tin, sulphur, and selenium. The properties of the material were initially explored in 1988 and the invented PV device with this material was first demonstrated in 1996 with an efficiency of 0.7%. The investigation was silent until 2004 with 5% of power conversion efficiency [7]. A remarkable efficiency of 14% was demonstrated in 2014 by IBM T. J. Watson Research Centre [8]. The benefits of CZTS solar cells are based on certain principles. The kesterite material as a photovoltaic absorber layer must satisfy two necessary conditions such as direct tuneable band gap and optimal cell thickness to absorb photons from sunlight. Because of high photon absorption coefficient greater than 10^4 cm^{-1}, CZTS absorber layer possesses the ability to achieve high photon absorption from sunlight with low cell thickness. Additionally, kesterite materials as absorber layers do not affect or damage generated photocurrent in solar cell. One of the key emerging aspects of kesterite solar cells is the ease-of-fabricating device for mass fabrication with low-cost techniques. In general, fabrication of kesterite solar cells is more attractive with solution-based process for deposition followed by low temperature annealing (500–600 °C). Research community has explored many different types of physical and chemical deposition techniques. Some of the leading techniques are given below which are frequently reported by the researchers globally as follows.

Growth techniques	References
Sputtering (Atomic beam)	[9]
Sputtering (RF magnetron)	[10, 11]

(continued)

(continued)

Growth techniques	References
Sulphurization (e-beam or electrodeposition method)	[12, 13]
Photochemical depositions	[14]
Spray pyrolysis	[15]
Pulsed laser deposition	[16, 17]
Solgel sulphurization	[18]
Chemical synthesis	[19, 20]
Screen printing	[21]
SILAR techniques	[22]
chemical vapour deposition (CVD)	[23]

The efficiency achieved by using different type of growth method is given below to convey the status of the CZTS solar cell.

Growth techniques	Conversion efficiency	References
Spin coating	11	[24]
Co-evaporation	9	[24]
Electrodeposition (Sequential)	7.3	[24]
Electrodeposition (One-step)	3.4	[24]
CVD	6	[24]
Sol–gel method	5	[24]
SILAR technique	1.85	[24]
Spray pyrolysis techniques	1.1	[24]
Doctor blade coating	6.8	[24]
Sputtering	6.7	[25]
Sputtering (Sequential)	4.6	[25]
Physical layer deposition	3.1	[25]
Reactive ion co-sputtering	3.4	[25]
Stacked deposition (sputtering along with evaporation)	6	[25]

3 Device Structure and Mathematical Model

The physical structure of bulk kesterite solar cell (AZO/ZnO/CdS/kesterite/Mo) and nanostructure of multijunction kesterite solar cell (AZO/ZnO/CdS/(CZTS/CZTSe)/Mo) [26, 27] are explored for the investigation of next generation solar cells in thin film technology. Figure 2 shows the structure of kesterite CZTSe or CZTS absorber layer based solar cell and kesterite

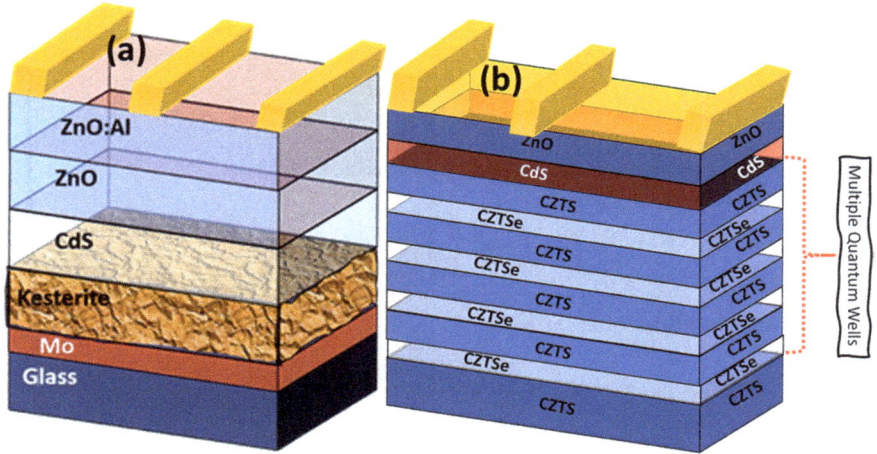

Fig. 2 Schematic of **a** bulk and **b** QW kesterite solar cell

nanostructured multiple quantum well (MQW) solar cells. Moreover, p-CZTS/p-CZTSe is used as the absorber layer whereas n-doped CdS film is incorporated as the buffer layer that forms the p–n junction, which establishes the fundamental carrier dynamics for the photovoltaic device. In this study, molybdenum (Mo) is used as a back contact to collect the carriers and the undoped ZnO or popularly i-ZnO is incorporated as a window layer. It is reported that most of the remarkable performance achieved from CZTS solar cell are fabricated using non-vacuum process.

The assumptions taken for modelling are [28]

- Dependence of biasing on the depletion width is neglected.
- Energy levels shift in quantum wells under electric field is neglected.
- Non-radiative recombination (NRR) and radiative recombination (RR) coefficients are assumed to be constant and independent of carrier density.
- Carrier distribution all through the structure is described by quasi-Fermi levels and non-degenerated Maxwell–Boltzmann statistics.
- Solar spectrum is projected as perpendicular to the device.
- Losses due to ohmic contacts are neglected

Let us consider a charged particle q, moving along $\overrightarrow{x}$ direction plunged into the electric field $\overrightarrow{\varepsilon}$. The particle comes under the force $\overrightarrow{F} = q \times \overrightarrow{\varepsilon}$ and potential energy $U = -F \times x$. Hence, the Schrodinger's wave equation can be written as [28]

$$\frac{\mathrm{d}^2 \psi}{\mathrm{d}x^2} + \frac{2m}{\hbar^2}(E + Fx)\psi = 0 \tag{1}$$

where E is total energy of the particle, m stands for mass of the particle, and ψ is called the complex-valued wave function/eigen function.

It can also be written as

$$\frac{d^2}{d\xi^2} + \xi\psi = 0$$

where ξ is the one-dimensional variable and can be written as,

$$\xi = \left(x + \frac{E}{F}\right)\left(\frac{2mF}{\hbar^2}\right)^{1/3}$$

The solution of this equation is

$$\psi(\xi) = N A_i(-\xi) + N' B_i(-\xi)$$

where A_i and B_i are the homogeneous Airy functions [28]. The shift in the energy levels in wells is due to the dominant effect of the electric field and is commonly known as the *Stark Effect*. It is observed that the shift in energy levels in well is very short (only few meV) and hence can be neglected here.

The current continuity and current density equations are the traditional drift–diffusion equations written for electron and holes as follows [28]:

$$\frac{dn_b}{dt} = G_b - U_b + \frac{n_{qw}}{\tau_{en}} - \frac{n_b}{\tau_{cn}} + \frac{1}{q}\frac{dJ_n}{dx} = 0 \tag{2}$$

$$\frac{dn_b}{dt} = G_b - U_b + \frac{p_{qw}}{\tau_{ep}} - \frac{p_b}{\tau_{cp}} + \frac{1}{q}\frac{dJ_p}{dx} = 0 \tag{3}$$

$$J_n = q\mu_n n_b \vec{E} + q D_n \frac{dn_b}{dx} \tag{4}$$

$$J_p = q\mu_p p_b \vec{E} - q D_p \frac{dp_b}{dx} \tag{5}$$

The current–voltage characteristics of kesterite solar cells without nanostructure can be written [28]

$$J(V) = J_0\left[\exp(qV/kT) - 1\right] - J_{PH} \tag{6}$$

where J_{PH} and J_0 are photogenerated current density and overall reverse saturation current density, respectively. Furthermore, k, V, q, and T are the Boltzmann's constant, terminal voltage, electron charge, and room temperature, respectively. The J_0 term can be found by integrating the RR rate over total width of absorber layer, which leads to the below equation:

$$J_0 = q W n_i^2 B$$

where W is the thickness of absorber, n_i stands for intrinsic concentration of carrier and furthermore B is radiative coefficient of CZTS or CZTSe material. The radiative recombination coefficient can be found as follows:

$$B = \frac{8\pi n_r^2}{c^2 h^3 n_i^2} \int_{E_g}^{\infty} \frac{\alpha_{\text{bulk}} E^2 dE}{\exp\left(\frac{E}{kT}\right) - 1}$$

where h is planck's constant, c is speed of light, n_r represents refractive index, α_{bulk} is absorption coefficient of bulk absorber layer and E_g is the energy bandgap of the material. More details investigation and mathematical models can also be found from T. Jiménez et al. [38].

Additionally, fabricated CZTS absorber layer is characterized by high density of traps and defects that, in turn, increase the radiative as well as non-radiative recombination of photogenerated carriers in the absorber layer. Hence, the modified J-V characteristics of kesterite solar cell can also be written as [28]:

$$J(V) = J_{0\_\text{rad}}\left[\exp(qV/kT) - 1\right] + J_{0\_\text{nonrad}}\left[\exp(qV/kT) - 1\right] - J_{PH} \quad (7)$$

where J_{0_rad} is reverse saturation current density because of RR, thermionic emission, and carrier diffusion, while J_{0_nonrad} represents the reverse saturation current density because of NRR, trap assisted tunnelling recombination and interface trap recombination.

The current density (reverse saturation) because of radiative recombination is defined as [26]:

$$J_{0\_\text{rad}} = q W_n n_{i-\text{CdS}}^2 B_{\text{CdS}} + q W_p n_{i-\text{kesterite}}^2 B_{\text{kesterite}} \quad (8)$$

where W_n and W_p are the space charge width towards CdS and kesterite material respectively.

Similarly, current density (reverse saturation) because of NRR and trap assisted tunnelling is defined as [26]:

$$J_{0\_\text{nonrad}} = q\left(\frac{W_n n_{i-\text{CdS}}(1 + \Gamma_{\text{CdS}})}{\tau_n} + \frac{W_p n_{i-\text{kesterite}}(1 + \Gamma_{\text{kesterite}})}{\tau_p}\right) \quad (9)$$

The Γ term defines effect of tunnelling due to captured carriers by traps and emission rate. τ_n and τ_p are the lifetime of non-radiative minority carriers.

The photo current density can be calculated as follows

$$J_{ph} = q \int_{\lambda_1}^{\lambda_2} F(\lambda) EQE(\lambda) d\lambda \quad (10)$$

where λ_1 and λ_2 are the limits of solar spectrum, $F(\lambda)$ is the intensity of solar spectrum. EQE is defined as external quantum efficiency of the solar cell and can be calculated as follows:

The contribution of photogenerated current in i-layer to quantum efficiency can be calculated as [28]

$$QE = [1 - R(\lambda)] \exp\left\{-\left(\sum \alpha_i z_i\right) \times [1 - \exp(\alpha_B W - N_w \alpha_w^*)]\right\} \quad (11)$$

where $R(\lambda)$ denotes the surface reflectivity spectrum of the anti reflecting layer. The α_i states the absorption coefficient and z_i stands for the width of (i) anti reflecting coating (ii) emitter layer (iii) space charge region from emitter layer. Similarly, α_B is the absorption coefficient of bulk barrier material and α_w^* defines the coefficient for quantum well absorption and is a dimensionless quantity.

Furthermore, the contributions from p region and n-region are calculated by considering the carrier transport mechanism for the minority carriers at room temperature followed by depletion approximation and is given as [28]

$$\text{IQE}(\lambda) = \frac{J_p(\lambda)}{qF(\lambda)[1 - R(\lambda)]} + \frac{J_n(\lambda)}{qF(\lambda)[1 - R(\lambda)]} + \frac{J_{dr}(\lambda)}{qF(\lambda)[1 - R(\lambda)]} \quad (12)$$

$$\text{EQE}(\lambda) = \text{IQE}(\lambda)[1 - R(\lambda)] \quad (13)$$

A quantum well solar cell (QWSC) with number of wells N_w and length of L_w each with an intrinsic region length of W, a barrier band gap of E_{gB} and well band gap of E_{gW} is proposed in this study. The doping concentrations are uniform inside the n-p regions. Under these conditions, the I-V relation of the MQWSC under radiative limit considering radiative and non-radiative effects is given by [28]

$$J_{\text{MQW}} = J_0(1 + r_R \beta) \times \left[\exp\left(\frac{qV}{kT}\right) - 1\right] + \alpha r_{NR}\left[\exp\left(\frac{qV}{2kT}\right) - 1\right] - J_{PH} \quad (14)$$

where q stands for electron charge, kT represents the thermal energy, V represents the terminal voltage, and α, β are constants that can be defined as [28]:

$$\alpha = qWA_B n_{iB}$$

$$\beta = \frac{qWB_B n_{iB}^2}{J_0}$$

where A_B represents non-radiative coefficient for barrier in the depletion region ($A_B = 1/\tau_B$) and $n_{iB} = g_B \exp[-E_B/2kT]$, E_B is the bandgap for base material, g_B stands for effective volume densities of states for barrier material, τ_B is the lifetime of barrier non-radiative mechanism.

B_B stands for radiative barrier recombination coefficient. r_R depicts the radiative enhancement ratio that signifies the fractional increase in radiative recombination in the intrinsic region because of addition in quantum wells and given as [28]

$$r_R = 1 + f_w\big[\gamma_B \gamma_{\text{DOS}}^2 \exp((\triangle E - qFx)/KT - 1)\big]$$

Here are the parameters

$$\gamma_B = B_w / B_B$$

$$\gamma_{\text{DOS}} = g_w / g_B$$

$$\Delta E = E_{g,B} - E_{g,w}$$

where, g_w and g_B are the effective volume densities of states for well and barrier material. f_w represents fraction of the absorber region filled with quantum wells.

γ_{DOS} and γ_B are the enhancement factors for density of states and the oscillator, respectively. r_{NR} is non-radiative enhancement ratio and given as follows [28]

$$r_{NR} = 1 + f_w\big[\gamma_A \gamma_{\text{DOS}}^2 \exp((\Delta E - qFx)/2KT - 1)\big]$$

The parameters are as given below

$$\gamma_A = \frac{A_w}{A_B}$$

$$A_w = 1/\tau_w$$

where τ_w is the barrier non-radiative lifetime, γ_A is the lifetime reduction factor. F is denoted as built-in field F and x is the position inside the well.

The influence of non-radiative recombination on MQW solar cells is evaluated by the theory of Shockley–Read–Hall (SRH) recombination. The low injection mechanism, trap states in band-gap, and inside the depletion region, a constant recombination rate for each material are assumed for the development of the model. The maximum recombination is achieved at a point where equal carrier concentrations are observed. Moreover, the recombination mechanism inside the depletion region is detailed by Courel et al. [26, 27]

The photogenerated current density J_{PH} is given by [28]

$$J_{PH} = q \int F(\lambda)\big[1 - \exp(-\alpha_b W - \alpha_w N L_w)\big]d\lambda$$

where $F(\lambda)$ states the number of photons per wavelength of solar spectrum. α_b and α_w denote to absorption coefficient of well and barrier of MQW solar cell respectively. The total number of wells are defined as N. The absorption coefficient for barrier and well and effective density of state (DOS) is calculated from expression reported in the literature [28]

Furthermore, the eq. considering only radiative limit can be written as:

$$J_{\mathrm{MQW}} = q W n_i^2 B \left\{ 1 + f_w \left[\gamma_B \gamma_{\mathrm{DOS}}^2 \exp\left(\frac{\Delta E_g - q F L_w}{kT} \right) - 1 \right] \right\}$$
$$\times \left[\exp\left(\frac{qV}{kT} \right) - 1 \right] - J_{PH} \tag{15}$$

where, the electric field at the absorber layer is represented by F and the well thickness by L_w.

The absorption coefficient of the barrier is estimated from the expressions mentioned below as reported in the literature:

$$\alpha(E) = \frac{\sqrt{2} q^2 E_g \left[m_h^* m_e^* / m_h^* + m_e^* \right]^{3/2}}{3 \pi n_r c \varepsilon_0 m_e \hbar^2} \times \frac{\sqrt{E - E_g}}{E}$$

where E_g is bulk material band-gap, $\hbar$ represents reduced Planck's constant, E is the photon energy, and ϵ_0 correspond to vacuum dielectric permittivity, m_e is the electron mass, $m_h^* and m_e^*$ are electron and hole effective masses, respectively.

Similarly, the absorption coefficient for the well material can be calculated by the expression given by Courel et al. [26]. The effective masses of carriers are calculated considering band offsets of conduction band and valence band, relative dielectric permittivity, and energy bandgaps. The detailed expression can be found in Courel et al. [26].

The escape mechanism of carriers from quantum wells are of two types: (i) tunnelling escape and (ii) thermionic escape. The thermionic escape time of the processes is expressed as [26]:

$$\frac{1}{\tau_{\mathrm{Thermonic}}} = \frac{1}{L_w} \sqrt{\frac{kT}{2\pi m_w}} \exp[(\Delta E - q F L_w)/KT - 1] \tag{16}$$

where, m_w is the effective mass in the well and L_w is thickness of the well. All other parameters are same as used in Eq. 6. Similarly, expression for tunnelling escape time can be found from literature [28].

The open circuit voltage for bulk solar cell is given as [28]

$$V_{oc,B} = \frac{KT}{q} \ln\left(\frac{|J_{SC,B}|}{J_0(1 + \beta)} \right) \tag{17}$$

whereas open circuit voltage for QW solar cell considering recombination is given by

$$V_{oc,QW} = \frac{KT}{q} \ln\left(\frac{|J_{sc,QW}| + J_0(1 + r_R \beta)}{J_0(1 + r_R \beta)} \right) \tag{18}$$

The open circuit voltage for QW solar cells considering both radiative and non-radiative recombination is given by [28]

$$V_{oc,QW} = \frac{KT}{q} \ln\left(\frac{|J_{sc,QW}| + J_0(1 + r_R\beta) + \alpha r_{NR}}{J_0(1 + r_R\beta) + 1.64\alpha r_{NR}}\right) \qquad (19)$$

where, $r_R = 1 + f_w[\gamma_B \gamma_{DOS}^2 \exp(\Delta E/2KT - 1)]$.

Hence, mathematically it is evident from the above equation that V_{oc} in QW solar cell is less as compared to bulk solar cells, which is due to higher denominator factor in Eq. (18). V_{oc} is reduced from 1 to 0.7 V because of the incorporation of low band gap material and recombination inside the well.

The material parameters used in the simulation are listed in Tables 1 and 2. The other parameters such as valence band, conduction band, DOS, hole thermal velocity, hole capture cross-section, series and shunt resistance, interface recombination speed, band offset, minority carrier life time, etc. are referred from literature [28].

Table 1 Different materials parameters [28]

Parameters	ZnO	CdS	CZTS	CZTSe
Thickness (nm)	200	100	10	5
Bandgap (eV) [22]	3.3	2.4	1.45	1.05
Effective mass of electron (m_e/m_o) [22]	0.275	0.25	0.18	0.10
Effective mass of holes (m_h/m_o) [22]	5	0.59	0.22	0.12
Electron Affinity (eV)	4.2	4.4	4.1	4.46
Permittivity [28]	9	9	7	9.1
N_c (cm^{-3}) [28]	2.2×10^{18}	1.8×10^{19}	2.2×10^{18}	2.2×10^{18}
N_v (cm^{-3}) [28]	1.8×10^{19}	2.4×10^{18}	1.8×10^{19}	1.8×10^{19}
e^- mobility (cm^2/V. s) [29]	100	100	60	40
h^+ mobility (cm^2/V. s) [30]	25	25	20	10
Radiative recombination coefficient [6]	–	1.02×10^{-10}	1.04×10^{-10}	5×10^{-9}
Doping conc. (cm^{-3}) [22]	1×10^{18}	1×10^{16}	5×10^{18}	–

Table 2 Interface trap density [28]

Parameters	CZTS/CZTSe bulk	Along interface
Energy states w.r.t V_B (eV)	0.6/0.5	−0.115
Total defect density, N_t (cm^3)	1×10^{15}	1×10^{13}
Electron capture area (cm^2)	3×10^{-15}	1×10^{-12}
Hole capture area (cm^2)	3×10^{-15}	1×10^{-15}

4 Results and Discussion

The carrier recombination in solar cells before collection is one of the major performance limiting factors. The presence of native deep defects in the band gap of material generates energetic intermediate defect states and behaves as epicentres for recombination. These deep defects lead to generation of shallow donor and acceptor states, and mid-gap states which are nothing but traps in the forbidden band gap of bulk material. These available intermediate states in the forbidden gap have the potential to exchange carriers with the valence and conduction bands. Additionally, the quaternary crystallography of kesterite materials and disorders in crystallographic structure such as dislocations are enhanced in presence of unwanted discrete energy levels within the band gap. The presence of these discrete levels largely contributes to the recombination mechanism. The associated energy of trap states drops in the forbidden gap can exchange charge with the valence and conduction bands. The presence of both bulk and interface trap states is quite possible, which further contributes to change in electric field in the depletion region and thereby, affects the J–V characteristics of the solar cell. Figure 3 shows the current–voltage characteristics of the kesterite solar cell. The influence of different types of recombination on J–V characteristics of kesterite solar cells is shown in Fig. 4. A remarkable current density is observed in CZTSe based solar cells. This may be anticipated due to lower band gap and higher material absorption coefficients than traditional CZTS cells. Additionally, CZTSe material can collect more photons due to lower bandgap than CZTS material. Open circuit voltage of CZTS cell is found to be 1 V whereas a lower open circuit voltage is observed for CZTSe cell. It is also clear that current density increases with only non-radiative recombination effect. On the other hand, photogenerated current density falls while increase in radiative recombination rate. Additionally, involvement of trap defects in bulk absorber layer decreases the current density. This may be because of presence of traps and defect centres along the band gap of bulk material which, in turn, create epicentre for recombination and decreases cathode current.

Figure 4 depicts a bar chart presentation of conversion efficiency in kesterite solar cells with different possible loss mechanisms. Achievable low conversion efficiency

Fig. 3 Comparison of JV curve in CZTS and CZTSe solar cell with different loss mechanisms

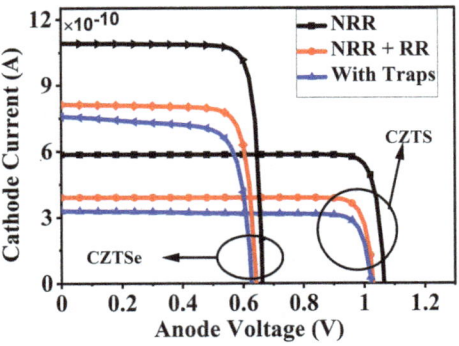

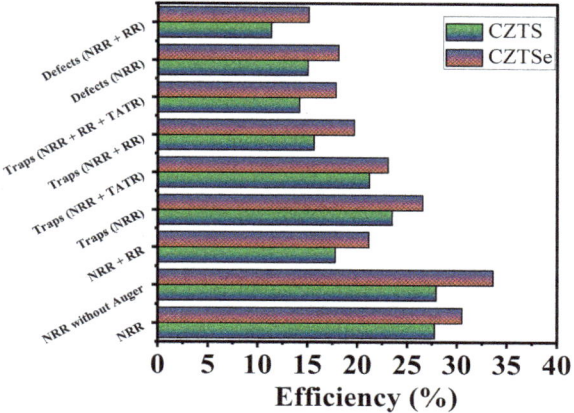

Fig. 4 Bar chart for efficiency in CZTS and CZTSe solar cell with different loss mechanisms

may be due to the presence of defect states leading to carrier recombination centres and are associated with radiative recombination. As discussed earlier, the recombination rates in kesterite solar cells are due to auger and radiative recombination near interfaces and bulk region of material.

The influence of radiative recombination can be optimized by implementing photon recycling techniques. However, non-radiative recombination mechanisms will be inevitably present. On the other hand, very fundamental auger recombination mechanisms cannot be fully avoided but the impact can be minimized to achieve high performance in solar cells as depicted in Fig. 5. The analysis shows a remarkable 40% reduction in performance of solar cell due to radiative recombination. The rest of the carrier recombination is primarily due of SRH recombination. Therefore, it is concluded that radiative recombination (RR) mechanism reduces the major performance of the thin film solar cells.

Figure 5 shows the J–V curve of 2, 4, 10 QW kesterite solar cells. It is evident that there is an increase in the current with increase in number of QWs. However, a reduction in voltage is observed with an increase in the number of QWs. Bulk CZTS

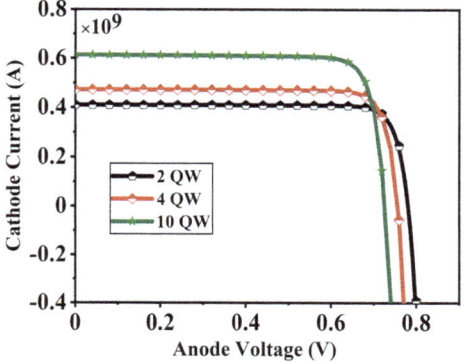

Fig. 5 J–V curve of two, three, and ten Quantum Well kesterite solar cell

Fig. 6 Efficiency of the
MQW solar cell with
increase in number of QWs

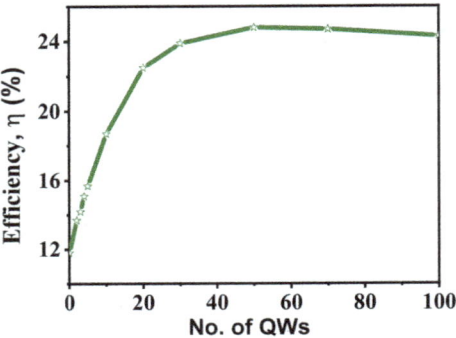

solar cell achieved V_{oc} near to 1 V, while open circuit voltage of MQW solar cell drops to around 0.7 V. This drop in voltage may be because of the incorporation of lesser bandgap well material. Additionall 5.y, interface recombination is also enhanced due to an increase in QWs as a greater number of layers share their interfaces with barrier materials which, in turn, enhance the recombination rate. Further explanation with supporting mathematical model can be referred from Eq. 19.

Figure 6 depicts the efficiency of the MQW solar cell with different quantum wells. It is evident that after reaching a threshold of increase in QW numbers, the efficiency of the solar cell started degrading as shown in the figure. A similar trend in the decrease in efficiency can also be found from the fabricated MQW solar cell by Bushnell et al. [29] and Lynch et al. [30]. The increase in efficiency is continued until 50 QWs. Efficiency starts degrading after the inclusion of 70 QWs. The decrease in V_{oc} begins to outweigh the enhancement in absorption from the QWs. So, an upper limit of QWs is optimized to 50 for achieving remarkable theoretical efficiency.

5 Conclusion

In this chapter, a theoretical study of bulk kesterite solar cell with CZTS and CZTSe as absorber layer is investigated and further extended to nanostructured QW solar cell. Effect of different recombination mechanism in bulk solar cell is mathematically modelled. The basic model is also developed for J_{sc}, V_{oc}, efficiency and quantum efficiency of the solar cell. In this device model, in addition to the bulk properties of CZTS and CZTSe, intrinsic native defects like trap centres and defect states are considered. Radiative recombination is considered as the most likely loss mechanism in CZTS and CZTSe based thin film solar cells. The effect of defects and traps on performance of solar cell is also discussed in detail with a concluding remark that lesser defects and traps are essential to enhance the performance of the solar cell. Moreover, application of nanostructures and quantum confinement to kesterite solar cells are also highlighted. It is concluded that on increasing the number of QWs, the J_{sc} of the cell can be tuned without much detrimental effect on its V_{oc}. Theoretically, an impressive efficiency of 24.8% and fill factor of 79.8% are reported with 50 number of QWs.

References

1. Routray SR, Lenka TR (2017) InGaN-based solar cell: a wide solar spectrum harvesting technology for 21st century. CSI Trans ICT 6:83–96
2. NASA. Image of vanguard 1 satellite. https://history.nasa.gov/sputnik/galleryvan.html (Online; accessed 01 Oct 2020)
3. Qi C, Marco ND, Yang Y(Michael), Song TB, Chen CC, Zhao H, Hong Z, Zhou H, Yang Y (2015) Under the spotlight: the organic–inorganic hybrid halide perovskite for optoelectronic applications. Nano Today 10:355–396
4. Green MA, Hishikawa Y, Dunlop ED, Levi DH, Hohl-Ebinger J, Ho-Baillie AW (2018) Solar cell efficiency tables (version 52). Prog Photovolt Res Appl 26:427–436
5. Jatosado. Schematic of an electron beam physical vapor deposition system. https://en.wikipe dia.org/wiki/Electron_beam_physical_vapor_deposition#/media/File:Electron_Beam_Depo sition_001.jpg. (Online; accessed 01 Oct 2020)
6. Barbalace K. Periodic table of elements 1995–2016. http://EnvironmentalChemistry.com/yogi/ periodic/In.html. (Online; accessed 01 Oct 2020)
7. Mitzi DB, Gunawan O, Todorov TK, Barkhouse DAR (2013) Prospects and performance limitations for Cu-Zn-Sn-S-Se photovoltaic technology. Philos Trans R Soc Lond A Math Phys Eng Sci 371:20110432
8. Wang W, Winkler MT, Gunawan O, Gokmen T, Todorov TK, Zhu Y, Mitzi DB (2014) Device characteristics of CZTSSe thin-film solar cells with 12.6% efficiency. Adv Energy Mater 4:1301465
9. Ito K, Nakazawa T (1988) Electrical and optical properties of stannite-type quaternary semiconductor thin films. Jpn J Appl Phys 27:2094
10. Seol JS, Lee SY, Lee JC, Nam HD, Kim KH (2003) Electrical and optical properties of Cu_2ZnSnS_4 thin films prepared by RF magnetron sputtering process. Sol Energy Mater Sol Cells 75:155
11. Katagiri H, Jimbo K, Maw WS, Oishi K, Yamazaki M, Araki H, Takeuchi A (2009) Development of CZTS based thin film solar cells. Thin Solid Films 517:2455
12. Katagiri H, Saitoh K, Washio T, Shinohara H, Kurumadani T, Miyajima S (2001) Development of thin film solar cell based on Cu_2ZnSnS_4 thin films. Sol Energy Mater Sol Cells 65:141
13. Katagiri H, Sasaguchi N, Hando S, Hoshino S, Ohashi J, Yokota T (1997) Preparation and evaluation of Cu_2ZnSnS_4 thin films by sulfurization of E-B evaporated precursors. Sol Energy Mater Sol Cells 49:407
14. Moriya K, Tanaka K, Huchiki H (2005) Characterization of Cu_2ZnSnS_4 thin films prepared by photo-chemical deposition. Jpn J Appl Phys 44:715
15. Kamoun N, Bouzouita H, Rezig B (2007) Fabrication and characterization of Cu_2ZnSnS_4 thin films deposited by spray pyrolysis technique. Thin Solid Films 515:5949
16. Moriya K, Tanaka K, Huchiki H (2007) Fabrication of Cu_2ZnSnS_4 thin-film solar cell prepared by pulsed laser deposition. Jpn J Appl Phys 46:5780
17. Moriya K, Tanaka K, Huchiki H (2008) Cu_2ZnSnS_4 thin films annealed in H_2S atmosphere for solar cell absorber prepared by pulsed laser deposition. Jpn J Appl Phys 47:602
18. Tanaka K, Oonuki M, Moritake N, Huchiki H (2009) Cu_2ZnSnS_4 thin film solar cells prepared by non-vacuum processing. Sol Energy Mater Sol Cells 93:583
19. Shinde NM, Dubal DP, Dhawale DS, Lokhande CD, Kim JH, Moon JH (2012) Low cost and large area novel chemical synthesis of Cu_2ZnSnS_4 (CZTS) thin films. J Photochem Photobiol A 235:14

20. Cao M, Li L, Zhang BL, Huang J, Wang LJ, Shen Y, Sun Y, Jiang JC, Hu GJ (2009) One-step deposition of Cu_2ZnSnS_4 thin film for solar cells. Sol Energy Mater Sol Cells 93:583

21. Zhou Z, Wang Y, Xu D, Zhang Y (2010) Fabrication of Cu_2ZnSnS_4 screen printed layers for solar cells. Sol Energy Mater Sol Cells 94:2042

22. Shinde NM, Dubal DP, Dhawale DS, Lokhande CD, Kim JH, Moon JH (2012) Room temperature novel chemical synthesis of Cu_2ZnSnS_4 (CZTS) absorbing layer for photovoltaic application. Mater Res Bull 47:302

23. Washio T, Shinji T, Tajima S et al (2012) 6% efficiency Cu_2ZnSnS_4-based thin film solar cells using oxide precursors by open atmosphere type CVD. J Mater Chem 22:4021

24. Abermann S (2013) Non-vacuum processed next generation thin film photovoltaics: towards marketable efficiency and production of CZTS based solar cells. Sol Energy 94:37

25. Delbos S (2012) Kesterite thin films for photovoltaics: a review. EPJ Photovoltaics 3:35004

26. Courel M, Andrade-Arvizu JA, and Vigil-Galan O (2014) Towards a CdS/Cu_2ZnSnS_4 solar cell efficiency improvement: a theoretical approach. Appl Phys Lett 105:233501

27. Courel M (2019) An approach towards the promotion of Kesterite solar cell efficiency: the use of nanostructures. Appl Phys Lett 115:12390

28. Sravani S, Routray SR, Pradhan KP (2020) Towards quantum efficiency enhancement of kesterite nanostructured absorber: a prospective of carrier quantization effect. Appl Phys Lett 117:13

29. Bushnell D, Tibbits T, Barnham K, Connolly J, Mazzer M, Ekins-Daukes N, Roberts J, Hill G, and Airey R (2005) Effect of well number on the performance of quantum-well solar cells. J Appl Phys 97:124908

30. Lynch M, Ballard I, Bushnell D, Connolly J, Johnson D, Tibbits T, Barnham K, Ekins-Daukes N, Roberts J, Hill G et al (2005) Spectral response and I-V characteristics of large well number multi quantum well solar cells. J Mater Sci 40:1445–1449

Experiment

Nano-Material Based Sensitized Solar Cells

Abhigyan Ganguly and Viranjay M. Srivastava

Abstract Quantum dot sensitized solar cells are emerging technology in photovoltaics that utilize nanoparticles in the device structure. Nanoparticles are used as sensitizing materials in the devices that help absorb a broader solar spectrum region during its working. This chapter will briefly explain the evolution of single crystal solar cells evolution to nano-material-based solar cells. The general fabrication techniques and characterization parameters for such solar cells will be discussed briefly. The chapter describes nano-structured solar cells working without going into the physics and mathematical details, focusing more on the technology's practical aspects. The modern techniques adapted by researchers to further enhance solar cells' efficient working, utilizing nanotechnology, have been discussed in the chapter. The technological limitations faced by photovoltaic researchers have also been discussed.

Keywords Quantum dot · Solar cell · Nano-material · Silicon · Nanotechnology · VLSI

1 Introduction

The world has witnessed tremendous growth in terms of technology and population within the last few decades. As a result, worldwide power consumption has also been increased by many folds. The primary source of fuel is coal and oil, both of which are exhaustible resources. By the end of the twentieth century, the world population quadrupled, and energy demand increased sixteen times. It was reported that the annual energy of 13 TW (Tera-Watts) is required for a population of 6.5 billion worldwide. It is estimated that by 2050, another 10 Tw of clean energy will be needed to sustain the current lifestyle [1].

A. Ganguly (✉) · V. M. Srivastava
Department of Electronic Engineering, Howard College, University of KwaZulu Natal, Durban 4041, South Africa

V. M. Srivastava
e-mail: viranjay@ieee.org

© The Author(s), under exclusive license to Springer Nature Singapore Pte Ltd. 2022
R. Goswami and R. Saha (eds.), *Contemporary Trends in Semiconductor Devices*,
Lecture Notes in Electrical Engineering 850,
https://doi.org/10.1007/978-981-16-9124-9_7

On the other hand, the sun provides about 10^4 times extra energy than our daily global energy consumption. Thus solar energy is often considered one of the best options available. Solar energy can be converted into different forms of energy, like electricity, heat, or gas. But the most direct method of converting solar energy into electricity is using a solar cell [2]. The solar cell is a photovoltaic device based on the physics and principle of operation of junctions in semiconductors, either composed of two or multiple layers on either side of the junction, which converts light incident photons into electrical energy. Based on their underlying technology and operation, photovoltaic cells can be classified into first, second, and third-generation devices. Let's briefly look into each solar cell type to understand better where the current technology stands today.

1.1 Silicon Single Crystal Solar Cell (1st Generation)

The first generation of the solar cell is a single crystal pn junction. The schematic representation of silicon solar is shown in Fig. 1. The n side of the silicon crystal is heavily doped and kept very thin so that light falling on the n side can easily penetrate the crystal to reach the junction. The p-side is kept lightly doped so that most of the depletion region lies in the p-region. The penetration of photons into the crystal depends on the absorption coefficient of the crystal's material. Because of the built-in potential and the electric field, electronic charge carriers, that is, electron and holes, are generated at the junction. The photogenerated electrons move toward the n-region, while the holes move toward the p-region. An external connection to the two terminals provides a path for the electron. Through the load, the excess electron travels to recombine with the holes. In a silicon solar cell, the shorter wavelengths are absorbed in the n-side while the p-region absorbs the longer wavelengths.

The "Single-crystal Si solar cells" are also known as conventional solar cells and have an efficiency in the range of 22 to 24%. The efficiency of a single crystal solar cell is dependent on the band-gap of the crystal material. At present, single or polycrystalline p–n junction silicon cells are the most common solar cells. But these conventional solar cells suffer from inevitable significant setbacks. High-purity silicon crystals are required, along with high fabrication temperatures. Si crystal of

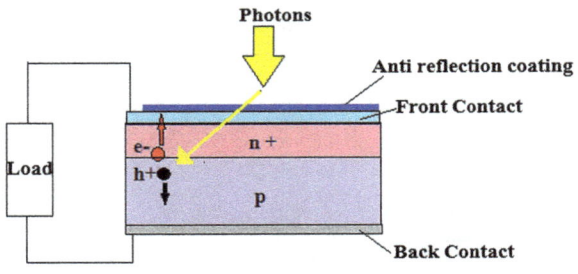

Fig. 1 Schematic representation of first-generation single crystal solar cells

such high purity is not easy to get, and also, a large amount of material needed for wafer-based cell results in significant cost issues [3]. Although they possess higher efficiencies, the cost-to-efficiency issue is a substantial setback for pn-junction solar cells.

1.2   Thin-Film Solar Cell (2nd Generation)

To overcome the high purity crystal requirement and cost issue in traditional solar cells, the second generation or Thin-film solar cells were invented. They are amorphous solar cells that utilize paste-like amorphous material layers instead of the crystalline semiconductor. The semiconductor material layer used in these devices has a thickness in the range of few micrometers only, hence the name. As the material requirement is significantly less and the fabrication process is relatively more straightforward; therefore the overall manufacturing cost of solar panels is low.

The materials used in second-generation solar cells are mainly amorphous silicon, Cadmium Telluride (CdTe), and Indium Gallium Di-Selenide (CIGS) [4]. At present, the second generation, that is, amorphous "thin-film solar cell" modules have reached the stage o commercialization. But compared to the first generation, these solar cells possess lower efficiency, but the main advantage is that they can be grown directly on glass substrates by sputtering and other techniques. Hence the overall cost of manufacturing is much lower. Further design improvements in a thin-film solar cell can result in enhancement inefficiency (Fig. 2).

A photon of energy ($=E_g$ eV) can excite only one electron from the lower energy valance band to the higher-energy conduction band, but if the photon has energy (h_v) $> E_g$, then excess energy will generate heat instead of generating extra EHPs. This limits overall efficiency called the "Shockley and Queisser limit," which specifies that maximum efficiency in a single crystal or single-layer solar cells, under standard illumination condition of AM 1.5G, cannot exceed 32.9% [5].

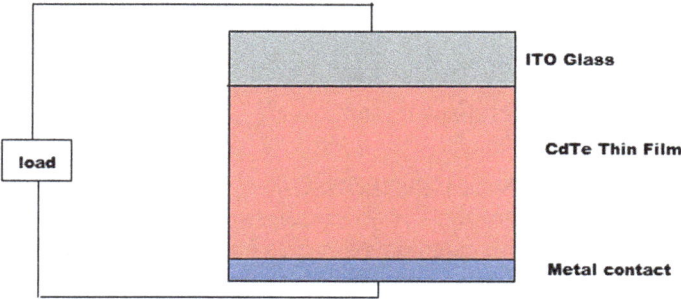

Fig. 2   Schematic diagram of the second-generation thin-film solar cell

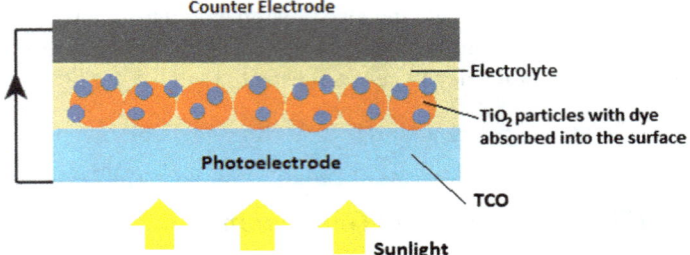

Fig. 3 Schematic diagram of the third-generation dye-sensitized solar cell

1.3 Dye-Sensitized Solar Cells (3rd Generation)

The idea of using light-absorbing material coating as a sensitizing layer in solar cells came from a leaf. A green leaf consists of chlorophyll pigment that has light-absorbing properties. Scientists came up with the idea, what if similar light-absorbing organic dyes be utilized in solar cells for better photonic absorption. The schematic representation of "third-generation solar cells" or, more commonly, "Dye-Sensitized Solar Cells" (DSSCs) is shown in Fig. 3. To overcome the Shockley- Queisser limit in single-layer solar cells, the third-generation devices utilize a tandem structure, that is, a multi-layer structure. An amorphous layer of titanium dioxide thin film is deposited on a Transparent Conduction Oxide (TCO) coated glass substrate. The oxide layer is then covered by organic dyes that have light-absorbing properties. Next, the photo-electrode is covered with an electrolyte (generally poly-sulfide electrolyte) and sandwiched with a Platinum-based counter electrode. Sunlight falling from the photo-electrode side passes through the transparent electrode to reach the organic dye layer, which excites electrons. The photogenerated electrons then flow into the titanium dioxide, where they travel through an external circuit for powering a connected load. But even this "dye-sensitized solar cells" suffers from drawback. The organic dyes are unstable toward water and oxygen and also expensive, as a result, organic dyes cannot be considered as a suitable sensitizing material [6, 7].

1.4 Quantum Dot Sensitized Solar Cell

Quantum Dot Sensitized Solar Cells (QDSSCs) or, more commonly, Quantum dot solar cells are derivatives of dye-sensitized solar cells [8]. Here the organic dyes are replaced by colloidal quantum dots to act as a sensitizing layer. The QDSSCs are an emerging field in low-cost photovoltaic research, as Quantum Dots (QDs) are easily synthesized and are stable toward water and oxygen. The quantum dots act as sensitizing layers in these devices and are deposited over a wide band-gap

Fig. 4 Schematic diagram of a quantum dot sensitized solar cell

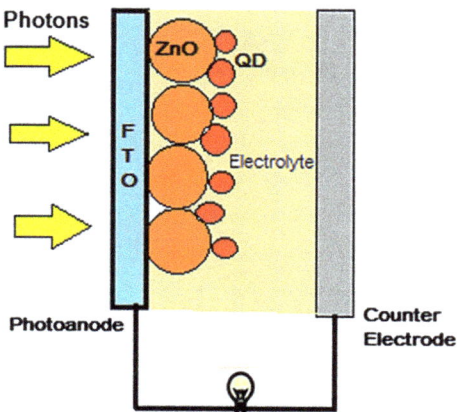

semiconductor oxide layer, typically TiO_2 or ZnO [9, 10]. A schematic representation of a typical QDSSC is shown in Fig. 4.

Quantum dots are considered zero-dimensional crystals with dimensions in the nanometer range (i.e., Billionth of a meter). Thus, quantum dots are systems where all dimensions are comparable to de Broglie's wavelength from a quantum mechanical viewpoint. Alternatively, they can be defined as three-dimensional confined systems of any material. Due to the size confinement, the electronic motion in a quantum dot is also restricted along all three dimensions. As a result of this quantum confinement effect, the optoelectronic properties in quantum dots are quite different from their bulk counterparts. Some of these properties are very useful in their use as sensitizing material in solar cells.

There are numerous advantages of using nano-materials/quantum dots as sensitizing material in solar cells. Due to the size-tunable band-gap in quantum dots, the band-gap of quantum dots can be adjusted to match with the band-gaps that of the material of the active oxide layer (TiO_2 or ZnO), on which they are to be deposited. Figure 5 shows how the electron can jump from the conduction band of quantum dots to that of the conduction band of oxide, in the case of the TiO_2/PbS-QD system, which is quite impossible in bulk PbS due to its smaller band-gap [11, 12].

Also, as the absorption in quantum dots ranges from the ultraviolet (UV) to that of the visible region, a wider range of solar spectrum ranging from UV to visible wavelengths can be utilized in photovoltaic cells. A larger Surface to Volume (S/V) ratio in nano-materials implies a larger surface area per unit volume for photon absorption. Multiple Exciton Generation (MEG) and a more significant extinction coefficient in QDs can be harnessed to improve solar cells' efficiency [13]. Doping and ion implantation of quantum dots with suitable transition metal ions can lead to a more significant current density in sensitized solar cells.

Fig. 5 Schematic
representation of electron
transfer from PbS quantum
dot to TiO$_2$

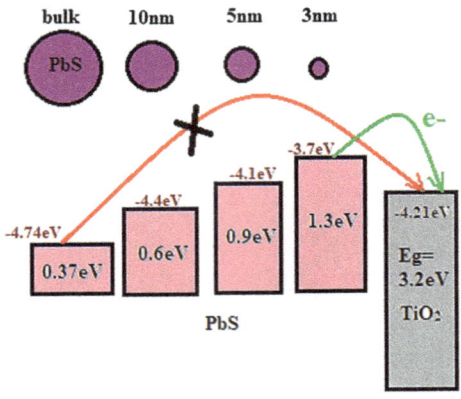

2   Fabrication and Working Principle of Quantum Dot Sensitized Solar Cells

The photo-electrode in a Quantum Dot Sensitized Solar Cell (QDSSC) is generally fabricated by adsorption and quantum dots' deposition on a wide bang gap oxide, typically TiO$_2$ or ZnO. This oxide thin film is sintered on Transparent Conducting Oxide (TCO) glass to form a mesoporous film. The most common and simplest method of deposition of oxide on TCO is by "Tape—template" and "Doctor blade technique" [14]. The thickness of the tape defines the thickness of the thin film deposited. The oxide deposited glass plates are heated at about 80 °C and then air annealed at around 450–500 °C. The primary purpose of heating the deposited layer is to harden the oxide ant and be better attached to the TCO glass surface. These oxide films act as electron conductors (or acceptors) and the transport layers [15].

Next, the quantum dots or nano-materials are deposited on the oxide film to act as the sensitizing layer. The simplest method of depositing the quantum dots on oxide is by dip coating, where the oxide deposited TCO glass plate is immersed in a solution of quantum dots or quantum dot precursors. This TCO/Oxide/QD coated glass plate would act as the photo-electrode, that is, an anode for the fabricated QDSSC. Few drops of the polysulfide electrolyte solution are then added to the photo-electrode using a dropper. They are attached with another conducting plate, forming a sandwiched structure with thin glass coverslip spacers to prevent any short between the two electrodes. This second conducting plate acts as a "counter electrode" or cathode of the device. The counter electrode can be held together with the transparent photo-anode using simple scotch tapes and paper clips. Two metallic probes were connected, each with the photo-anode plate and the cathode plate to act as connections. The detailed fabrication method of QDSSC is explained in Fig. 6, using a flow diagram and schematics. A cross-sectional Scanning Electron Microscopic (SEM) image of a quantum dot sensitized solar cell is shown in Fig. 7a, and a practical lab-made QDSSC is demonstrated in Fig. 7b.

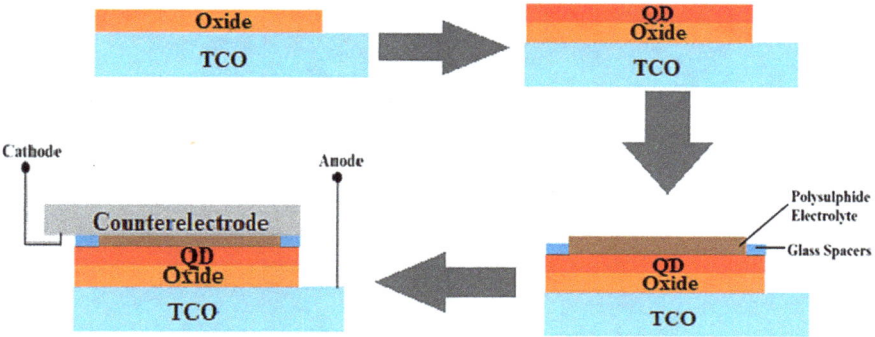

Fig. 6 Schematic diagram showing fabrication steps of a QDSSC

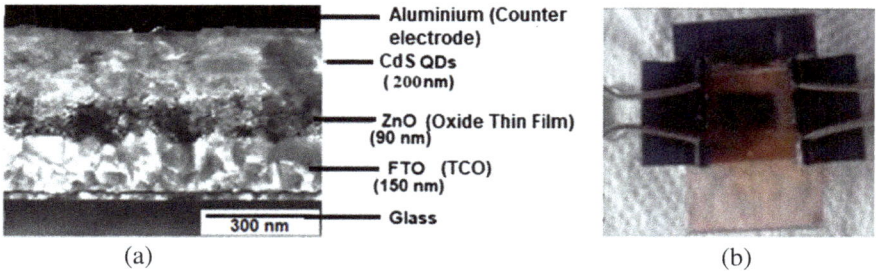

Fig. 7 **a** Cross-sectional SEM image of a QDSSC, **b** QDSSC prepared in a laboratory

During operation, light falling through the transparent conducting (TCO) electrode is captured by QDs, resulting in the generation of Electron–Hole Pairs (EHPs) [16]. The electrons are then separated from the holes, and they jump toward the oxide film, while redox couples in the electrolyte release the holes. Figure 8 shows the photon-induced charge transfer mechanism in an S_2/S_n^{2-} type redox couple system. Firstly, the charge is injected from the photo-excited QD into the oxide layer, and the photogenerated electrons move to the photo-electrode. Simultaneously, the hole gets transferred to the redox couple, thus, regeneration of the redox couple. The final step is the recombination of electrons from the QD and the oxidized form of the redox couple [17, 18].

The performance of solar cells is measured by obtaining and plotting the Current–voltage relation of the device. Photo-current is generally measured under the artificial AM 0.5G light illumination, which is the ideal condition spectrum for sunlight reaching the earth's surface after crossing the atmospheric layers. The two important parameters that are obtained from the Current Density (J) versus Voltage (V) plot are short circuit current density (J_{sc}) and Open-Circuit Voltage (V_{oc}). The other two critical parameters are the Fill Factor (FF) and Efficiency (η), which are calculated as follows [19, 20]:

Fig. 8 Working of a QDSSC

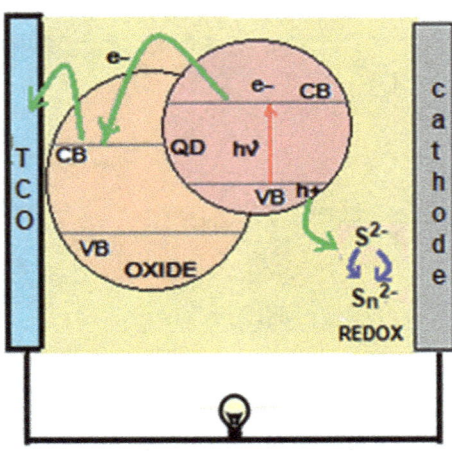

$$FF = \frac{J \max . V \max}{J_{sc} . V_{oc}} \quad \text{and} \quad \eta = \frac{V_{oc} \times J_{sc} \times FF}{P_{in}}$$

3 Components of a Quantum Dot Sensitized Solar Cell and Techniques to Improve Their Efficiency

With the basic understanding of structure and working of a primary nano-material-based sensitized solar cell, each of the components in a nano-material-based solar cell have been discussed in detail. It will also show the modifications and techniques adapted by modern researchers in each of the components to increase the solar cells' working efficiency even further.

3.1 Photo-Electrode

The photovoltaic performance of a Quantum dot sensitized solar cell is highly dependent on the satisfactory and proper assembly of quantum dots on the photo-electrode. Thus, oxide's structure and morphology play a vital role in the efficient working of quantum dot sensitized solar cells. The more the oxide layer's surface area, the more number of quantum dots can be deposited on them; thus, more photons can be harvested by them. However, solar cells' recombination process depends on the electrode surface area, which affects the open-circuit voltage (V_{oc}). The other factor influencing the overall performance of a quantum dot sensitized solar cell is the transport properties of generated careers, that is, electron–hole pairs, through the photo-electrode. Theoretically, as experimentally, it has been observed that electron transport is much faster for a higher degree of nano-material in photo-electrode.

The higher degree nano-material means nano-rods (or nanowires) and nano-tubes [21–24]. As the name suggests, nano-rods are rod-like elongated structures of nanometer dimension, and if the insides of these nano-rods are hollowed to form tube-like structures, they are called nano-tubes. Figure 9a shows the nano-rod arrays, and Fig. 9b shows a single nano-rod system.

In order to achieve higher photovoltaic efficiency in quantum dot sensitized solar cells, the active oxide is often replaced by the ZnO or TiO$_2$ nano-rod or nanotube-like structures instead of the thin film. Quantum dots deposited on a nano-rod surface is shown in Fig. 10. This improves the electron transport mechanism, and also, being elongated structures provides a higher surface area. It is replacing thin films with nano-structures increase the surface to volume ratio by many folds. As a result of which, the nono-structured oxide in photo-electrode provides more surface for quantum dot deposition. The more the number of sensitizing quantum dots, the more

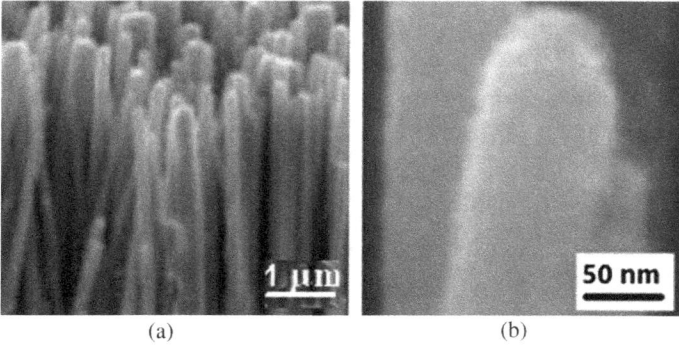

(a)   (b)

Fig. 9 High-Resolution Transmission Electron Microscope (HRTEM) image of **a** ZnO nano-rod array, **b** closer view of ZnO nano-rod

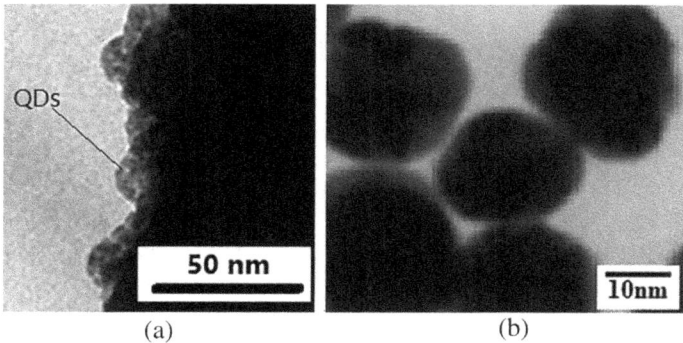

(a)   (b)

Fig. 10 **a** Quantum dots deposited in nano-rod surface, TEM of PbS quantum dots

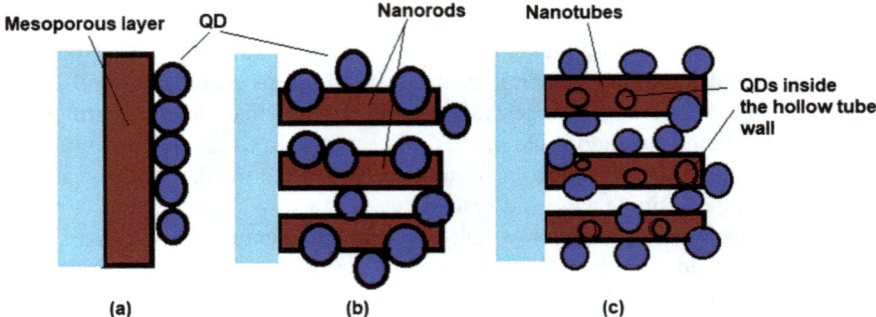

Fig. 11 The different morphologies of photo-electrodes for the QDSSCs are **a** mesoporous, **b** nano-rods or nanowires, and **c** nano-tubes

is the photon absorption, enhancing the overall efficiency of sensitized solar cells. In Fig. 11, it can be observed how the surface area in nano-tubes is higher than that of nano-rods, which has a higher surface area for quantum dot absorption than thin films. Also, the nano-tubes surface area is even higher than that of nano-tubes because of the internal hollow structure. Also, a faster electron transport reduces the chance of recombination. Therefore, to achieve better efficiency, it is necessary to balance recombination and photon harvesting. Recently, scientists have achieved even higher efficiency by using even advanced nano-material oxide structures such as nano-cones or nano-spikes [25].

Other methods that have been recently utilized to improve quantum dot sensitized solar cells' efficiency involve using multiple oxide layers or using doped oxide nano-structures. It has been experimentally observed that using multiple ZnO mesoporous layers in the photo-anode resulted in enhanced solar cell efficiency compared to single-layer devices [26]. Using copper and aluminum-doped ZnO nano-rods in sensitized solar cells has been proven to be more efficient than undoped nano-structured solar cells [27, 28].

3.2 Sensitizing Materials and Methods

As discussed earlier, using quantum dots as sensitizers in solar cells provides a series of advantages due to their unique properties. One of the significant advantages is the size dependency of light absorption property, due to which they can cover the whole solar spectrum. The most commonly used quantum dots are Lead Sulfide (PbS), Cadmium Sulfide (CdS), and Cadmium Selenide (CdSe). Other than this, other materials such as Lead Selenide (PbSe), Cadmium Telluride (CdTe), Zinc Sulfide (ZnS), Indium Arsenide (InAs), Indium Phosphide (InP), Bismuth (III) Sulfide (Bi_2S_3), Silver Sulfide (Ag_2S), and Antimony Trisulfide (Sb_2S_3) have also been used in the form of quantum dots as a sensitizer in solar cells. The proper coverage of the whole photo-electrode surface and adequate assembly of the quantum dots are vital for the cell's adequate working [3, 29].

The basic, benchtop method of deposition of quantum dots in-situ into the photo-anode is Chemical Bath Deposition (CBD) and Successive Ionic Layer Adsorption and Reaction (SILAR) [30]. In CBD, both the cationic and anionic precursors are present in a single solution, and the photo-anode is dipped into the solution for deposition. Whereas in SILAR, the anionic and cationic precursors are prepared in two separate containers. The oxide-coated photo-anode is dipped successively, first into the cationic solution and then into the anionic solution. The quantum dots are grown insitu on the oxide surface. Both CBD and SILAR are generally used to deposition sulfide type quantum dots on the photo-electrode, but recently they have been used in the case of selenides and tellurides, too [31, 32]. Figure 12 shows a schematic illustration of CBD and SILAR.

Another method involves ex-situ synthesis of the quantum dots and then depositing them into the photo-electrode surface either directly or via a linker. The electron transfer is relatively faster in the immediate attachment process, but due to quantum dots' aggregation, recombination occurs. Due to this, the sensitization process becomes slow. Thus, for improved functioning of QDSSC, a bifunctional molecular linker is often used. The nature of the bifunctional linker molecule has an essential role in the charge separation process in a solar cell, thereby affecting the conversion efficiency in a QDSSC. As shown in Fig. 13, a bifunctional linker

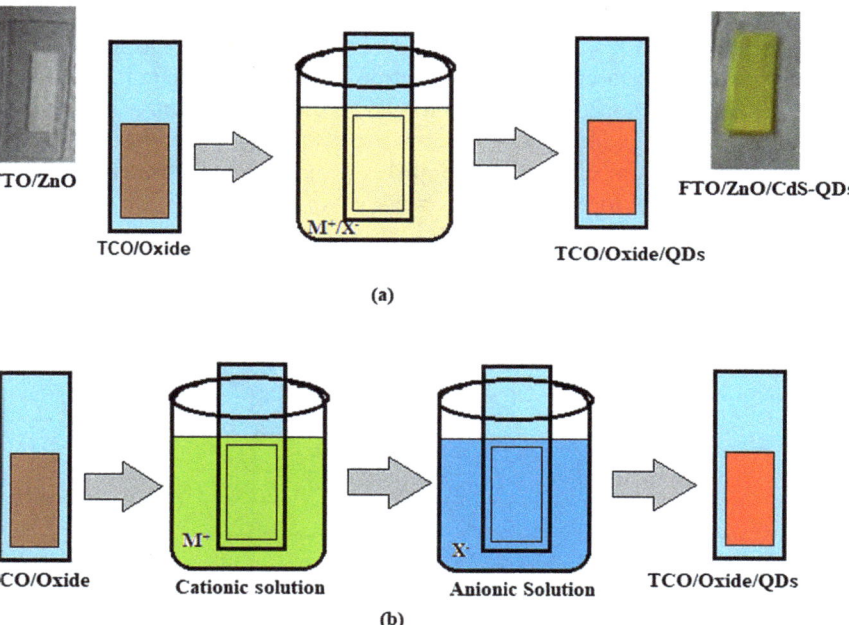

Fig. 12 Schematic illustration of **a** the CBD and **b** SILAR

Fig. 13 Schematic diagram
of a linker molecule

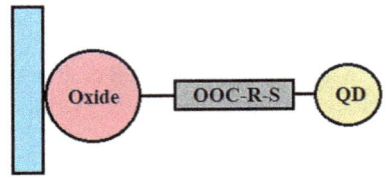

consists of a (COO)–R–SH structure [33]. The "R" in the molecule is the organic
core of the linker. While the carboxyl group gets attracted to the photo-anode oxide,
the thiol part is free to attach itself with the quantum dot. The linkers are first grown
on the oxide surface to modify the photo-electrode surface, and then they are simply
immersed into the QD dispersion solution. As the charge separation process depends
on the linker molecule, they also control the efficiency of generated photo-electrons
in solar cells. Hence optimization of the sensitization process and photo-electron
transfer is required for the proper working of QDSSC.

One of the recent methods for increasing the efficiency in QDSSC is by doping
the sensitizing quantum dots by transition metal ions [34]. The reason for choosing
transition metal ions as dopant lies in their unique electronic configuration, that is,
([Ar] ns^x $(n-1)d^y$). When excited by an external photon in transition metal ions,
the ion loses the electron in the s-orbital leaving behind an unpaired electron in
d-shell. This results in electron transition from d to d orbital in transition metal
ions. This photon-assisted d-d transition phenomenon increases current density (J)
in transition metal ion-doped quantum dot sensitized solar cells [35]. This increase in
current density, in turn, results in higher photovoltaic efficiency in QDSSCs. Figure 14
shows a schematic representation of an electron transfer mechanism in an undoped
QD on ZnO, and Cu doped (or implanted) QD on the ZnO system. More electrons
get transferred to the oxide because of extra d-d electronic transitions in Cu doped
(or implanted) QDs compared to the undoped system. Intermediate energy states are
created in between the forbidden energy band-gap of quantum dots on doping. This

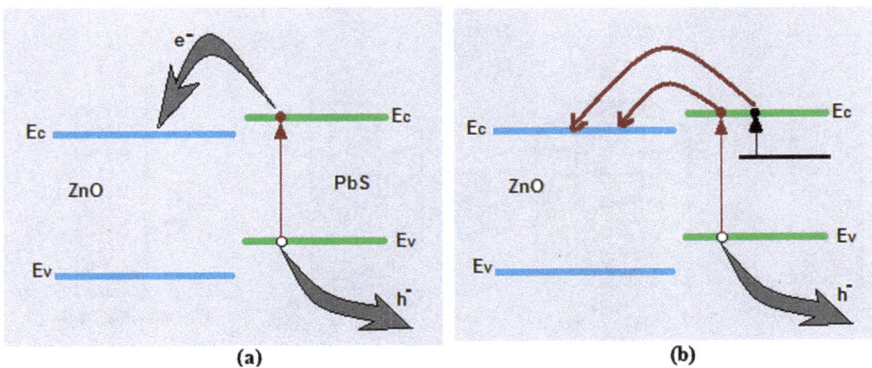

Fig. 14 Energy band diagram of **a** ZnO/PbS and **b** ZnO/PbS: Cu quantum dots

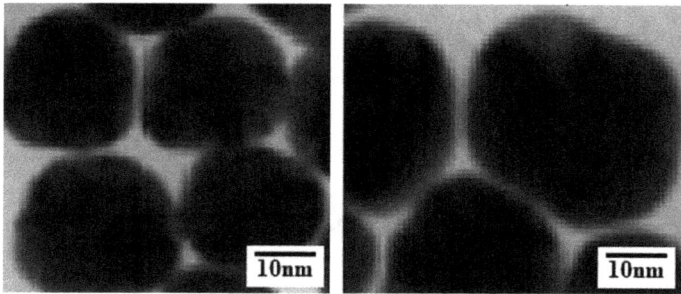

Fig. 15 HRTEM images of Ni SHI irradiated ZnS quantum dots at a higher irradiation level

facilitates the easier transfer of a higher number of electrons from PbS quantum dots to ZnO oxide, as illustrated. But in this process, the quantum dots cannot be doped beyond a specific limit. On excess doping, the dopant ions create excess midgap states that cause trapping photo-electrons instead of moving them to the oxide. This affects the charge transfer and hence the overall efficiency adversely [36].

Another improvised and a more sophisticated method of utilizing the property described above of transition metal ions in QDSSCs is ion irradiation. Recently, quantum dots irradiated with swift heavy ions of a transition metal have been incorporated as a sensitizer in solar cells. As ion irradiation of quantum dots results in better dopant ion implantation into the quantum dot crystal than that of chemical doping, irradiated quantum dot gives better results than chemically doped quantum dots in solar cells. In the case of irradiation, better doping gives rise to comparatively higher current density (J_{sc}), which enhances the overall efficiency in irradiated QD sensitized solar cells [37, 38]. But for a higher dose, the ion-implantation process is adversely affected by the formation of crystal defects, unwanted ion penetration, and agglomeration of closely spaced quantum dots, resulting in a fall in the current density [39]. This can be observed in the HRTEM image of irradiated quantum dots in Fig. 15.

In the second image, quantum dots' size due to quantum dots' agglomeration can be observed. Due to excess heat produced because of prolonged radiation treatment, quantum dots can fuse into one another. Thus, it is necessary to keep the radiation level low for the optimum function of QDSSCs. The current density versus voltage ($J–V$) characteristics of pristine, chemically doped, and iron irradiated QDSSCs are shown in Fig. 16. The corresponding solar cell parameters are shown in Table 1.

3.3 The Electrolyte and the Holes Transport Media

The redox potential in a sensitized solar cell can be adjusted by altering the concentration ratio of electrolytes. The electrolyte concentration also affects the charge transfer kinetics in the interface of QD and electrolyte. The most common electrolyte used

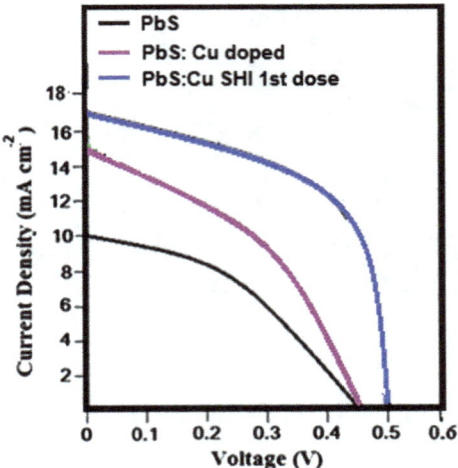

Fig. 16 Current density–voltage characteristics of undoped PbS, Cu doped PbS, and Cu SHI irradiated PbS quantum dot sensitized solar cell

Table 1 Photovoltaic parameters of pristine PbS, Cu doped PbS, and swift heavy ion of Cu^{+2} irradiated PbS quantum dot sensitized ZnO solar cells

Samples	V_{oc} (V)	J_{sc} (mA cm^{-2})	FF	H (%)
PbS	0.46	10.0	0.45	2.07
PbS:Cu doped	0.46	15.2	0.51	3.52
PbS:Cu irradiated	0.5	16.9	0.52	4.39

in QDSSCs is a polysulfide electrolyte, which is a solution of 2M of sodium sulfide (Na_2S) and 3M of sulfur (S) [40]. A Fe_{+3}/Fe_{+2} type electrolyte has also been utilized in CdS quantum dot-based sensitized solar cells [41]. Recently a new type of redox couple consisting of cobalt complex has been proposed to be used in PbS, CdSe, and CdS based QDSSCs [42].

But overall, it can be stated that the choice of electrolyte for QDSSCs is limited. The reason being, in most cases, the counter-electrode metal reacts with the electrolyte material, thus poisoning the counter-electrode. The chemistry of redox reaction in sensitized solar cells is complex, and as a result, it remains a challenge to prepare a proper electrolyte for the efficient working of sensitized solar cells. In dye-sensitized solar cells, I_3/I_2 type organic electrolyte works fine and is quite effective. But no such suitable electrolyte is still available for QDSSCs. Further research is very much needed in this direction.

Various researchers are trying to replace liquid electro-type with solid-state material, as handling a solid-state electrolyte is much easier, long-lasting, and more efficient. One such material recently used is spiro-MeOTAD and CuSCN [43]. However, the solid-state electrolyte used in QDSSCs has a significant drawback. The electrolyte material, such as MeOTAD, often penetrates the quantum dot sensitized nano-structured oxide film. This reduces the thickness of the mesoporous layer, resulting in incomplete light absorption in the solar cell. Thus this aspect requires the attention of future nano-material-based photovoltaics researchers.

3.4 The Counterelectrode

Often another TCO is used as a cathode in quantum dot sensitized solar cells. Platinum (Pt) electrodes are often used as a counter electrode, but Pt often results in low fill factor with polysulfide electrolyte. Gold (Au) may be used as an alternative material in QDSSCs, but Pt and Au are costly. Aluminum (Al) has also been used as counter-electrode material, which, although not highly efficient, is a cost-effective choice [37–39]. Recently, metal sulfides of Cobalt (Co), Copper (Cu), and lead (Pb) have been tested as counter-electrode material [44]. But it was observed that copper sulfide and cobalt sulfide both poisons the electrode surface on reacting with polysulfide electrolyte. Thus, researchers are still looking for suitable counter-electrode material for nano-material based sensitized solar cells.

4 Few More Techniques to Improve Efficiency

In the previous sections, we have already seen what modifications and techniques have been adapted by scientists to improve nano-material-based sensitized solar cells' efficiency. Various techniques have been adapted from nano-rods or nano-tubes instead of oxide, doping oxide or the sensitizing quantum dots, or ion irradiating quantum dots. Alternate materials for electrolyte and counter electrode have also been tested worldwide. Apart from this, let us look into some additional techniques that nanotechnology and photovoltaic researchers have utilized to date and have successfully achieved enhanced efficiency in QDSSCs.

Deposing a multi-layer porous ZnO layer instead of a single oxide layer has higher efficiency [45]. An additional compact layer of oxide (TiO_2) is often deposited first on the TCO, before depositing the mesoporous TiO_2 layer. The compact oxide layer, also called the blocking layer, prevents back transfer of an electron from TCO to polysulfide, thus increasing efficiency [46]. Post synthesis annealing the PbS quantum dot deposited ZnO-based anode at low temperature up to 140 °C has resulted in better efficiency [47]. Very recently, it has also been reported that passivation with molecular halide improves the working of PbS quantum dot-based solar cells [48]. Depositing an additional CdSe quantum dot buffer layer between the oxide and PbS quantum dot sensitizing layer has demonstrated better performance in sensitized solar cells [49]. Using quantum dot combination, such as PbS-CdS, as a sensitizer often increases efficiency in the solar cell [50].

5 Summary and Future Work

Nano-material based solar cells, more specifically QDSSCs, have attracted considerable attention for photovoltaic applications. The perfect crystallinity and size-tunable photonic absorption property in quantum dots make them a practical choice as sensitizing material in solar cells.

Although suitable material for wide band-gap oxide and quantum dot sensitizers is important, we have seen that various other factors can also be considered. Modern-day researchers strive to achieve higher efficiency in QDSSCs by following methods such as replacing nano-rods in place of thin-film, doping the oxide, or doping the quantum dots, ion irradiating quantum dots, using co-sensitized quantum dots, by annealing, and so on. Polysulfide electrolyte has shown good performance in quantum dot-based sensitized solar cells. However, scientists are yet to find a suitable material for counter-electrodes that can provide satisfactory and long-term stable performance. Alternative redox systems are also currently under investigation.

The nanotechnology-based sensitized solar cells hold promises for further improvements and analyses. They are a prospective candidate in the field of low-cost alternate photovoltaic devices. However, the efficiency achieved is still quite low and must be developed further before using them as a daily source of alternate power.

References

1. Kamat PV (2007) Meeting the clean energy demand: nanostructure architectures for solar energy conversion. J Phys Chem C 111(7):2834–2860
2. Ruhle S, Shalom MA, Zaban A (2010) Quantum-dot- sensitized solar cells. Chem Phys Chem 11: 2290–2304
3. Chen LY, Yin Y, Ho Th et al (2014) Sensitized solar cells via nano-materials a recent development in quantum dots-based solar cells. IEEE Nanotech Mag 16–21. https://doi.org/10.1109/MNANO.2014.2314182
4. Semonin OE, Luther JM, Beard MC (2012) Quantum dots for next-generation photovoltaics. Mat Today 15:508–515
5. Shockley W, Queisser HJ (1961) Detailed balance limit of efficiency of p-n junction solar cells. J Appl Phys 32(3):510–519
6. Ruhle S, Shalom M, Zaban A (2010) Quantum dot sensitized solar cells. Chem Phys Chem 1:2290–2304
7. Oregan B, Gratzel MA (1991) Low-cost, high-efficiency solar-cell based on dye-sensitized colloidal TiO_2 films. Nature 353:737–740
8. Nozik AJ (2002) Quantum dot solar cells. Phys E 14(1–2):115–120
9. Zaban A, Micic OI, Gregg BA et al (1998) Photosensitization of nanoporous TiO_2 electrodes with InP quantum dots. Langmuir 14:3153–3156
10. Lee HJ, Kim DY, Yoo JS et al (2007) Anchoring cadmium chalcogenide quantum dots (QD) onto stable oxide semiconductor for QD sensitized solar cells. Bull Korean Chem Soc 28:953–958
11. Moreels I, Lambert K, Smeets D et al (2009) Size-dependent optical properties of colloidal PbS quantum dots. Acs Nano 3:3023–3030
12. Graetzel M, Janssen RAJ, Mitzi DB, Sargent EH (2012) Materials interface engineering for solution-processed photovoltaics. Nature 488:304–312

13. Kamat PV (2013) Quantum dot solar cells- the next big thing in photovoltaics. J Phys Chem Lett 4:908–918
14. Ganguly A, Nath SS, Gope G et al (2018) A back illuminated solar cell using PbS quantum dots an sensitizers. Int J Nanoparticles 10(3):218–224
15. Kamat PV (2008) Quantum dot solar cells. Semiconductor nanocrystals as light harvesters. J Phys Chem C 112(48):18737–18753
16. Tada H, Fujishima M, Kobayashi H (2011) Photodeposition of metal sulfide quantum dots on titanium(IV) dioxide and the applications to solar energy conversion. Chem Soc Rev 40:4232–4243
17. Lee YL, Lo YS (2009) Highly efficient quantum-dot-sensitized solar cell based on co-sensitization of CdS/CdSe. Adv Funct Mater 19(4):604–609
18. Zhang QF, Chou TR, Russo B, Jenekhe SA et al (2008) Aggregation of ZnO nanocrystallites for high conversion efficiency in dye-sensitized solar cells. Angew Chem Int Ed 47:2402–2406
19. Ganguly A, Srivastava VM (2020) Synthesis and characterization of Fe doped CdS quantum dots. In: 2nd international conference on VLSI device, circuit and system, 5–8
20. Chou TP, Zhang QF, Fryxell GE et al (2007) Hierarchically structured ZnO film for dye-sensitized solar cells with enhanced energy conversion efficiency. Adv Mater 19:2588–2592
21. Seol M, Kim H, Tak Y, Yong K (2010) Novel nanowire array based highly efficient quantum dot sensitized solar cell. Chem Commun 46:5521–5523
22. Seol M, Ramasamy E, Lee J, Yong K (2011) Highly efficient and durable quantum dot sensitized ZnO nanowire solar cell using noble-metal-free counter electrode. J Phys Chem C 115:22018–22024
23. Kongkanand A, Tvrdy K, Takechi K et al (2008) Quantum dot solar cells. Tuning photoresponse through size and shape control of CdSe-TiO$_2$ architecture. J Am Chem Soc 130:4007–40015
24. Tian J, Cao G (2013) Semiconductor quantum dot sensitized solar cells. Nano Rev 4:22578. https://doi.org/10.3402/nano.v4i0.22578
25. Liu H, Zhang G, Sun W, Shen Z, Shi M (2015) ZnO hierarchical nanostructure photoanode in a CdS quantum dot sensitized solar cell. PLoS ONE 10(9):e0138298
26. Chen H, Li W, Liu H et al (2011) CdS quantum dots sensitized single- and multi-layer porous ZnO nanosheets for quantum dots-sensitized solar cells. Electrochem Comm 13:331–334
27. Raja M, Muthukumarsamy N et al (2015) Enhanced photovoltaic performance of quantum dot-sensitized solar cell fabricated using Al-doped ZnO nano-rod electrode. Superlatt Microstruct 80:53–62
28. Poornime K, Krishnan KG, Lalitha B et al (2015) CdS quantum dots sensitized Cu doped ZnO nano-structured thin films for solar cell. Superlatt Microstruct 83:147–156
29. Ganguly A, Nath SS (2019) Nickel doped ZnS quantum dots for sensitization in solar cell. J Nanoelectron Optoelec 14(2):286–290
30. Yang Z, Zhang Q, Xi J, Park K, Xu X, Liang Z et al, CdS/CdSe co-sensitized TiO$_2$ solar cell prepared by jointly using the successive ion layer absorption and reaction (SILAR) method and chemical bath deposition (CBD) process. Sci Adv Mater 4: 1013–1017
31. Gorer S, Hodes G (1994) Quantum size effects in the study of chemical solution deposition mechanisms of semiconductor films. J Phys Chem 98(20):5338–5346
32. Lee H, Wang M, Chen P (2009) Efficient CdSe quantum dot-sensitized solar cells prepared by an improved successive ionic layer adsorption and reaction process. Nano Lett 9(12):4221–4227
33. Robel I, Subramanian V, Kuno (2006) Quantum dot solar cells. Harvesting light energy with CdSe nanocrystals molecularly linked to mesoscopic TiO$_2$ films. J Am Chem Soc 128(7):2385–2383
34. Ganguly A, Nath SS (2020) Mn-doped quantum dots as sensitizers in solar cells. Mater Sci Eng B 225:114532
35. Santra PK, Kamat PV (2012) Mn-doped quantum dot sensitized solar cells: a strategy to boost efficiency over 5%. J Am Chem Soc 134:2508–2511
36. Ganguly A, Nath SS, Choudhury M (2018) Effect of Mn doping on multilayer PbS quantum dots as sensitized solar cell. IEEE J Photovoltaics 8(6):1656–1661

37. Ganguly A, Nath SS, Srivastava VM (2021) Comparative analysis of ZnO quantum dots synthesized on PVA and PVP capping matrix. Nanosyst Nanomater Nanotechnol 19(2):337–345
38. Ganguly A, Srivastava VM (2020) Enhanced efficiency in swift 100MeV Ni ion irradiated ZnS quantum dot sensitized solar cells. Chalcogenide Lett 17(10):487–493
39. Ganguly A, Nath SS, Choudhury M (2019) Effect of Cu doping and ion irradiation of PbS quantum dots and their applications in solar cells. IET Optoelectron 13(3):113–117
40. Radich J, Peeples N, Santra P et al (2014) Charge transfer mediation through CuxS. The hole story of CdSe in polysulfide. J Phys Chem C 118:16463–16471
41. Tachibana Y, Akiyama HY, Ohtsuka Y et al (2007) CdS quantum dots sensitized TiO$_2$ sandwich type photoelectrochemical solar cells. Chem Lett 36(1):88–89
42. Lee HJ, Chen P, Moon SJ et al (2009) Regenerative PbS and CdS quantum dot sensitized solar cells with a cobalt complex as hole mediator. Langmuir 25(13):7602–7608
43. Lee HJ, Yum JH, Leventis HC et al (2008) CdSe quantum dot-sensitized solar cells exceeding efficiency 1% at full-sun intensity. J Phys Chem C 112(30):11600–11608
44. Chen P, Yum JH, Angelis FD et al (2009) High open-circuit voltage solid-state dye-sensitized solar cells with organic dye. Nano Lett 9(6):2487–2492
45. Haining C, Li W, Liu H (2011) CdS quantum dots sensitized single- and multi-layer porous ZnO nanosheets for quantum dots-sensitized solar cells. Electrochem Comm 13:331–334
46. Kim J, Choi H, Nahm C et al (2011) The effect of a blocking layer on the photovoltaic performance in CdS quantum-dot-sensitized solar cells. J Power Sourc 196:10526–10531
47. Wanga H, Yangb S, Wangb Y et al (2017) Influence of postsynthesis annealing on PbS quantum dot solar cells. Org Electron 42:309–315
48. Lan X, Voznyy O, Kiani A et al (2016) Passivation using molecular halides increases quantum dot solar cell performance. Adv Mat 28(2):299–304
49. Zhao T, Goodwin ED, Guo J et al (2015) Advanced architecture for colloidal PbS quantum dot solar cells exploiting a CdSe quantum dot buffer layer. ACS Nano. https://doi.org/10.1021/acsnano.6b03175
50. Speirs MJ, Balazs DM, Fang HH et al (2015) Origin of the increased open circuit voltage in PbS–CdS core–shell quantum dot solar cells. J Mater Chem A 3:14501457

Lateral Straggle Parameter and Its Impact on Hetero-Stacked Source Tunnel FET

K. Vanlalawmpuia and Brinda Bhowmick

Abstract The strict power restrictions of integrated circuits and the non-scalability of the subthreshold slope in a metal-oxide semiconductor field-effect transistor (MOSFET) present major challenges to the continuous scaling of field-effect transistors. Henceforth, tunnel field-effect transistors (Tunnel FETs) have been deemed an optimistic contender to substitute the well-known conventional MOSFET as they possess a steep subthreshold slope lower than 60 mV/decade along with its fast switching characteristics at a low operating power supply voltage. Despite numerous advantages, the performance of the tunnel FET primarily relies on the accuracy in the manufacturing procedure. Ion implantation techniques have been employed to comprehend the variations in non-zero tilting angle. This extends the dopants from the source as well as drain regions into the channel region and affects the performances of tunnel FET significantly. Using Technology Computer Aided Design (TCAD) simulations, detailed examination on the impact of variations in lateral straggling parameters (σ) for a hetero-stacked source tunnel FET (HSS-TFET) is carried out. The chapter focuses on the investigation of Analog/RF figure of merits of the HSS-TFET due to lateral straggling variations. A mixed-mode HSS-tunnel FET is analyzed by the implementation of a digital HSS-tunnel FET inverter for variations in the straggling parameter.

Keywords Ion implantation · Lateral straggle · Tilt · Tunnel FET

1 Introduction

One of the most significant aspects of emphasis in the semiconductor technologies, particularly in semiconductor devices is the continuous downscaling of the device dimensions. Researchers have made numerous advancements in this process—from the massive vacuum tubes leading to large-room computers to the first metal-oxide semiconductor field-effect transistor (MOSFET) with a device channel length of 300

K. Vanlalawmpuia (✉) · B. Bhowmick
Department of Electronics and Communication Engineering, National Institute of Technology Silchar, Assam, India

micro-meters to the field-effect transistor (FETs) having a channel length of ~14 nano-meters or even lesser which resulted in integrated circuits (ICs) comprising of billions of transistors in it. Downscaling the size of MOSFETs has several advantages, in addition to increasing the number of transistors in an IC. Reduction in gate length of the MOSFET contributes to a lower gate capacitance, which increases the circuit switching speed. In addition, the voltage scaling which is essentially associated with device miniaturization even lowers the device power consumption. But the off-state power consumption of the MOSFET had become serious issue, when device sizes were minimized to ~50 nm along with power supply to 0.5 V [1–3]. Thermionic emission is the operating mechanism in MOSFETs which controls the drain current. The potential barrier at the source/channel junction decreases as the gate bias increases, resulting in an increment in the drive current which triggers two major issues: a substantially high off-state current as a consequence of subthreshold conduction as well as a high subthreshold swing. The subthreshold swing (SS) of MOSFET is well-defined as the amount of gate voltage (V_G) that is essential to lower the drain current (I_D) by one order of magnitude. With the intention of maintaining a higher on-state current to the off-state current (I_{ON}/I_{OFF}), a steeper subthreshold slope is required and thus will result in a reduced power dissipation in the off-state condition. The SS of the metal-oxide semiconductor FET, that is the reciprocal of the subthreshold slope is determined by:

$$SS = \left(\frac{d\,(\log_{10} I_D)}{d\,V_G}\right)^{-1} \approx \ln\,(10)\frac{kT}{q}\left(1 + \frac{C_d}{C_{ox}}\right) \tag{1}$$

where the depletion capacitance and the oxide capacitance are denoted by C_d and C_{ox}, respectively. The least possible subthreshold swing value for MOSFETs is 2.3 kT/q, that is 60 mV per decade at 300 K as seen from Eq. 1. An alternative method and device design are required to resolve certain fundamental limitations of large off-state leakage currents and a high subthreshold slope of the MOSFET.

Tunnel field-effect transistor (TFET) is among the device structure implementations that can be a feasible substitute to the conventional MOSFET. Owing to a fundamental difference in the current control mechanism from MOSFETs, TFETs can have a subthreshold swing of below 60 mV/decade. The drain current in MOSFETs depend on thermionic emission of charge carriers over the potential barrier between the source and channel region. On the other hand, the current in Tunnel FETs conversely rely on the charge carriers tunneling via a potential barrier from the valence band (E_V) of the source to the conduction band (E_C) of the channel. Since this potential barrier in tunnel FET is exceptionally large in the off-state, the transistor exhibits a substantially low off-state currents (I_{OFF}) in addition to a low average subthreshold swing [4–6]. Apart from the steep subthreshold slope and the low leakage currents, tunnel FETs possess a greater immunity to numerous short channel effects (SCEs) for instance drain-induce barrier lowering (DIBL), threshold voltage roll-off, charge sharing due to leakage currents, etc. as compared to MOSFETs [7]. It should be noted that tunnel FET differs solely in the kind of source doping from MOSFET.

Hence, the fabrication procedure of the tunnel FET is quite similar as the fabrication process of MOSFET.

2   Conventional MOSFET and Tunnel FET Structures and Their Operating Mechanism

In simple term, the tunnel FET is a gated reverse-biased p-i-n structure that works based on the principle of the quantum mechanical band-to-band tunneling (BTBT) of carriers from the valence band of the source region to the conduction band of the channel region through the forbidden energy bandgap. Tunnel FETs can be implemented both for p-type and n-type. However, this chapter will focus only on n-type TFET as the principle of operation is similar for both types (p-type injection into the channel from the source is holes, whereas it is electrons for n-type). For an n-type tunnel FET, the source region is doped with p^+, whereas the drain region is doped with n^+. The source and the drain are both doped with n^+ in n-type MOSFET, whereas the channel region is p-type. The channel is usually intrinsic or lowly doped semiconductor material of either type for tunnel FETs. Conventional structures for both n-type MOSFET and n-type tunnel FET are shown in Fig. 1.

In n-channel MOSFET, where the operating mechanism is based on thermionic emission, when a sufficient positive gate-to-source voltage (V_{gs}) is given to the gate electrode, the barrier potential in the channel region gets reduced thereby lowering the barrier height from the source to the channel region. In this manner, electrons can easily surmount this barrier height and are swept from the channel region to the drain terminal due to the positive bias at the drain side. However, in the case of n-type tunnel FET where quantum mechanical band-to-band tunneling (BTBT) is the conduction mechanism, when a positive gate-to-source voltage (V_{gs}) is given to the gate electrode, it induces a band bending at the source and channel junction resulting in tunneling of electrons from the valence band of the source region to the conduction band of the channel region. These electrons which are tunneled through the channel region are then collected to the drain terminal due to the positive drain to source voltage (V_{ds}). The tunnel FET is then considered to be in the on-state when

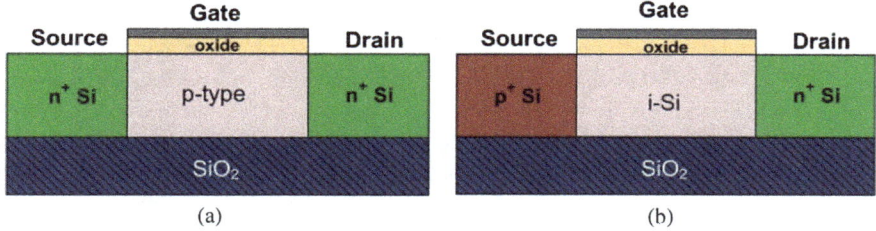

(a)   (b)

Fig. 1 Schematic of n-type **a** MOSFET and **b** tunnel FET

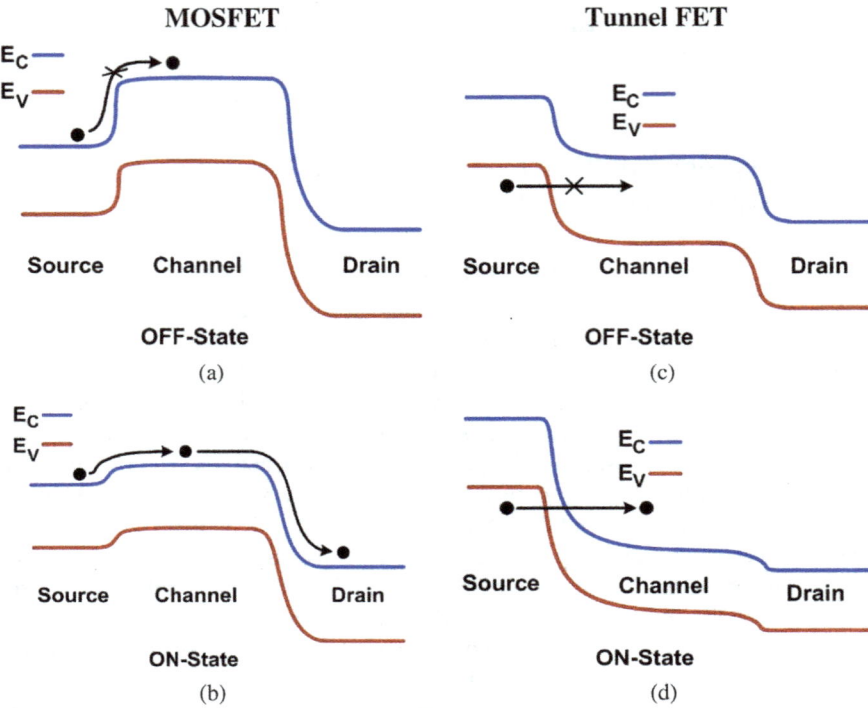

Fig. 2 Energy band diagram of conventional MOSFET and tunnel FET

tunneling of electrons occur, whereas when the gate bias is zero and no tunneling takes place from the source into the channel region, it is considered to be in the off-state. Figure 2 shows the energy band profile of both conventional n-channel MOSFET and n-type tunnel FET for both their on-state as well as off-state, respectively.

Here, the tunnel FET is in its off-state condition once the drain bias, $V_{DS} > 0$ V as well as the gate bias, $V_{GS} = 0$ V, as in the case of MOSFETs. Whatever charge carrier is there in the channel region's conduction band tends to get drifted to the drain side then accordingly produce a current in the off-condition in tunnel FETs. Be that as it may, since the source is made up of p-type semiconductor material, there are not many free electrons present in the conduction band, subsequently only few electrons are possible to be infused into the channel region which entails an insignificant off-state leakage current. It is also worth noting that on account of MOSFET, the source region and drain region are of n-types and possess free electrons in the E_C. Some of these free electrons could be infused to the channel region through the potential barrier across the source/channel region via thermionic emission which corresponds to a higher off-state current in MOSFETs in comparison with tunnel FETs.

However, when the gate voltage V_{GS} is enhanced to a large extent, the channel's energy band varies in relation to the source. At a certain V_{GS} value, the valence band of the source region is aligned with the conduction band of the channel region allowing electrons to tunnel from the valence band of the source into the conduction band of the channel through the potential barriers formed by the forbidden energy bandgap. In tunnel FETs, the gate bias where the source's valence band aligns with the channel's conduction band is termed as the commencement of the on-state. As the gate bias is enhanced further, energy band in the channel area is further suppressed resulting in a decrement in the tunneling barrier distance (also known as the tunneling length), more electrons possibly will tunnel into the conduction band of the channel, that resulted in a sharp enhancement in the drive current. It should also be noted that the tunneling probability given by Wentzel–Kramers–Brillouin (WKB) approximated equation from Eq. 2 enhances with increasing gate voltage, resulting in increased current [8, 9]. Thus, increasing V_{gs} raises not just the amount of electrons capable of tunneling but also the chances of tunneling. As a result, the current will also vary substantially when V_{gs} is varied.

$$T_{WKB} \approx \exp\left[-\frac{4\lambda\sqrt{2m^*E_g^3}}{3\,q\hbar(E_g + \Delta\Phi)} \right] \tag{2}$$

where, q symbolizes the electron charge, the effective carrier mass is m^*, E_g is the forbidden energy bandgap of the semiconductor material, λ is the screening tunneling length that defines the spatial range in the transition region at the source/channel edge which is reliant on the transistor geometric arrangements. The energy range over which tunneling takes place is denoted using $\Delta\Phi$ while $\hbar$ signifies the reduced Planck's constant. Here, higher on-current will be attained for higher tunneling probability, which should be unity for best-case scenario. From Eq. 2, it is clear that for the purpose of increasing the on-state current of the tunneling field-effect transistor, modification in the structure and materials or in other words, optimized design of the tunnel FET is required.

3   Subthreshold Swing in MOSFETs Versus Tunnel FETs

Again, looking at one of the main premises on why we are looking for a replacement for MOSFETs is due to possibility of a better steeper subthreshold slope attainable by the tunnel FETs [10]. Although MOSFETs have a thermal limitation of 60 mV/decade on the SS, Tunnel FETs can reach a subthreshold swing lower than 60 mV/dec since they work on dissimilar operating mechanisms i.e., tunneling of electrons in comparison to the MOSFET which is the thermionic emission. That being said, with the discrepancy in the current conduction mechanism in the subthreshold region, the characterization of the subthreshold swings are also different for MOSFET and tunnel FET. In the subthreshold region, MOSFETs conduct current by the mechanism of diffusion of carriers across the channel which is given by Eq. 3 [11]:

$$I_{DS} = \mu_{eff}\frac{W}{L}\sqrt{\frac{\varepsilon_{Si}\, q\, N_A}{4\,\psi_B}}\left(\frac{kT}{q}\right)^2$$
$$\exp^{q(V_{GS}-V_{th})/mkT}\left(1 - \exp^{-q\,V_{DS}/kT}\right) \qquad (3)$$

where ψ_B denotes the fermi potential of the body and the other terms have their usual meaning. The MOSFET's subthreshold swing (SS) is determined as:

$$SS = \frac{d\,(\log I_{DS})}{d\,V_{GS}} \qquad (4)$$

Be that as it may, tunnel FETs conduct currents by the mechanism of band-to-band tunneling in the subthreshold region as well. Hence, with the gate voltage variation, the subthreshold swing also fluctuates and fails to attain a fixed value in the case of MOSFETs. However, in tunnel FETs, there are two kinds of subthreshold swing. They are defined as: (1) point subthreshold swing and (2) average subthreshold swing.

The point subthreshold swing (SS_{pnt}) of a tunnel field-effect transistor is determined as the minimum value of the subthreshold swing anywhere on the transfer characteristics graph, usually measured at the steepest point in the curve. It is given by:

$$SS_{pnt}\,(V_{GS}) = \frac{d\,(\log\,(I_{DS}(V_{GS})))}{d\,V_{GS}} \qquad (5)$$

While the average subthreshold swing (SS_{AVG}) of a tunnel FET is specified across a range of the gate voltage values, it is expressed as:

$$SS_{AVG} = \frac{V_{TH} - V_{OFF}}{\log(I_{DS}(V_{TH}) - I_{DS}(V_{OFF}))} \qquad (6)$$

where V_{TH} designates the threshold voltage in tunnel FETs and V_{OFF} signifies the gate voltage at the off-state i.e., $V_{GS} = 0$ V. $I_{DS}(V_{TH})$ refers to the drain current at the threshold voltage and $I_{DS}(V_{OFF})$ is the value of the drain current at zero gate voltage. It is usually measured using constant current method. Nevertheless, the average subthreshold swing is more important in determining the device characteristic as relying exclusively on the reduction of SS_{pnt} is inadequate for ultra-low voltage operation. Therefore, from this point onward, the subthreshold swing in this chapter will refer to the average subthreshold swing, unless mentioned otherwise.

4   Enhancing the On-State Current of Tunnel FET

From time to time, we keep mentioning that the main reason why tunnel field-effect transistors ought to be regarded as a substitution for conventional MOSFET for future low-power applications owing to the SS_{AVG} being lower than the 60 mV/dec at room temperature. It is worth mentioning, however, that tunnel FETs should attain a subthreshold slope of lesser than 60 mV/dec for at least four decades of the drain current in becoming a legitimate contender for the substitution of MOSFETs. Hence, this is a very challenging task for researches [12–14]. The most vital limitation of TFETs is the low on-state currents (I_{ON}), primarily owing to the band-to-band tunneling that supports low current levels for silicon and other semiconductor materials possessing silicon-like bandgap. Until and unless the on-state in the tunnel FET is raised to a range comparable to that of MOSFETs, the tunnel FET cannot be implemented in the conventional complementary metal-oxide semiconductor (CMOS) circuit. Hence, so as to improve the on-state currents in tunnel FETs, many non-conventional tunnel FET device structures have been proposed by researchers. However, in this section, only few of the techniques and implementation relevant to this chapter are discussed here.

Several notable tunnel FET device structures proposed include the silicon-on-insulator (SOI) tunnel FET shown in Fig. 3a in which the whole thin silicon layer gets depleted and an oxide layer buried prevents all the source-to-drain leakages pathway via the bulk area, the device performance is increased by offering higher transconductance, enhanced electrostatic channel control thereby attaining higher on-state current [15, 16]. Double gate (DG) tunnel FET which comprises of two gates, one of those at the top (which is known as the front gate) and another one which is situated at the bottom (known as the back gate) which enhances the electrostatic controlling of the gate over the channel region is depicted in Fig. 3b. In contrast with a

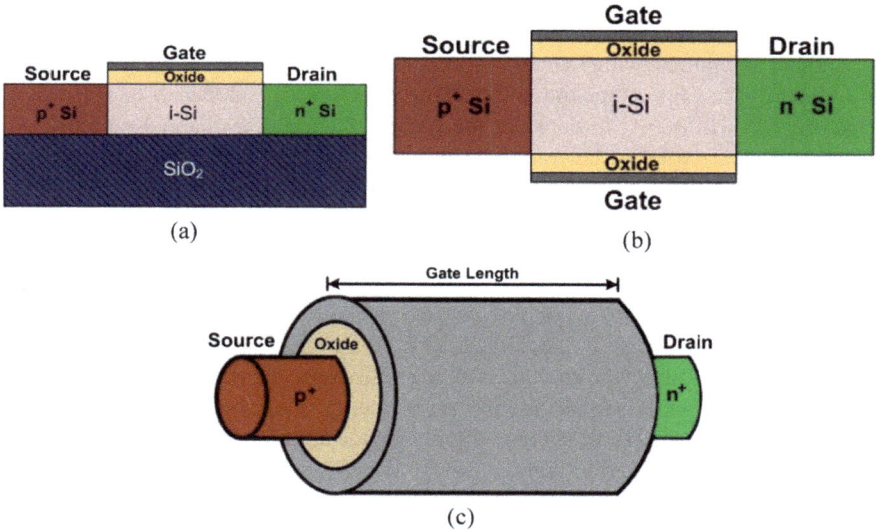

(a)   (b)

(c)

Fig. 3 **a** Silicon-On-Insulator (SOI) TFET, **b** Double Gate (DG) TFET and **c** Gate-All-Around (GAA) TFET

single gate tunnel FET, the on-current also improves as the TFET will benefit from the additional gate [17, 18]. This enhanced gate control seen in a double gate tunnel FET can be further enhanced in a Gate-All-Around (GAA) tunnel FET. GAA tunnel FETs shown in Fig. 3c can obtain a higher amount of electrostatic controlling by the gate, since the electric field lines emerging from the drain region get terminated at the gate without substantial intrusion toward the channel resulting in steeper subthreshold slope and reduction in SCEs thereby enhancing the on-state current [19–21].

Another well-known technique to boost the on-current is gate oxide engineering in which the thickness of the gate insulator (t_{ox}) is reduced or the dielectric constant (κ) of the gate material is enhanced by utilizing another dielectric material. As t_{ox} and κ of the gate insulator are exponentially associated with the tunneling current, the on-state current is increased significantly due to gate oxide engineering in tunnel FETs. However, it should be pointed out that in case of conventional MOSFETs, the on-current increases linearly by the reduction of the gate insulator thickness or enhancing the dielectric constant of the gate material. Tunnel FETs are thus predicted to benefit even more in the on-current due to gate dielectric engineering than the improvement made with the same technique in MOSFETs. Since the dielectric gate thickness cannot be reduced to a certain limit as direct tunneling of carriers through the oxide (which is undesirable) restricts further reduction in dielectric thickness of the gate. The technique of using high-κ gate dielectric material having a higher permittivity instead of the popular silicon dioxide (SiO_2) owns an additional advantage of further decreasing the oxide capacitances thereby enhancing the controlling of the channel by the gate which is necessary to achieve a higher on-current as well as improving the subthreshold swing in tunnel FETs. High-κ materials based on hafnium oxide (HfO_2) with a dielectric constant of ~22 have been widely studied and used successfully which enhance the device performance substantially [22–25].

Another technique without making changes to the material is to add a pocket doping region i.e., for n-type tunnel FET where the source is doped with p^+, pocket-doped region will be n^+) in between the source and intrinsic channel regions. The doping region is depleted, leaving ionized positive charge. These depletion charges participate in an alternative electric field element to the intrinsic field of the channel and this larger field across the tunnel junction tends to enhance the electron band to band generation rate, hence increasing the tunneling current. Simultaneously, the extra field often ensures that the current gets enhanced subsequently when the inter-band tunneling is activated. For this reason, the minimum subthreshold swing is additionally decremented. In the pocket-doped tunnel FET as shown in Fig. 4a, the drain voltages only affect the channel intrinsic field excluding impacting the depletion of the pocket region. All in all, the impact of the drain on the tunnel junction gets reduced and hence less short channel effects are realized [26–28].

One alternative material that is being actively and commonly pursued by researchers for implementing in tunnel FET is germanium (Ge) material. Germanium is chemically identical to silicon and is generally compatible with existing CMOS process to a huge extent. Germanium has higher carrier mobility and a lower bandgap (0.66 eV) in comparison with silicon and it has drawn the interest of device

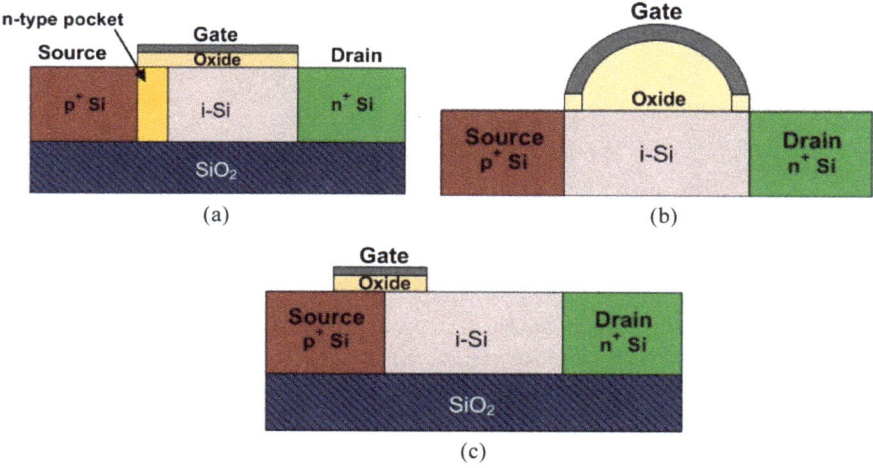

Fig. 4 **a** Pocket doped TFET, **b** circular gate TFET and **c** gate on source/channel SOI TFET

engineers for the last few decades. But, because of the smaller bandgap in germanium, the leakage current is quite high [29]. However, this smaller bandgap is a beneficial property for the employment of germanium for tunnel FET applications. The on-current is in excess of a significant degree higher in germanium-based tunnel FET than in silicon-based tunnel FET for a similar structure as germanium by its very nature endorse higher band-to-band tunneling [30]. Nevertheless, conventional tunnel FET based solely on germanium does have a high and undesirable off-state leakage current. Different methods have been implemented to keep the off-current under control and to make use of the higher band-to-band tunneling. For instance, employing short gate arrangement in which the gate on the drain end is shortened to prevent the high channel carrier density from extending to the drain can suppress the tunneling on the drain side. It was shown using simulations that the off-current could be diminished by 3–5 orders using short gate arrangement germanium-based tunnel FET [31]. Another way is to use lower doping concentrations at the drain side, where the ambipolar currents can be decreased by up to two orders of magnitude by decreasing the drain doping by five times [32]. Another commonly implemented technique is the use of germanium combined with silicon in a heterojunction where the smaller bandgap in germanium is utilized at the source region and the large bandgap of silicon material is employed in the channel and drain regions which holds the off-state leakage current within necessary limits [33, 34]. Other noteworthy tunnel FET structure which uses different techniques for improving the overall device performance includes the circular gate tunnel FET as shown in Fig. 4b [35], silicon germanium channel tunnel FET [36] and Gate on source/channel SOI tunnel FET as depicted in Fig. 4c [37]. A hetero-stacked source tunnel FET in which the germanium material is stacked on top of a silicon material in the source region has also been implemented to boost the on-current to some extend while maintaining a low subthreshold swing [38]. However, the hetero-stacked source device structure has been further optimized in terms of the transistor dimensions and will be discussed later in detail in Sect. 6.

5 Lateral Straggle Due to Ion Implantation

Any structural modification in tunnel FET for the improvement in performance including those mentioned in Sect. 4 exclusively relies on the process of fabrication. As numerous stages in the fabrication sequence of MOSFETs device as well as tunnel FETs require precise control of doping in designated region of the wafers, ion implantation is employed in this regard. In 1954, William Shockley decided to submit a patent that describes "Forming of Semiconductor Devices by Ionic Bombardment". The original patent outlines this same area that is being practiced till date and describes the application of implantations for chemical doping as well as electrical activations [39]. However, due to the advancements in modern technologies, the complexity and the fabrication steps have become much simpler and about 10–15 steps of them are different ion implantation steps. This section focuses on the one particular problem i.e., lateral straggle that can occur from ion implantation process.

Ion implantation can be explained as the injection of electrically charged particles onto targets with sufficient energies to infiltrate past the surface layer [40, 41]. The surface of the semiconductor is exposed to a beam of fixed-energy dopant ions, usually in the range of 10–100 kilo electron Volt (keV). The depth of the penetration is decided by the intensities of the incident ions. The main benefit of ion implantation is the ability to precisely regulate the quantity of ions added. These implants are employed to form the source region and drain region, as well as to control the threshold voltage, enhance latch-up immunity and reduce the formation of hot carriers in MOS devices. Advanced design and novel integrated circuits demand an accuracy in controlling the dopant concentration gradients and profiles. However, only precise controlling the dose of the dopants is insufficient, another degree of freedom provided by ion implantation, for instance, regulation of the beam energies and the angle of incident are effectively used [41]. Implants with zero-degree tilting angle between the trajectory of the beam and normally to the wafer substrate during fabrication are employed to prevent channeling effects that impact the transistor performance [42]. This is done with the ion beam striking perpendicularly to the wafer substrate, or by rotating the wafer at various angles throughout the implantation (Fig. 5).

On the contrary, with the decrease in device dimensions to sub-100 nano-meter regime, tilt in the wafer substrate in reference to the beam trajectory contributes a meaningful part. So far, the influence of non-zero tilt angle is difficult to accurately evaluate during ion implantation techniques of manufacturing process.

Figure 6 shows the impact of non-zero tilt angle which can cause lateral doping extension. The lateral spreading is distinguished by a non-zero tilt angle, which will be referred to as lateral straggle parameter (σ). With Very Large Scale Integration

Fig. 5 Schematic representation of the tilt angle [43]

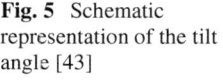
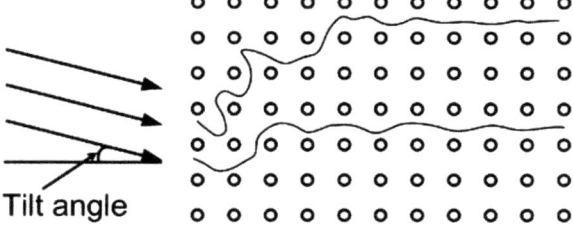

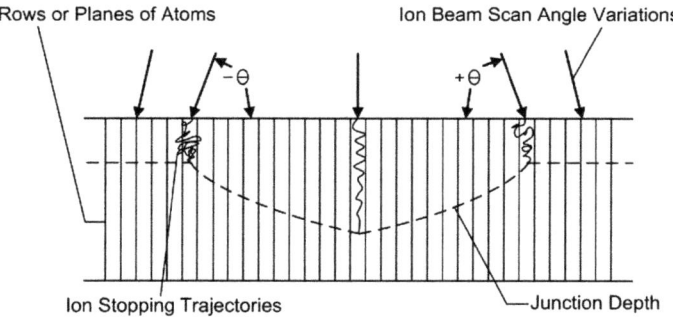

Fig. 6 Influence of non-zero tilt angle beam [47]

and Ultra Large Scale Integration circuits, sizes of the transistors are shrinking drastically into the nano-scale regime and the lateral straggling should be taken into considerations. As semiconductor devices dimensions are more reduced, the effect of the straggling turns out to be more and more crucial. The lateral doping extension caused by the non-zero tilt angle of the source and drain regions will alters the geometric length of a short channel and result in intrusion of source and drain regions into the channel region [44–46]. The impact of tilting angle is acknowledged utilizing Gaussian profiles at the source/channel and drain/channel junctions and is mathematically modeled using Eq. (7):

$$N_{SD}(x) = (N_{SD}(x))_{peak} \exp(-x^2/\sigma^2) \tag{7}$$

where $N_{SD}(x)$ is the source and drain doping concentrations at x, $N_{SD}(x)_{peak}$ determines the peak doping concentrations of the source and drain regions to the channel junction while σ defines the fluctuations in the ion beam [48]. Lateral straggle of impurity concentrations of injected ions has a massive impact on the performance of devices. This effect would prompt declining in the effective channel length (L_{effCh}) that thus expanded the short channel effects of the device, yet because of the decrease in L_{effCh}, the on-state current would also improve. This non-abrupt lateral expansion of the source and drain dopants into the channel can affect the performances of a transistor in many ways. Hence, later on, in the chapter, this influence of variations in lateral straggle parameter (σ) and how it affects the transistor characteristics will be studied briefly on a hetero-stacked source tunnel field-effect transistor. However, before that, firstly, the operation principle and the hetero-stacked source tunnel FET structure will be studied in detail in the next section.

6 Hetero-Stacked Source Tunnel FET and Simulation Details

As stated in the previous sections, even though tunnel FETs can attain the sub-60 mV/decade subthreshold swing at low operating voltage, unfortunately conventional tunnel FET still suffers from high SS_{avg} and low on-state current. In this regard, an optimized geometrical configuration hetero-stacked source tunnel field-effect transistor (HSS-TFET) is employed and a brief study of the structure and the operating principle is discussed here.

The hetero-stacked source tunnel FET (HSS-TFET), whose schematic is shown in Fig. 7 comprises of two materials in the source region stacked on top of each other. The upper layer of the source consists of silicon material whose energy bandgap is of 1.12 electron volt (eV) at room temperature and the lower (underlying) layer uses germanium material having a smaller energy bandgap of 0.66 eV at 300 K. The p$^+$ silicon and germanium in the stacked-source region are both highly doped with doping concentrations of 1×10^{20}/cm^3. In order to ensure low leakage currents, the wider bandgap silicon material is implemented in the channel region along with the drain region where the channel region is lowly doped with a doping concentration of 1×10^{16}/cm^3 and the n$^+$ drain region is moderately doped with 1×10^{18}/cm^3 for suppressing the ambipolar current. The rest of the materials used and the geometrical dimensions of the HSS-tunnel FET are as listed: The gate length (L_G) is of 30 nanometer (nm) and uses a metal that possesses a work-function of $\phi_M = 4.05$ eV, the upper layer of the stacked-source thickness is 5 nm while the underlying layer is of 10 nm thick. This makes the total thickness of the stacked source to $t_{Ch} = 15$ nm, i.e., the same thickness for both the intrinsic channel and drain regions. The gate dielectric oxide employed hafnium oxide (HfO$_2$) of thickness 1.5 nano-meter. A buried oxide (t_{Box}) of silicon dioxide material with thickness of 15 nm is used.

Fig. 7 Hetero-stacked source tunnel FET

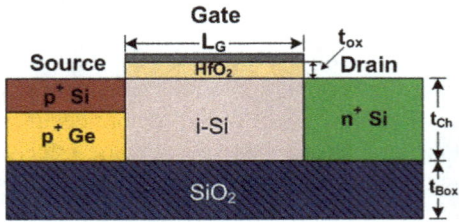

To implement and analyze the HSS-TFET, Sentaurus Technology Computer Aided Design (TCAD) tool is used [49]. In the Sentaurus tool, the following physics models are considered: Shockley–Read–Hall (SRH) model is employed for activating the generation and recombination. As bandgap plays a major part in tunnel FETs, the bandgap narrowing model is enabled. Doping-dependent mobility models have been utilized to map the impact of doping concentrations on carrier mobility. Fermi–Dirac statistics have been utilized taking into consideration degenerate dopings exhibited in the stacked-source and drain regions. As numerous numerical methodologies must be employed for obtaining an answer in calculating the tunneling probability, non-local band-to-band tunneling (NLBTBT) model that gives a precise assessment of the current in the device is used where the parameter values $A_{path} = 1.63 \times 10^{14}$/cm^3 s^1, $B_{path} = 1.47 \times 10^7$ V/cm, $P_{path} = 0.0567$ eV along with reduced mass $(m_r) = 0.033$ are adjusted to follow the experimentally validated results in [50]. Figure 8 illustrates the non-local BTBT model calibration against the experimental results and shows decent coordination between them authorizing for the validation of the TCAD simulations.

The energy band profile of the hetero-stacked source tunnel FET which is portrayed in Fig. 9 is explained here. During the transition period, as the gate-to-source voltage (V_{gs}) is raised slightly, the channel's conduction band begins to decline, the band-to-band tunneling occurs as the channel's conduction band tends to get lower than the valence band of the stacked source. For smaller gate-to-source voltage, say at 0.2 V as shown in Fig. 9a, the intrinsic channel region showcases stronger electrostatic at the surface (the p$^+$ silicon upper layer taken at A – A$'$) than in depth (the p$^+$ germanium lower layer taken at B – B$'$), hence the tunneling barrier width of the silicon stacked-source layer, $\lambda_{A-A'}$ is significantly shorter and thus the band-to-band tunneling is more dominating at this point than the tunneling width of the germanium underlying stacked-source layer, $\lambda_{B-B'}$. However, as seen from Fig. 9b, the tunneling width of the germanium layer $\lambda_{B-B'}$ is further decreased and becomes quite similar to the upper layer of the dual-stacked source $\lambda_{A-A'}$ with further increased in the gate-to-source voltage. Hence, the band-to-band tunneling probability of the underlying germanium layer would greatly enhance. Given that the tunneling probability of the material of the lower underlying layer is higher than the

Fig. 8 Non-local BTBT model calibration against experimental data in [50]

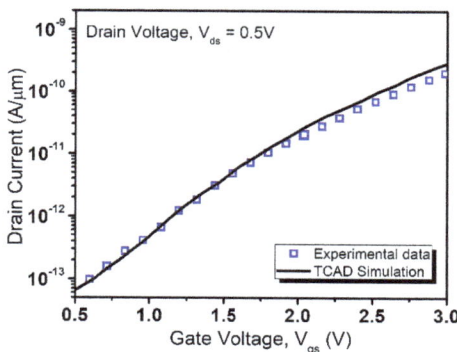

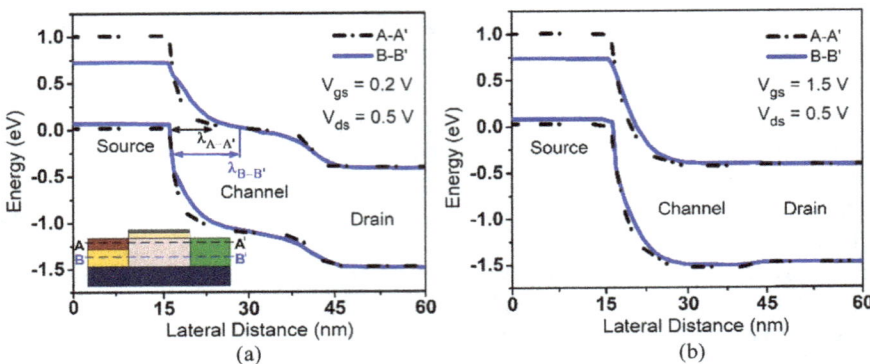

Fig. 9 Energy band diagram of the HSS-tunnel FET taken separately beneath the channel surface at a distance of 1 nm (A – A') and a distance of 6 nm (B – B') at **a** gate-to-source voltage (V_{gs}) = 0.2 V and **b** V_{gs} = 1.5 V both for drain voltage (V_{ds}) = 0.5 V

upper layer stacked-source material, the underlying germanium layer will then have provided increased tunneling probability in the HSS-TFET for higher gate-to-source voltage as compared to conventional tunnel.

The transfer characteristic of the silicon and germanium dual source stacked tunnel FET and a conventional tunnel FET is illustrated in Fig. 10. With the introduction of the silicon and germanium hetero-stacked source, the overall characteristic of the device is improved. As the germanium underlying layer of the dual-stacked source is a smaller band-gap material, higher efficiency in the electron BTBT is observed which increases the drain current (I_D) at high gate-to-source voltage. Additionally, this greatly progresses the subthreshold characteristic of the HSS-tunnel FET and resulted in a steeper SS in comparison with conventional tunnel FET. Due to the employment of silicon material, a larger bandgap in the upper layer of the source, the channel and drain regions, the HSS-TFET also exhibits substantially low off-state current. The I_{ON}/I_{OFF} current ratio of the HSS-TFET turned out to be 4.58×10^{10} while for conventional tunnel FET $I_{ON}/I_{OFF} = 6.98 \times 10^8$. The subthreshold swing

Fig. 10 Transfer characteristic comparison of conventional tunnel FET and hetero-stacked source tunnel FET

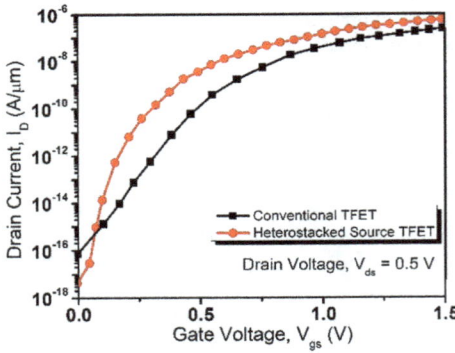

of the HSS-TFET is 28.61 mV/decade, whereas the $SS_{AVG} = 42.32$ mV/decade for the Si-based conventional tunnel FET.

7 Impact of Lateral Straggle on Hetero-Stacked Source Tunnel FET

The operating mechanism and structure of the hetero-stacked source tunnel FET were briefly studied in the previous section. In this section, the influence of the aforementioned lateral straggling parameter (σ) variations on different DC and Analog/RF parameters of the HSS-tunnel FET will be discussed.

The lateral straggle is implemented in the HSS-TFET by using a Gaussian doping profile at the Silicon–Germanium hetero-stacked source region with peak density of 1×10^{20} /cm^3 and a peak density of 1×10^{18} /cm^3 at the drain, both with dopant gradients of 0, 1, 3, 5 and 7 nm/decade. The lateral straggling parameters variability from the non-zero tilt alongside the channel region is obtained from the variations of σ at 0, 1, 3, 5 and 7 nm as depicted in Fig. 11. It merits referencing that for this particular HSS-TFET structure, the use of the ultra-low doped channel decreases the probability of fluctuation in the intrinsic parameters and mitigates impurity scattering.

The transfer characteristic curve and the electron band-to-band tunneling rate for the effect of different lateral straggle parameter of the HSS-tunnel FET are illustrated in Fig. 12. The gate-to-source voltage (V_{gs}) is varied from 0 to 1.5 V, whereas the drain-to-source voltage (V_{ds}) is maintained fixed at 0.5 V. The values of σ are varied from 0 to 7 nm in steps of odd numbers. It is apparent from Fig. 12a that the on-state current (I_{ON}) of the HSS-TFET rises considerably with the increment in the lateral straggling parameters. The rationale is that the quantity of dopants infiltrating the channel region from the Silicon–Germanium hetero-stacked source and the drain regions get incremented as the lateral straggling parameters are incremented, thereby considerably decreasing the effective channel length (L_{effCh}) of the transistor. This reduced effective channel length at higher σ (say $\sigma = 7$ nm) enhances the electron

Fig. 11 Influence of lateral straggling on doping concentrations of hetero-stacked source tunnel FET

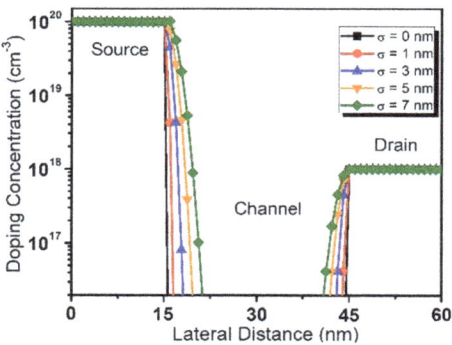

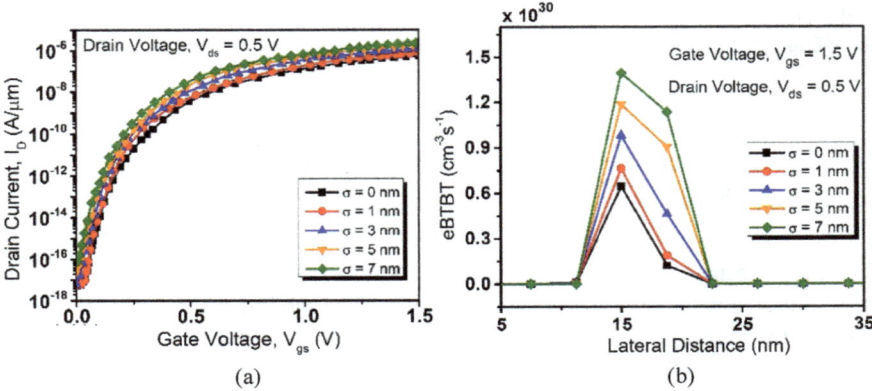

Fig. 12 Effect of lateral straggle variations on **a** transfer characteristic and **b** electron band-to-band tunneling rate of the HSS-tunnel FET

band-to-band tunneling in addition to the lateral electric field (V/L_{effCh}) in the intrinsic channel for fixed value of $V_{\text{ds}} = 0.5$ V. In this regard, higher values of lateral straggling parameters resulted in enhanced I_{ON}.

The electron band-to-band tunneling rate for various straggle parameters is depicted in Fig. 12b to help understand this rise in the drain current attributed to the increased in the straggling parameter. It can be seen that the generation rate of the electron band-to-band tunneling increments through the increase in straggling parameters, hence the electrons could tunnel easily from the stacked-source region's valence band to the conduction band of the channel region at higher straggling parameter, thereby increasing I_D of the HSS-TFET. However, with the decrement in L_{effCh} at higher straggling parameter, the subthreshold slope in addition to the off-state leakage currents of the HSS-tunnel FET deteriorate as evident from Fig. 12a.

For the output characteristics in tunnel FETs, at first, the potential from the gate gets more prominent as compared to the channel potential, that pinpoints the channel on the drain potential. Hence, when the drain potential is enhanced, potential at the channel also enhances, prompting a substantial increment in I_D. When V_{DS} is close to V_{GS}, the channel potential becomes independent of the drain potential. As a consequence, the current stays almost consistent as the drain bias increases. As channel length modulation (CLM) has an insignificant influence on I_D in tunnel FETs, the output resistance in the saturation region is much higher for tunnel FETs as compared to MOSFETs. Furthermore, as varying the drain bias does indeed have negligible influence on the channel potential in the saturation region, by virtue of which I_D as the source/channel tunneling remains constant. The effect of variations in straggling parameter on the output characteristics of the Silicon–Germanium hetero-stacked source tunnel FET for an overdrive voltage of $V_{OV} = 0.5$ V ($V_{gs} - V_{TH}$) is shown in Fig. 13. As stated earlier, reducing the straggling parameter enhances the effective length of the channel. Therefore, the carriers could not flow easily from the dual-layered stacked source to the drain region. As a result, the drain current gets

Fig. 13 HSS-tunnel FET
output characteristics
(I_D–V_ds) variation due to
straggling parameters

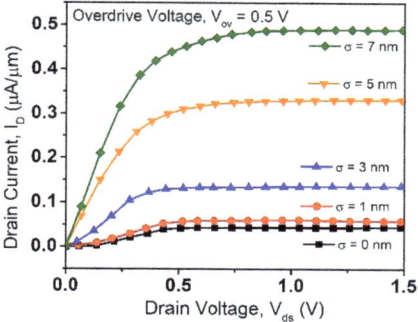

saturated at high drain voltage at a low value of lateral straggling parameters. On the contrary, I_D gets saturated at low drain voltage for high values of straggle parameters owing to a reduction in channel length.

The bar graph showing the effect of variations in the lateral straggling parameters on the $I_\mathrm{ON}/I_\mathrm{OFF}$ ratio, the subthreshold swing and threshold voltage of the dual-stacked source tunnel FET are portrayed in Fig. 14. These are perhaps the most significant electrical parameters in devices. These parameters are extracted from the transfer characteristics of the HSS-tunnel FET from Fig. 12a. As observed from the $I_\mathrm{ON}/I_\mathrm{OFF}$ ratio graph (Fig. 14a) that the current ratio degrades as the straggling parameter is increased. This is due to the fact that, even though I_ON is enhanced for higher σ, the deterioration in I_OFF due to short channel effects as a consequence of reduced L_effCh has a tremendous effect on the current ratio for higher σ, this has led to the decrement in the $I_\mathrm{ON}/I_\mathrm{OFF}$ ratio of the hetero-stacked source tunnel FET. The subthreshold swing of the dual-stacked source tunnel FET is calculated using Eq. 6. Since the decrement in the effective channel length owing to high value of the straggling parameter leads to more SCEs, the subthreshold swing of the transistor also gets degraded for higher lateral straggle that can be justified from Fig. 12b. Since tunnel FET does not possess a specific description of the threshold voltage (V_TH) as MOSFET, however, here we adopted the concept that the tunnel FET's threshold voltage is specified as V_gs for which the energy bandgap narrowing begins to get saturated. For this, a constant current method at a drain current $I_\mathrm{D} = 10^{-8}$A is employed to specify V_TH of the Silicon–Germanium hetero-stacked source tunnel FET. Due to the improvement in I_ON as a result of higher lateral straggling, the threshold voltage is also improved for higher value of σ. The actual values of the current ratio, SS and V_TH are presented in Table 1.

Transconductance (g_m) which is an important parameter in devices is defined as the electrical attributes that correlate the current through a device's output to the voltage through the device's input. For a given drain voltage (V_ds), it is mathematically defined as:

$$g_\mathrm{m} = \left(\frac{\partial I_\mathrm{D}}{\partial V_\mathrm{gs}}\right)_{V_\mathrm{ds}} \tag{8}$$

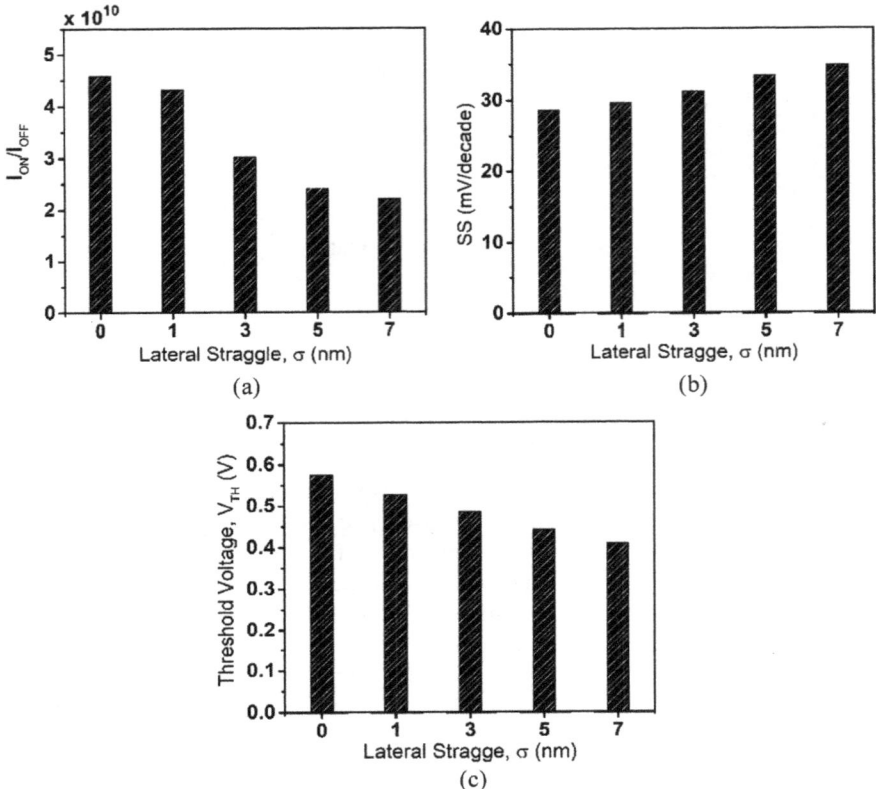

Fig. 14 Influence of lateral straggling on **a** I_{ON}/I_{OFF} ratio, **b** Subthreshold swing and **c** Threshold voltage of the HSS-TFET

Table 1 Electrical parameters for different straggling parameters of the hetero-stacked source tunnel FET

Lateral straggle parameter (σ) nm	I_{ON}/I_{OFF} ratio	SS (mV/decade)	Threshold voltage (V)
0	4.581×10^{10}	28.61	0.575
1	4.312×10^{10}	29.74	0.527
3	3.057×10^{10}	31.25	0.485
5	2.404×10^{10}	33.46	0.441
7	2.305×10^{10}	34.82	0.408

The tunnel FET's intrinsic parameters, which are important for Analog circuits such as transconductance and output conductance (g_d) are heavily dependent on the precise implementation of the tunnel FET. The transconductance (g_m) in the subthreshold region in tunnel FETs could be determined using [51]:

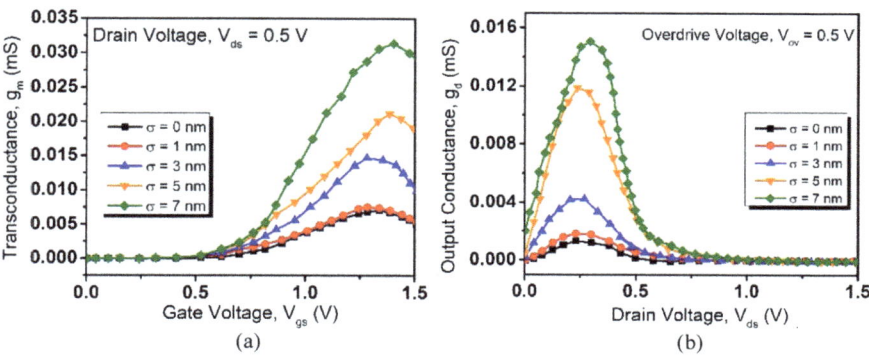

Fig. 15 Transconductance (g_m) and output conductance (g_d) of the HSS-TFET due to variations in lateral straggling parameter

$$g_m = \frac{\partial I_D}{\partial V_{gs}} = \frac{\ln(10) I_D}{SS_{pnt}} \tag{9}$$

where SS_{pnt} is the point subthreshold swing as specified in Eq. 5. The SS_{pnt} of the tunnel FET changes with I_D: it is minimal at very low I_D and progressively rises as I_D is enhanced. The output conductance is expressed as the fraction of the drain current to the drain to source bias:

$$g_d = \frac{\partial I_D}{\partial V_{ds}} \tag{10}$$

Figure 15a, b illustrate the variations in transconductance and output conductance of the HSS-tunnel FET due to lateral straggling parameters, respectively. It is apparent from Fig. 15a that the variations in the lateral straggling have greatly affected the transconductance of the HSS-TFET substantially. The transconductance is increased with an increment in the gate bias yet declines at higher gate voltages. Be that as it may, if the gate-to-source is not high, the transconductance rises with the increment in the straggling parameter owing to an increment in I_D for higher σ as a result of reduction in the L_{effCh}. This reduced L_{effCh} increased the lateral electric field for higher value of the straggling parameter. The tunneling influence of the higher σ originating from the Silicon–Germanium hetero-stacked source to the drain is also greatly increased due to an increased electric field relative to lower straggling parameter for the same gate-to-source voltage value. The transconductance curve declines for all the lateral straggling parameters after a certain gate voltage (say $V_{gs} = 1.25$ V). This attributes to the rise in the electric lateral fields at a higher gate bias, contributing to mobility degradations [52]. However, the output conductance (g_d), as seen from Fig. 15b, showed a decrement at low straggling parameter when the drain voltage is low and this results in increased output resistance ($R_o = 1/g_d$). This indicates the HSS-tunnel FET driving capacity is enhanced for lower straggling parameter values. The driving capabilities determine how a system or circuit regulates other devices. The output conductance reduces with the low values of the straggling parameter resulting in an increased output resistance. Conversely, the driving capabilities of the dual-source tunnel FET is diminished due to a higher values of lateral straggle parameter.

In tunnel FETs, under all bias conditions, the total gate capacitance (C_{gg}) is subjugated by the gate-to-drain capacitance (C_{gd}) instead of the gate-to-source capacitance (C_{gs}) [53, 54]. When the tunnel FET is in the off-state, the gate-to-drain capacitance is much greater than that of gate-to-source capacitance since this resistance between channel and drain is considerably smaller than that between channel and source. There seems to be only few potential drops within the channel and the drain while the tunnel FET is in the on-state condition, contributing to a stronger C_{gd} in both linear region and saturation region [55]. This behavior is opposite to a conventional MOSFET in which the C_{gs} and C_{gd} contributions are almost similarly equal in the linear region, and the C_{gs} is dominant in the saturation region. As a consequence of the large amount of gate-to-drain capacitance in a tunnel FET, a better coupling occurs between the input pin (Gate) and the output pin (Drain) in tunnel FET circuits and has significant repercussions for the circuit's actions owing to the Miller effect. However, the total gate capacitance ($C_{gs} + C_{gd}$) in TFET depends heavily on the properties of the material used. For instance, the total capacitance of a silicon-based tunnel field-effect transistor is noticed to be greater than the total capacitance of the conventional MOSFET [56].

In this case of Fig. 16a, the extraction of C_{gg} of the HSS-TFET is calculated by measuring the changes in charge at the gate with regards to minor variations in the terminal voltage at each bias point. It is witnessed that C_{gg} is slightly increased for higher lateral straggling. The gate-to-source capacitance and gate-to-drain capacitance are composed of the inner fringing capacitance (C_{if}), outer fringing capacitance (C_{of}) and the overlap capacitance (C_{ov}). An enhancement in the C_{gs} and C_{gd} is also perceived for higher straggling parameters from Fig. 16b. This rise in capacitances on account of increasing straggling parameter attributes to the increased in the quantity of dopants intruding into the channel region close to stacked-source/drain

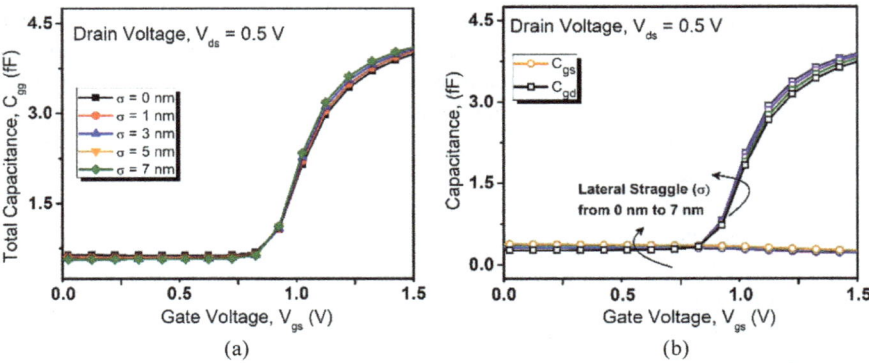

Fig. 16 The influence of lateral straggling parameters variation on **a** The total gate capacitance (C_{gg}) and **b** Gate-to-drain capacitance (C_{gd}) and gate-to-source capacitance (C_{gs}) of the HSS-TFET

Fig. 17 Consequence of lateral straggling parameter on the cut-off frequency (f_T) of HSS-tunnel FET

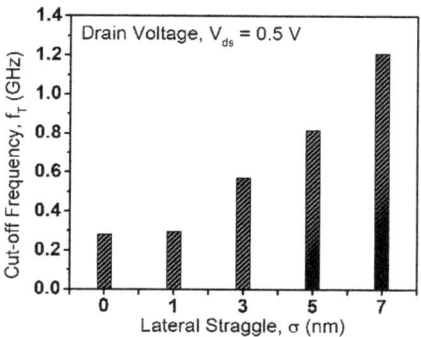

regions. As a result, charges in the channel region close to the stacked-source / drain regions increase, this enhances the intrinsic capacitances. In this manner, the overall capacitances C_{gs} and C_{gd} are enhanced.

The combined impact of the transconductance and capacitances may be taken into consideration in the unity gain cut-off frequency (f_T) of tunnel FET, which can be determined as:

$$f_T = \frac{g_m}{2\pi \left(C_{gs} + C_{gd} \right)} \qquad (11)$$

As the transconductance and the total capacitances alter with I_D, f_T too varies along with the drain current. Typically, the peak obtainable cut-off frequency in tunnel FET is normally less than a MOSFET. In Fig. 17, the f_T is depicted as a function of the variation in straggling parameters that characterize the device's high-frequency output. As discussed earlier, since the transconductance in addition to the total gate capacitance ($C_{gs} + C_{gd}$) increases for increasing lateral straggle parameters, giving rise to the cut-off frequency at higher lateral straggling parameter values for the hetero-stacked source tunnel FET.

8 Mixed-Mode Hetero-Stacked Source Tunnel FET Due to Lateral Straggle Parameters

In order to apply tunnel FETs to CMOS-type circuits, it is essential that the features of the n-tunnel FET and the p-tunnel FET correspond to those of the CMOS-type circuits. The circuit diagram of HSS-tunnel FET inverter is shown in Fig. 18a. For conventional MOSFET, the driving capabilities of the p-MOS and n-MOS transistors are balanced by an appropriate scaling of the width (W) and length (L) ratios of the transistors. That being said, in the case of the n-tunnel FET and the p-tunnel FET, there are other variations in electrical properties, such as the subthreshold swing, additionally to the mismatch in the drive forces. To enhance the on-state current,

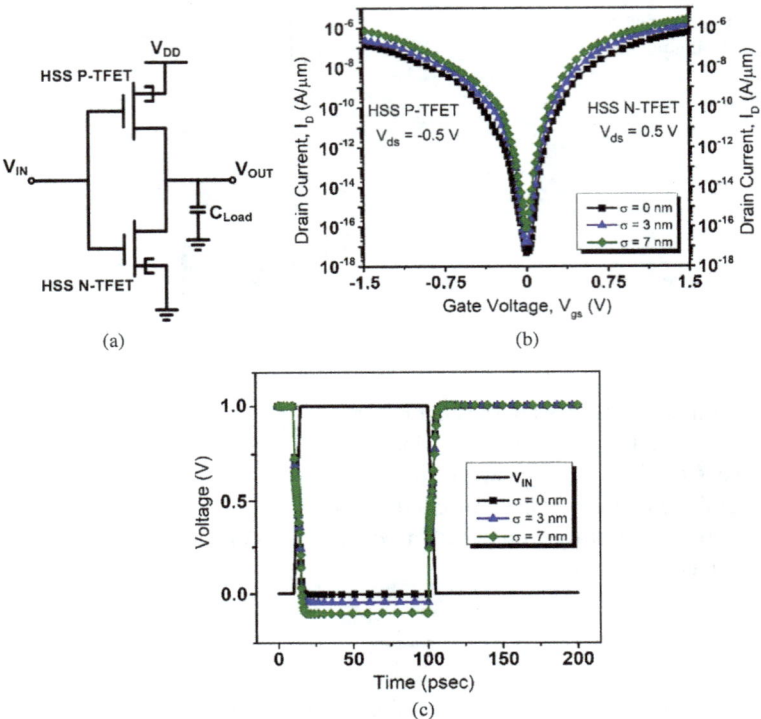

Fig. 18 **a** Circuit diagram of HSS-tunnel FET inverter, **b** transfer characteristic curve of HSS-TFET for both P-type and N-type for different lateral straggling parameters, **c** transient behavior of HSS-tunnel FET inverter showing the overshoot and the undershoot

the source doping concentration should be high in tunnel FETs. The source doping in the p-tunnel FET is n⁺ material, while the source doping in the n-tunnel FET is p⁺ material. Owing to the low density of states in E_C, for equivalent doping concentrations for n-type and p-type materials, the fermi energy level (E_F) shifts further toward E_C for n-type material as opposed to E_F shifting into E_V for p-type semiconductor material. Hence, a p-tunnel FET typically has a lower I_{ON} as well as slight deterioration in subthreshold slope than an n-tunnel FET. This discrepancy in the n-tunnel FET and p-tunnel FET characteristics has an important influence in the energy consumption and delay characteristics of tunnel FET-controlled CMOS-type circuits. Hence, this discrepancy needs to be minimized by optimizing the device. After optimizing the n and p-type HSS-TFET along with the work function (ϕ_M = 4.05 eV and 5.23 eV for n-type and p-type HSS-tunnel FET, respectively), the transfer characteristic is depicted in Fig. 18b with the variations due to lateral straggling. It is visualized that the hetero-stacked source n-tunnel FET and p-tunnel FET have an almost identical values of the current ratio for various straggling parameters. The transient response of the inverter realized using the hetero-stacked source tunnel

FET for different lateral straggling parameters of 0, 3 and 7 nm is illustrated in Fig. 18c. An input voltage (V_{IN}) of ramp 0–1.0 V is introduced while keeping the load capacitance (C_L) fixed at 3.15 fF. The transient response of a tunnel FET inverter normally portrayed an undershoot transient and overshoot transient. This undershoot and overshoot along with long settling time in tunnel FET inverter diminishes the circuit's operating speed, yet in addition raises the energy utilization along with yielding a distorted output waveform. The HSS-TFET inverter shows an increase in the undershoot (when transitioning from high to low) by a substantial amount for increasing lateral straggling parameters, however, the overshoot varies negligibly. This voltage undershoot is because of the high Miller capacitance, that comes from large coupling between the V_{IN} and V_{OUT} of the HSS-tunnel FET inverter due to the high C_{gd}. The voltage undershoot ($V_{Undershoot}$) is evaluated bearing in mind the conservation of charges which is determined by:

$$V_{Undershoot} = \frac{C_M}{(C_M + C_L)} V_{DD} \tag{12}$$

where C_M denotes the Miller capacitance, C_L is the load capacitance and V_{DD} signifies the supply voltage. From Eq. 12, the percentage of the voltage undershoot can be calculated which indicates further degradation for high straggling parameters of $\sigma = 3$ and 7 nm. This is attributed to the higher gate-to-drain capacitances for higher lateral straggle parameters. The average delay time parameter (τ_p) is also determined using Eq. 13 where τ_{PHL} is the high to low voltage delay time and τ_{PLH} is the low to high voltage delay time. The average delay parameter has shown a reduce delay time for higher straggling parameters and is listed in Table 2.

$$\tau_p = \frac{\tau_{PHL} + \tau_{PLH}}{2} \tag{13}$$

Since there is an enhancement in both the on-state current as well as the total gate capacitance at higher lateral straggling parameters for the hetero-stacked source tunnel FET as witnessed from Figs. 12a and 16a, respectively, the average delay has improved significantly due to more substantial enhancement in the on-state current than the total gate capacitance.

Table 2 Delay parameters of the HSS-tunnel FET inverter due to lateral straggle parameters

Delay parameters (ps)	$\sigma = 0$ nm	$\sigma = 3$ nm	$\sigma = 7$ nm
τ_{PHL}	0.21	0.143	0.101
τ_{PLH}	0.34	0.22	0.13
τ_p	0.275	0.181	0.115

9 Summary

In this chapter, we explored the difference between conventional MOSFET and tunnel FET as regards to both the working mechanism and subthreshold characteristics. The problems that are faced in the conventional tunnel FET, in particular the issue of low on-state current is stressed. Different techniques that are utilized to enhance the tunnel FET drive current are discussed. The problem encounters during the fabrication techniques of transistors, especially the ion implantations and the impact on device characteristics is addressed. A hetero-stacked source tunnel FET structure, its operating principle and some important parameters and how it has lowered the subthreshold slope have been studied. The impact on the variations in lateral straggling on different parameters of the Silicon–Germanium hetero-stacked source tunnel FET is analyzed. A mixed-mode tunnel FET for the HSS-TFET structure is implemented where the average delay parameters are calculated. It is concluded that for attaining a good device performance, proper optimization in the straggling parameters of the Gaussian doping is required.

References

1. Kahng D (1976) A historical perspective on the development of MOS transistors and related devices. IEEE Trans Electron Devices 23:655–657
2. Brinkman WF, Haggan DE, Troutman WW (1997) A history of the invention of the transistor and where it will lead us. IEEE J Solid-State Circ 32:1858–1865
3. Mack CA (2011) Fifty years of Moore's law. IEEE Trans Semicond Manuf 24:202–207
4. Zhang Q, Zhao W, Seabaugh A (2006) Low-subthreshold-swing tunnel transistors. IEEE Electron Device Lett 27:297–300
5. Ionescu AM, Riel H (2011) Tunnel field-effect transistors as energy-efficient electronic switches. Nature 479:329–337
6. Lu H, Seabaugh A (2014) Tunnel field-effect transistors: State-of-the-art. J. Electron Devices Soc 44–49
7. Koswatta SO, Lundstrom MS (2009) Performance comparison between p-i-n tunneling transistors and conventional MOSFETs. IEEE Trans Electron Devices 56:456–465
8. Knoch J, Appenzeller J (2005) A novel concept for field-effect transistors: The tunneling carbon nanotube FET. In: Proceedings of 63rd DRC. IEEE, pp 153–156
9. Knoch J, Appenzeller J (2008) Tunneling phenomena in carbon nanotube field-effect transistors. Physica Status Solidi (a) 205(4):679–694
10. Choi WY, Park BG, Lee JD, Liu TJK (2007) Tunneling field-effect transistors (TFETs) with subthreshold swing (SS) less than 60 mV/dec. IEEE Electron Device Lett 28:743–745
11. Taur Y, Ning TH (1998) Fundamentals of modern VLSI devices. Cambridge University Press, Cambridge, UK
12. Khatami Y, Banerjee K (2009) Steep subthreshold slope n- and p-type tunnel-FET devices for low-power and energy-efficient digital circuits. IEEE Trans Electron Devices 56:2752–2761
13. Lee M, Jeon Y, Son KS, Shim JH, Kim S (2012) Comparative performance analysis of silicon nanowire tunnel FETs and MOSFETs on plastic substrates in flexible logic circuit applications. Physica Status Solidi (a) 209(7):1350–1358
14. International Technology Roadmap for Semiconductors (ITRS (2013) Emerging research devices, 2013 edn. http://www.itrs.net

15. Verhulst AS, Vandenberghe WG, Leonelli D, Rooyackers R, Vandooren A, Pourtois G, Gendt SD, Heyns MM, Groeseneken G (2010) Boosting the on-current of Si-based tunnel field-effect transistors. ECS Trans 33(6):363–372
16. Bhushan B, Nayak K, Rao VR (2012) DC compact model for SOI tunnel field-effect transistors. IEEE Trans Electron Devices 59(10):2635–2642
17. Toh EH, Wang GH, Yeo GSYC (2007) Device physics and design of double-gate tunneling field-effect transistor by silicon film thickness optimization. Appl Phys Lett 90(263507)
18. Liu L, Mohata D, Datta S (2012) Scaling length theory of double-gate interband tunnel field-effect transistors. IEEE Trans. on Electron Devices 59(4):902–908
19. Verhulst AS, Sorée B, Leonelli D, Vandenberghe WG, Groeseneken G (2010) Modeling the single-gate, double-gate, and gate-all-around tunnel field-effect transistor. J Appl Phys 107:024518-1-024518–6
20. Zhan A, Mei J, Zhang L, He H, He J, Chan M (2012) Numerical study on dual material gate nanowire tunnel field-effect transistor. In: International conference on electron devices and solid-state circuit (EDSSC), Bangkok, Thailand, 3–5 Dec 2012
21. Vishnoi R, Kumar MJ (2014) Compact analytical drain current model of gate-all around nanowire tunneling FET. IEEE Trans Electron Devices 61(7):2599–2603
22. Bohr MT, Chau RS, Ghani T, Mistry K (2007) The High-k Solution. IEEE Spectr 44(10):29–35
23. Boucart K, Ionescu AM (2007) Double-gate tunnel FET with high-κ gate dielectric. IEEE Trans Electron Devices 54:1725–1733
24. Schlosser M, Bhuwalka K, Sauter M, Zilbauer T, Sulima T, Eisele I (2009) Fringing-induced drain current improvement in the tunnel field-effect transistor with high-κ gate dielectrics. IEEE Trans Electron Devices 56:100–108
25. Anghel C, Chilagani P, Amara A, Vladimirescu A (2010) Tunnel field effect transistor with increased ON current, low-κ spacer and high-κ dielectric. Appl Phys Lett (122104)
26. Jhaveri R, Nagavarapu V, Woo J (2011) Effect of pocket doping and annealing schemes on the source-pocket tunnel field-effect transistor. IEEE Trans Electron Devices 58:80–86
27. Verreck D, Verhulst AS, Kao KH, Vandenberghe WG, Meyer KD, Groeseneken G (2013) Quantum mechanical performance predictions of p-n-i-n versus pocketed line tunnel field-effect transistors. IEEE Trans Electron Devices 60:2128–2134
28. Abdi BD, Kumar MJ (2014) In-built N$^+$ pocket p-n-p-n tunnel field-effect transistor. IEEE Electron Device Lett 35:1170–1172
29. Agopian PGD, Martino MDV, Santos SD et al (2015) Influence of the source composition on the analog performance parameters of vertical nanowire-TFETs. IEEE Trans Electron Devices 61:16–22
30. Toh EH, Wang GH, Samudra G, Yeo YC (2008) Device physics and design of germanium tunneling field-effect transistor with source and drain engineering for low power and high performance applications. J. Applied Physics 103(104504)
31. Verhulst AS, Vandenberghe WG, Maex K, Groeseneken G (2007) Tunnel field-effect transistor without gate-drain overlap. Appl Phys Lett 91(053102)
32. Alam K, Takagi S, Takenaka M (2014) A Ge ultrathin-body n-channel tunnel FET: effects of surface orientation. IEEE Trans Electron Devices 61:3594–3600
33. Chander S, Baishya S (2015) A two-dimensional gate threshold voltage model for a hetero-junction SOI-tunnel FET with oxide/source overlap. IEEE Electron Device Lett 36(7):714–716
34. Kim SH, Jacobson ZA, Liu TJK (2010) Impact of body doping and thickness on the performance of germanium-source TFETs. IEEE Trans Electron Devices 57(7):1710–1713
35. Goswami R, Bhowmick B (2017) An analytical model of drain current in a nanoscale circular gate TFET. IEEE Trans. on Electron Devices 64(1):45–51
36. Saha R, Bhowmick B, Baishya S (2019) Impact of WFV on electrical parameters due to high-k/metal gate in SiGe channel tunnel FET. Microelectron Eng 214:1–4
37. Mitra SK, Bhowmick B (2018) A compact interband tunneling current model of gate-on source/channel SOI-TFET. J Comput Electron 17(4):1557–1566
38. Wu C, Huang Q, Zhao Y et al (2016) A novel tunnel FET design with stacked source configuration for average subthreshold swing reduction. IEEE Trans Electron Devices 63(12):5072–5076

39. Shockley W (1957) Forming Semiconductive devices by ionic bombardment. US Patent 2,787,564, 2 April 1957
40. Mayer JW, Eriksson L, Davies JA (1970) Ion implantation in semiconductors. Academic Press, New York
41. Ziegler JF (1992) Ion implantation technology. North Holland Amsterdam
42. Kranti A, Armstrong GA (2007) Source/drain extension region engineering in FinFETs for low-voltage analog applications. IEEE Electron Device Lett 28(2):139–141
43. Rimini E (1995) Ion implantation: Basics to Device Fabrication. Springer Science+Business Media, New York
44. Gosh S, Kolen K, Sarkar CK (2015) Impact of the lateral straggle on the analog and RF performance of TFET. Microelectron Reliab 55:326–331
45. Vanlalawmpuia K, Saha R, Bhowmick B (2018) Performance evaluation of heterostacked TFET for variation in lateral straggle and its application as digital inverter. Appl Phys A: Mater Sci Process 124(10):701. https://doi.org/10.1007/s00339-018-2121-4
46. Saha R, Vanlalawmpuia K, Bhowmick B, Baishya S (2019) Deep insight into DC, RF/analog, and digital inverter performance due to variation in straggle parameter for gate modulated TFET. Mater Sci Semicond Process 91:102–107. https://doi.org/10.1016/j.mssp.2018.11.011
47. Ziegler JF (1992) Handbook of ion implantation technology. Elsevier Science Publications, p 119
48. Kranti A, Armstrong GA (2007) Design and optimization of FinFETs for ultra-low-voltage analog applications. IEEE Trans Electron Devices 54(12):3308–3316
49. Sentaurus Device Users' Manual (2013) Synopsys Inc. Mountain View, CA, USA
50. Biswas A, Dan SS, Royer C, Grabinski W, Ionescu AM (2012) TCAD simulation of SOI TFETs and calibration of non-local band-to-band tunneling model. Microelectron Eng 98:334–337
51. Trivedi AR, Carlo S, Mukhopadhyay S (2013) Exploring tunnel-FET for ultra-low power analog applications: a case study on operational transconductance amplifier. In: Proceedings of the 50th annual design automation conference, IEEE, p 109
52. Hoyniak D, Nowak E, Anderson RL (2000) Channel electron mobility dependence on lateral electric field in field-effect transistors. J Appl Phys 87:876–881
53. Yang Y, Tong X, Yang LT, Guo PF, Fan L, Yeo YC (2010) Tunneling field-effect transistor: capacitance components and modeling. IEEE Electron Device Lett 31:752–754
54. Dagtekin N, Ionescu AM (2015) Impact of super-linear onset, off-region due to uni-directional conductance and dominant C_{GD} on performance of TFET-based circuits. IEEE J Electron Devices Soc 3:233–239
55. Mookerjea S, Krishnan R, Datta S, Narayanan V (2009) On enhanced Miller capacitance effect in interband tunnel transistors. IEEE Electron Device Lett 30:1102–1104
56. Mallik A, Chattopadhyay A (2012) Tunnel field-effect transistors for analog/mixed-signal system-on-chip applications. IEEE Trans. on Electron Devices 59:888–894

Fabrication of ZnO and ZnO Heterostructures for Gas-Sensing Applications

Argha Sarkar and Santanu Maity

Abstract Researchers in the past have reported improvement in sensitivity for gas sensors by using nanostructured materials or catalysts; however, less work has been discussed on the improvement of sensitivity of gas sensors based on both nanostructured materials and composites. This is the motivation to fabricate highly sensitive gas sensors. The discussion of this chapter is motivated toward the physical property which is depending on the crystal shape, size, arrangement, alignment, and aspect ratio. Specific traditional metal oxide gas sensors, which are mainly, based on alumina substrate, importantly used for sensing hydrocarbons (like CH_4) are inflammable in nature and other hazardous gas (like CO). However, it faces the two problems, viz. (1) their comparatively high activation temperature and (2) huge power dissipation. The use of silicon microelectromechanical system (MEMS) structure to host active metal oxide sensing platform could be a suitable option to design microheater to achieve the required temperature with minimum power dissipation. Using nanocrystalline ZnO or its ZnO heterosructures as a sensing layer may successfully lower the operational temperature less than 300 °C providing better sensitivity. So synthesis of zinc oxide or its heterostructures and the modification through doping in a feasible way are of great challenge.

Keywords Zinc Oxide · Heterostructures · Gas Sensor · Interdigited Electrode Position · Sensitivity

1 Introduction

ZnO is a promising material which is substantially having high gas sensitivity, minimal cost, comparatively low operational temperature as well as consuming less power [1–6]. The main bases of ZnO gas sensors are changes of electrical resistivity,

A. Sarkar (✉)
Vishnu Institute of Technology, Bhimavaram 534202, India

S. Maity
Indian Institute of Engineering Science and Technology, Shibpur, India

and it is significantly depending on its size, shape, and the defect state present in a crystal [7]. If there is any stoichiometric change during ZnO synthesis, then crystal defects may appear [8]. Hence, a simple and a workable way of synthesis are desired, and utmost care must be taken during the synthesis because it is having remarkable impact on the elementary properties and thus sensitivity of the gas-sensing device. Several gas sensors based on several nanostructured oxides (e.g., ZnO, TiO_2, etc.) have been investigated [9]. It is reviewed from several reports that thin films made of graphene flakes or graphene oxides are gaining much more attention as sensing layer. Actually, the physical and electrical properties of those nanostructured flake or oxide are much responsive to adsorbates on surfaces because of the target analyte [10–12]. The high mobility of carrier (at room temperature) and large surface to volume ratio make graphene a potential candidate for sensing layer in gas sensors [13]. The transition metal oxide nanostructures are not suitable in this regard because it provides desired sensitivity only at high temperatures. Graphene is having much higher adsorptive capacity than solid-state surfaces reported so far [14, 15]. Extensive experiments have been carried out to fabricate reduced doped graphene oxide (rGO)/Mg:ZnO heterostructure.

The prime objectives of the chapter are as follows:

1. To optimize the responsivity of gas sensor device through the structural improvement of ZnO nanorods and minimizing of defect states

   (a) Formation of high sensitive hexagonal ZnO nanorod with large surface to volume ratio.
   (b) Even distribution and uniformity in shape and size of the nanorods.
   (c) The study of positioning interdigited electrode below/above sensing layer.
   (d) Study of performance and sensitivity with respect to sensing layer, device performance (applied voltage with temperature).

2. To study the behavior of graphene oxide/metal oxide heterojunction to enhance the sensing performance

   (a) Reduced graphene oxide (rGO) synthesis and doping of zinc oxide (ZnO)/zinc magnesium oxide (ZnMgO) on rGO.
   (b) Investigation of thin film rGO for CH_4 sensing properties.
   (c) Enhanced sensitivity due to the existence of magnesium (Mg) in the ZnO/rGO crystalline.

This chapter is subdivided into seven different sections. Section 1 introduces the topic of research. Section 2 discusses the hexagonal ZnO nanorod synthesis for gas sensor fabrication. Section 3 starts with fabrication of GO/Mg:ZnO heterostructure gas sensor devices. Section 4 is based on experimental observation of synthesized ZnO nanorod. Electrical and optical characterization of GO/Mg:ZnO heterostructure ZnO nanorod is detailed in Sect. 5. Further enhancement of gas-sensing performance is elaborated in Sect. 6. The conclusion of the chapter is drawn in Sect. 7 converging the individual summary of each section.

2 Hexagonal ZnO Nanorod Synthesis for Gas Sensor Fabrication

Optimization of Device

Most of the MOS gas-sensing platform operates nearly 300 °C temperature, so a microheater remains important [16]. An easiest way of batch fabrication steps which has been discussed which is applied here to optimize the device is given in Fig. 1. RCA I and RCA II are considered for cleaning p-type silicon (Si) substrate (reduced resistive 0.01 Ω-cm (100)). In silicon dioxide (SiO$_2$) layer, one micrometer thickness is developed on both sides of the substrate for passivation of IDE and heater. During the process, pyrogenic oxidation is followed (is better grown procedure) with controlled temperature of 1100 °C. To do the back-side patterning, photolithography

Fig. 1 Block diagram of fabrication technique to form IDE and microheater for gas-analyzing setup (© 2017 Elsevier Ltd.)

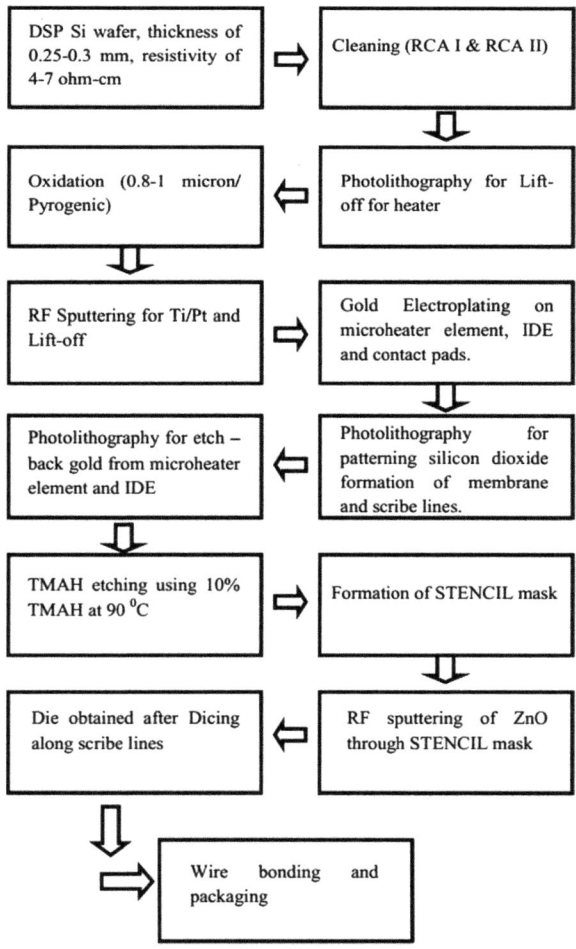

process (using positive photoresist: SPR 700 1.8 and manufacturer MF319, with positive photomasking) is followed, and micromachining technology (utilizing chemical etchant, TMAH) is considered for reducing the silicon substrate thickness so that the device outcome is increased. Both side alignment technique is followed for positioning the IDE and heater as a function of reverse side micro-machined window. By magneto sputtering system, deposition of Ti/Pt (20/100 nm) stag is done on front portion of the silicon substrate to achieve higher temperature (resistive performance). Photo lithography and lift-off are carried out for transforming the heater and IDE. Gold (Au) electroplating is done on IDE, heater element, and contact pads. Electroplating and masking are done for reducing the contact resistance of input pad (heater) and output pad (IDE) [17].

Stencil mask is used to deposit sensing material. ZnO and rGO/Mg: ZnO are synthesized for sensing layer, and the preparation of sensing layer is discussed in 'optimization of sensing layer' section. Finally, wire bonding technique is followed for electrical characterization.

Optimization of Sensing Layer

Two distinct techniques are considered for preparing sample S1, sample S2, sample S3, and sample S4. 200 nm ZnO layers are synthesized for making seed layer. As a precursor, 0.15 M zinc-acetate-dehydrate [$Zn(CH_3COOH)_2 \cdot 2H_2O$] is dissolved in ethanol (C_2H_5OH) and de-ionized water. Using magneto bit stirrer, the solution is continuously mixed for 1 h at 600 °C. The solution is then aged for 30 h at room temperature and deposited on IDE through stencil mask (S1 and S3 samples taken from this following this technique 1). Hydrothermal technique is followed for growing ZnO nanorods (for preparing S2 and S4). Zinc-acetate-hexahydrate is added with hexamethamine in a ratio of 20:10 mM to prepare the chemical solution to obtain desired arrangement of uniform nanorods. Several temperature is fixed using normal heater to observe the effect of temperature on growth process. Isopropyl alcohol (IPA) is used to control the smoothness of ZnO surface. Table 1 describes about the different sample for the position of IDE and fabrication outcomes 17].

Table 1 Sensor based on different samples as a function of size, shape, distribution, and IDE position [17]

Sensors	State of nanorods	IDE position
S1	Non-uniform size, shape, and scattered	Top of the nanorods
S2	Uniform size, shape, and scattered	Top of the nanorods
S3	Non-uniform size, shape, and uniformly distributed	Below the nanorods
S4	Uniform size, shape and uniformly distribution	Below the nanorods

3 Fabrication of GO/Mg:ZnO Heterostructure Gas Sensor Devices

Synthesis of Graphene

Iron oxide crystal of 0.5 g by mass is put in a quartz glass tube with an electrical heating system. To reduce the targeted oxide to iron particles, a chemical process named in situ reduction (ISR) is followed by injecting hydrogen. Methane and nitrogen are mixed with 30:20 volumetric ratios of methane and nitrogen. At a rate of 25 ml/min, the mixture is flowed to iron catalyst (obtain from ISR) with methane as a source of carbon. After the thermocatalytic decomposition of methane, graphene sheets are repeatedly yielded at different reaction temperatures (800, 700, and 600 °C) at ambient pressure. The temperature varied during the thermocatalytic decomposition of methane is 600–800 °C; based on this temperature variation, the respective yields are also examined (Fig. 2).

Few carbon nanofibers are seen in sample A (i.e., simple reduced GO sample), and it is treated at 600 °C. However, self-decomposition of methane happens at 800 °C. It results in carbon nanospheres. Graphene yield rates for different time and temperatures are given in Fig. 2.

The yield of graphene sheets shows that the characteristic varies with reaction temperatures. After attaining temperatures (i.e., at 800 °C), the yield rate of graphene monotonically decreases with increasing reaction duration. Initially at 700 °C, the production rate is increased with reaction time, but after a certain temperature it starts to show opposite behavior and starts decreasing, and noticeably, the production rate first rises linearly at low temperature. The observed pattern of the curve indicates that, due to fast reaction at high temperatures, the iron surfaces are covered by graphene layers, thereby it reduces the yield (Fig. 3).

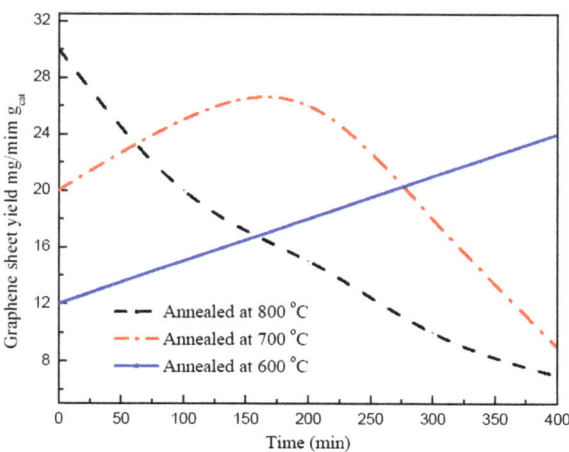

Fig. 2 Production yield graphene sheet for varying temperature [18]

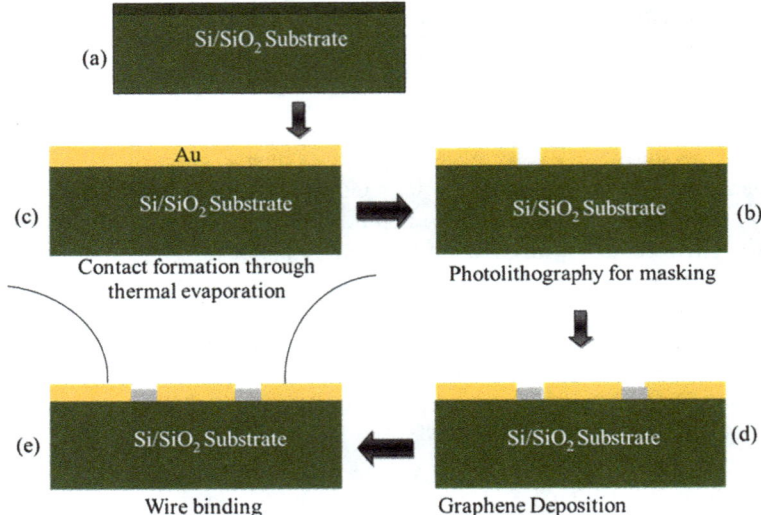

Fig. 3 Diagram presenting the fabrication process of graphene-based sensor: **a** base substrate is Si/SiO₂, **b** deposition of Au on Si/SiO₂, **c** photolithography for IDE formation, **d** deposition of graphene layer, and **e** formation of contact for resistance measurement [18]

IDE Formation

Silicon <100> with 0.001 Ω-cm² is considered during fabrication process. A pyrogenic oxidation process is followed to grow the silicon dioxide (SiO₂) layer of 1000 nm thickness. Considering the thermal evaporation technique, a gold (Au) layer (100-nm-thick) is deposited for fabrication of the IDE. A pattern is generated by photolithography technique. A positive photoresist (SPR 700 1.8) is used while fabricating IDE. A process of making the layout of IDE is already discussed showing (1) development of silicon substrate with silicon dioxide SiO₂ layer by pyrogenic oxidation, (2) Au (500 nm) deposition by thermal evaporation, (3) photolithography to form IDE, (4) deposition of doped graphene oxide, and (5) formation of contact for measurement 18].

Mg:ZnO/Graphene Layer Optimization

A mixture of 1–3 mg/ml Mg-doped ZnO (i.e., Mg:ZnO) ethanolic colloidal solution and an aqueous solution is prepared with a solute rGO (0.1–0.7 mg/ml rGO), and deposition is followed sequentially on gold electrodes by a spraying method. Annealing of thin film samples is carried out at 800 and 900 °C with nitrogen flow. Pure rGO film is made following the same process. Two different samples are synthesized, reduced graphene oxide (sample A) and reduced graphene oxide with Mg:ZnO (sample B). Both the samples (sample A and sample B) are taken into consideration for measuring the dynamic response. Investigations of electrical as well as optical properties are done for Sample B. Figure 4 describes the preparation of the ZnO/ZnMgO:graphene for sensing layer [18].

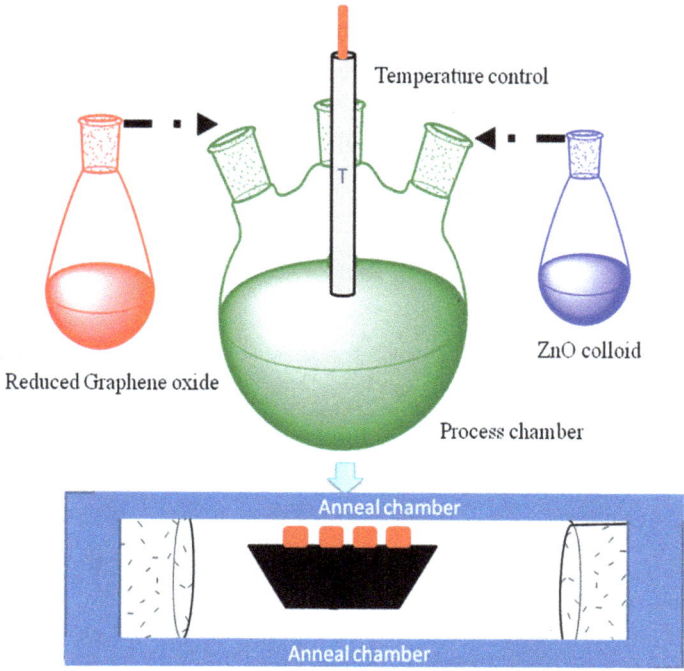

Fig. 4 Schematic diagram showing the deposition of ZnO/ZnMgO over reduced graphene oxide layer [18]

4 Experimental Observation of Synthesized ZnO Nanorod

The effects of varying duration of reaction, solution concentrations, growth, alignment monitoring, and morphology are emphasized in the experimental procedure. Characterization of each sample is done by SEM to know the structure of ZnO nanorods. X-ray photoelectron spectroscopy (XPS), X-ray diffraction (XRD), and photoluminescence (PL) provide appropriate results owing to the generation of hexagonal nanorod. All samples are experimented and observed with methane gas (CH_4) in moisture-free air by an experimental gas-analyzing system. In this chapter, discussion about the characterization of the synthesized samples and the corresponding results from gas-analyzing setup in dry air ambient are presented.

Synthesized ZnO nanorods are modified to get better surface morphology by doing chemical treatment. Depending on the duration of the chemical treatment, distinct samples are prepared, and scanning electron microscope (SEM), XRD, XPS, and photoluminescence (PL) are done for characterization of these samples.

Scanning Electron Microscopy (SEM):

The SEM microimages of the hydrothermally synthesized ZnO nanorods are recorded considering principle issues like solution concentration and reaction time.

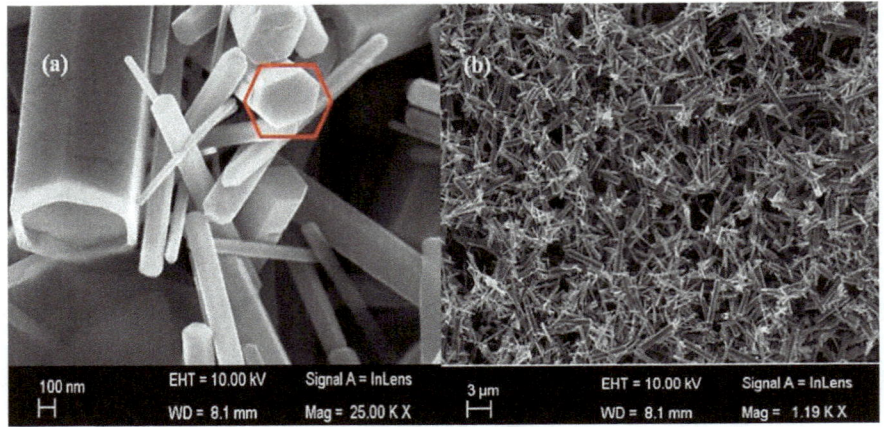

Fig. 5 SEM images showing **a** formation of hexagonal ZnO nanorods and **b** distributions of ZnO nanorods (© 2017 Elsevier Ltd.) [17]

A cautious observation of Fig. 5 (for synthesis procedure 1) tells the development of hexagonal-structured nanorods and formation of more desirable surface quality. It is presumed that surface area and sensitivity will be increased for gas adsorption, but it is observed that the alignment is poor as well as uniformity is also low. Solution concentration can change the density and diameter of the particle [19]. Changing the solution by varying the solution concentrations, the desired shape, size, and proper distribution of particles are obtained, and it becomes much more aligned and uniformly distributed in comparison with the previous sample which is viewed in Fig. 6a. If the reaction time crosses a certain limit, then ZnO becomes over etched, and it causes damaged ZnO nanorods that are observed in Fig. 6b.

X-ray diffraction (XRD):

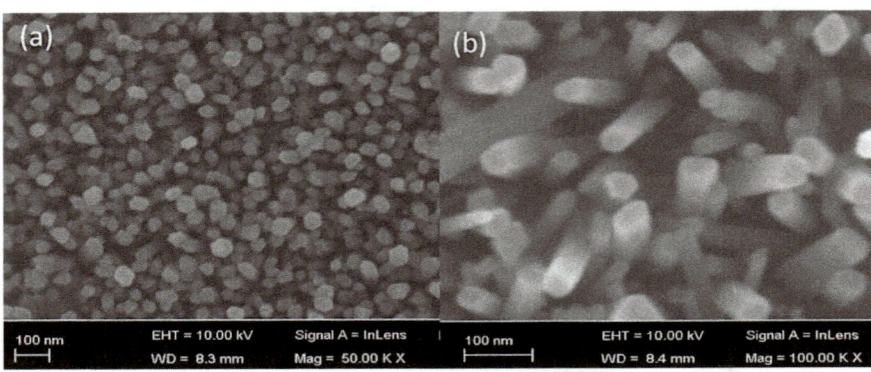

Fig. 6 SEM micrographs: **a** homogeneous hexagonal ZnO nanorods and **b** damaged ZnO nanorods (© 2017 Elsevier Ltd.) [17]

Fig. 7 Comparative XRD patterns for as grown sample, S3 non-uniform nanorods, and S4 uniform nanorods (© 2017 Elsevier Ltd.) [17]

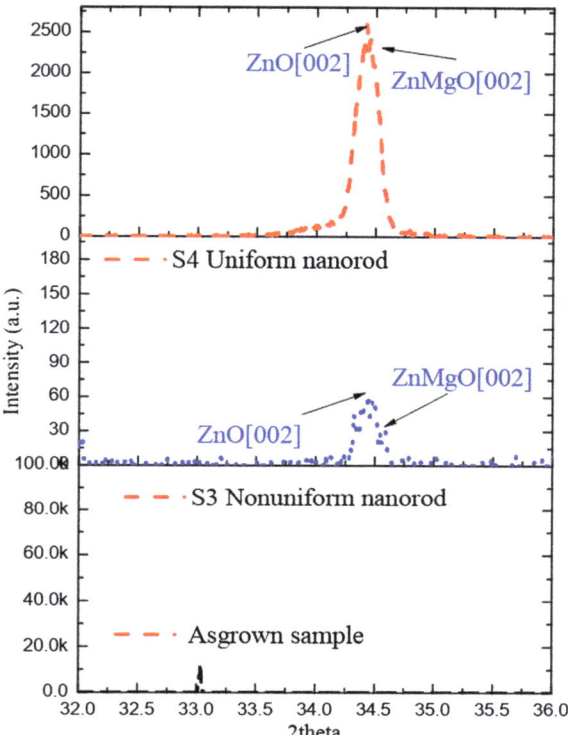

XRD technique is considered to authenticate the crystal structure in fabricated ZnO nanorods. The presence of small nanocrystals is assured by the peak broadening which is apparently observed in the XRD pattern. Figure 7 demonstrates X-ray scattering spectrum, illustrating the development of hexagonal morphology and good crystal configuration of the synthesized nanorods. In S3 and S4 samples (according to JCPDS card no 36–1451), it is detected that ZnO has a higher intensity peak (002) at 34.30°. It is recognized from Fig. 7 that S4 sample has improved crystallinity (002) at 34.30° than other samples and also reveals that ZnO is distinctively developed on the c-axis. No other peaks are observed by XRD which clearly tells about the absence of other impurities and free from the bulk remnant material. It also indicates the generation of only hexagonal-structured ZnO lattice. Moreover, if the XRD image ((002) peak for S3, S4) is enlarged, then a minor discontinuity is seen, as shown in Fig. 7. It assures that magnesium is favorably integrated into ZnO. The calculation of crystal size is done according to the diffraction peak broadening at (002) using the Debye–Scherrer Eq. (1) [19].

$$D = \frac{K\lambda}{\beta_{hkl}\mathrm{Cos}\theta} \tag{1}$$

where β_{hkl} implies integral half width, $\lambda = 0.1540$ nm implies the incident X-ray wavelength, K is having a constant value (0.90), D is the size of the crystal, and the Bragg angle is denoted by θ. The size of the particle obtained becomes 27.36 nm. The presence of defects states in the lattice is called dislocation density (δ), i.e., the dislocation lines length with respect to per unit volume of the lattice. It can easily be evaluated from the formula given below (2) [20].

$$\delta = \frac{1}{D^2} \tag{2}$$

Here, the crystallite size is D. The calculated value of dislocation density (δ) is 13.35×10^{-4} nm^{-2}.

The ZnO bond length (L) is calculated from the Eq. (3) [21] and found to be 0.2610 nm.

$$L = \sqrt{\left(\frac{a^2}{3} + \left(\frac{1}{2} - u\right)^2 c^2\right)} \tag{3}$$

where a and c are the lattice constants and u denotes amount of displacement of each atom with respect to the next across 'c' axis and termed as positional parameter.

The magnitude of a and c is mathematically determined from the equation given below (4) [22].

$$\text{Sin}^2\theta = \frac{\lambda}{4a^2}\left[\frac{4}{3}(h^2 + k^2 hk) + \left(\frac{a}{c}\right)^2 l^2\right] \tag{4}$$

Taking first-order approximation $n = 1$, a and c are measured as 0.3269 nm and 0.5222 nm, respectively, which resembles with ZnO JCPDS data.

X-ray photoelectron spectroscopy (XPS):

In accordance with the working rule of sensor, gas detection mainly depends on the reactivity of gas with chemisorbed O_2 species (O^-, O_2^-, O_2^-) on top of the MOS. It results in significant change in the magnitude of material resistance. The sensor response may be upsurged for electron donor. This occurs when a large number of oxygens are chemisorbed and the ionization occurs on top of the sensing layer [23–25]. So, the XPS is performed to see the constitutional properties of the hexagonal ZnO nanorods surface. XPS shows the formation of several peaks which correspond to different bands ranging from 0 to 1100 eV binding energy.

The existence of Zn, Mg, C, and O is confirmed from Fig. 8, which also resembles with the XRD investigation. C peak is formed mainly on account of hydrocarbon contaminants. Mg peak tells about the doping of Mg. $Zn_{2p3/2}$ peak formed at 1021.20 eV is highly intense, and Fig. 8 reveals that the peak is symmetrical in shape. And it confirms the existence of only zinc state. The O1s curve consists of three important parts. In case of O1s spectrum, it is in O_2^- ions in hexagonal structure lattice at minimal binding energy side. In ZnO matrix, the occurrence of O_2^- ions in

Fig. 8 XPS spectra of hexagonal-structured ZnO nanorods (S4 sample) (© 2017 Elsevier Ltd.) [17]

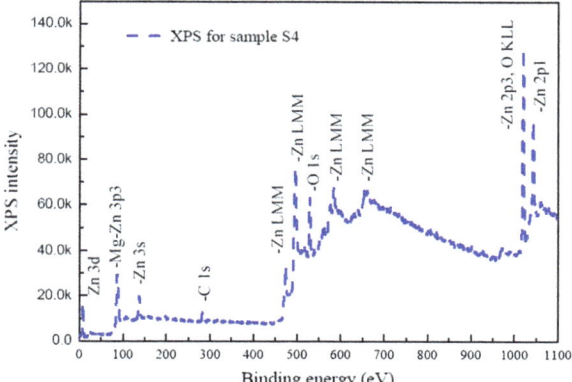

O_2 limited zone of O1s spectrum is proved by the moderate binding energy. In O1s, high-valued binding energy is associated with O_2 which is the chemisorbed over the top of sensing layer. Thus, an enhanced sensing is desired while creating sensing layer by S4 ZnO nanorod. XPS (Fig. 9) indicates the symmetrical peak of $Zn_{2p3/2}$ spectrum positioned at 1021.20 eV with a, showing that the existence of Zn only in oxide state.

Photoluminescence (PL):

Photoluminescence (PL) is done to examine the formation of ZnO nanorod. The variations of PL spectrum of ZnO (S4) nanorods (pre- and post-annealing done at 800 °C) are shown in Fig. 10. By photoluminescence, the comparison of excitons and emission depending on UV and visible band of synthesized nanorods samples is carried out. Free excitons are recombined, and it creates highly intensed (UV band: near band emission) peak at 380 nm. The reason behind deep-level emission (visible band) lies on interstitials of Zn and O_2 deficiencies. In case of S4 sample, the ratio of intensity for UV and visible bands and also the development of highly intensed peak at 380 nm assures about the defect of ZnO nanorod at material surface, and enhanced surface to volume ratio is obtained [26, 27]. It results in increased sensitivity.

It is already described in the previous chapter that in ZnMgO, it is deposited on ZnO seed layer following technique 2. This bilayer creation improves crystallinity and uniformity of nanoparticles (as depicted in Fig. 6). Figure 11 illustrates that compared to simple ZnO, better intensity peak is found when magnesium is doped with S4. Device sensitivity is also improved for minimized defect state and uniform configuration.

Performance of Gas Sensor

The detection is based on the variation of resistance of the semiconducting sensing platform on account of chemisorbed oxygen from the ambient air over the top of ZnO/ZnMgO. When ZnO/ZnMgO-based platform is placed in the atmosphere, electrons present in the conduction region of semiconducting platform are captured by

Fig. 9 Narrow scanned XPS spectrum of Zn2p3 (© 2017 Elsevier Ltd.) [17]

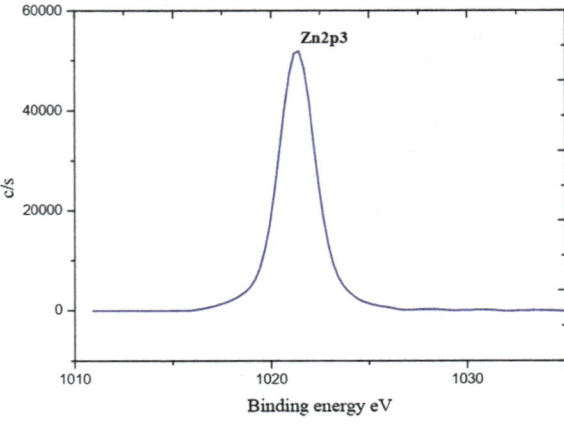

Fig. 10 Photo luminescence spectrums of S4 sample before and after annealing treatment (800 °C) (© 2017 Elsevier Ltd.) [17]

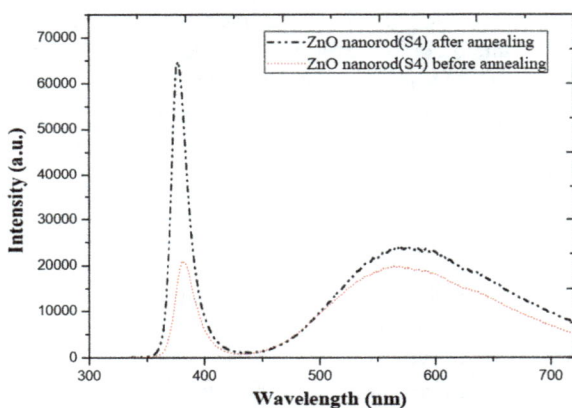

Fig. 11 Better intensity peak (PL's response curve) of sample S4 (Mg doped) (© 2017 Elsevier Ltd.) [17]

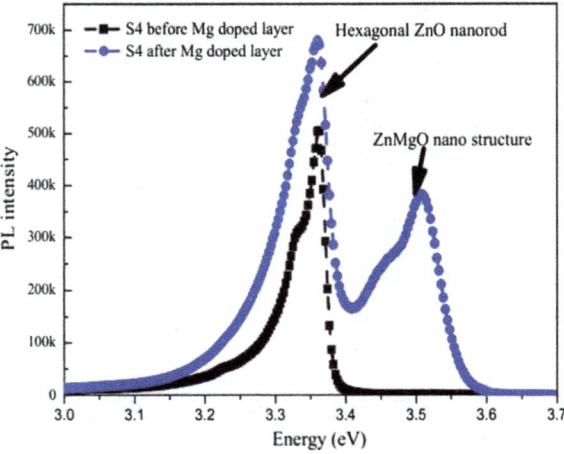

chemisorbed oxygen, and negatively charged species are formed. Creation of (O^-, O_2^-, O_2^-) depends on the temperature [28], and it results in the resistance change.

As a result, the space charge layer is formed [29] in the ZnO, and resistance (Ra) grows up. When the required operating temperature is supplied and the sensor makes contact with methane, CH_3 and H are formed by breaking CH_4. It starts reaction with ionized oxygen which is already chemisorbed, and it leads to an upsurge in electron density. Due to this procedure, resistance (Rg) significantly becomes smaller. If the difference between Rg and Ra is higher, better value of responsivity can be obtained. The schematic representation of sensing mechanism is depicted in Fig. 12. Methane sensing properties are investigated depending on the synthesized samples and IDE positioning. ZnO exposed at different activation temperatures is examined. According to Fig. 13, it is seen that the sensor response (sample S4 as a sensing layer) starts increasing from 100 °C temperature, and sensor response attains its optimum value at 200 °C.

The sensor response again starts decreasing with rise in temperature. It is because of excessive scattering of electron, de-absorption and contact resistance. If the operating temperature is below 200 °C, methane will not be able to make reaction with chemisorbed O_2 due to insufficient thermal energy. At 200 °C temperature, methane gets sufficient energy to activate, which may turn into satisfactory response. If the activation temperature is greater than 200 °C, it becomes difficult, and the exothermic reaction (CH_4 with oxygen) and also CH_4 desorption start happening increasingly

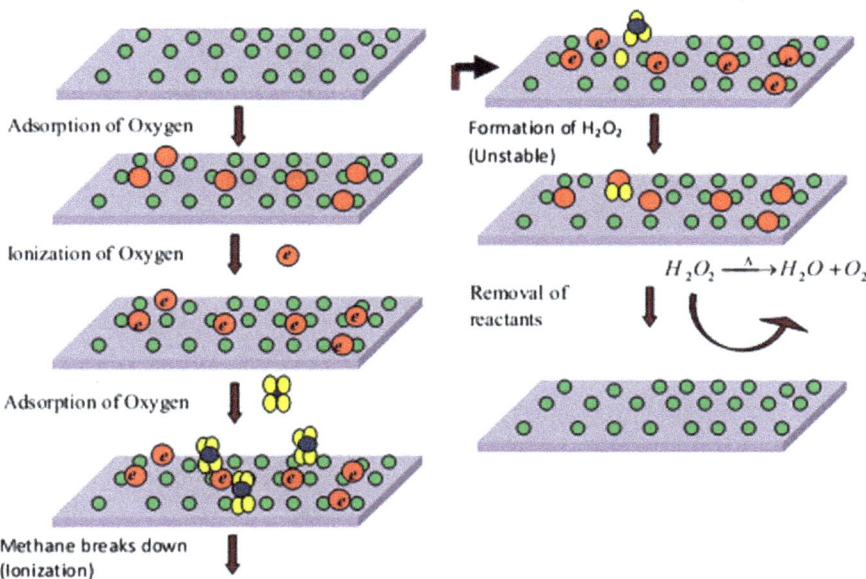

Fig. 12 Schematic representation of CH_4 sensing mechanism on sensing layer (© 2017 Elsevier Ltd.) [17]

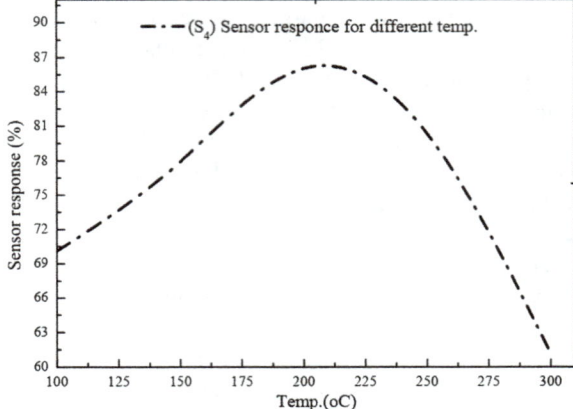

Fig. 13 Sensitivity of S4 sample as sensing material for different temperature (methane as target gas) (© 2017 Elsevier Ltd.) [17]

from the top of ZnO layer. General principle for all the metal oxide semiconductor is that there is a specific operational temperature, and it is also observed for ZnO [30].

Leveling the operational temperature (200 °C), four distinct samples are made for sensors as per Table 1.

Figure 14 is schematic diagram which depicts the position of interdigited electrode (above and below sensing layer). Figure 15 is the comparison of the responses for

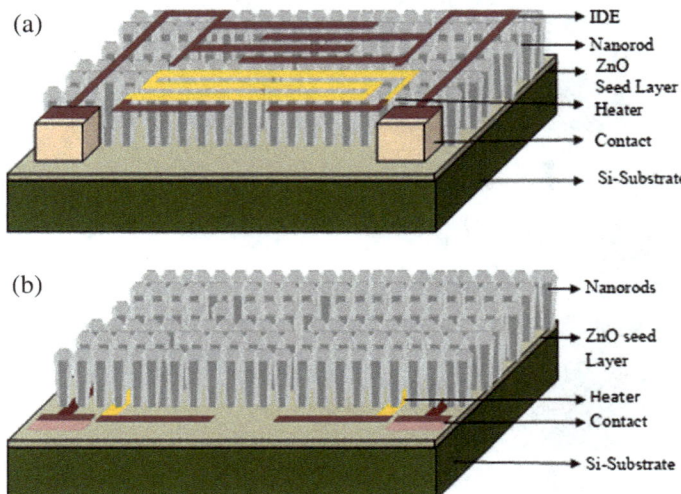

Fig. 14 Schematic diagram of position of interdigited electrode **a** IDE above and **b** below the nanorods [17]

Fig. 15 Sensor responses (for S1, S2, S3, and S4) (© 2017 Elsevier Ltd.) [17]

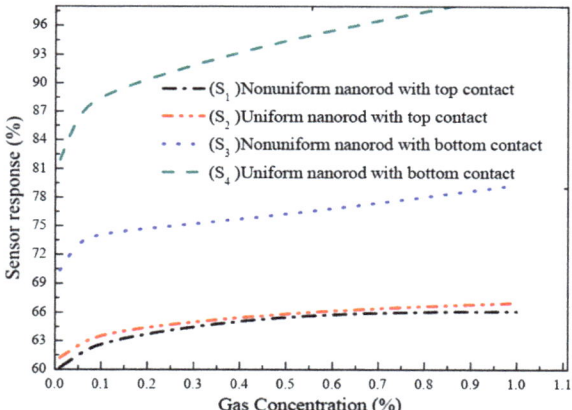

different samples S1, S2, and S3 with respect to S4. In S4, IDE is above the hexagonal ZnO nanorod, and the nanorods are uniform in shape, size, and distribution.

Theoretically, gas sensitivity, $S_m(\% \text{ ppm}^{-1})$ can be evaluated from Eq. (5) [31] where $\Delta R/R_i$ (%) signifies relative resistance change, ΔR implies the difference between the magnitudes of steady-state resistance (R_f) on exposure to analyte and in air (R_i), and C_j is taken as concentration of gas.

$$S_m(\%\text{ppm}^{-1}) = \frac{1}{n}\sum_{j=1}^{n}\frac{\left[\frac{\Delta R}{R_i}\right]_j}{C_j} \qquad (5)$$

The S4-based sensor proves that sensitivity enhances only because of uniform size, shape, and arrangement because of the electron depleted situation in lattice structure. Bottom contact has also been found to be an important issue for increased sensitivity toward CH_4 gas. The sensor response using S4 sample for varying concentration of methane at 200 °C temperature is studied, and it is shown in Fig. 16. 0.01–1.0% of CH_4 gas is exposed on the surface of the sensing material by using gas-analyzing system. When the gas starts flowing, sensing layer collects electron, and automatically resistance decreases, and when the gas flow is stopped, the resistance increases.

5   Electrical and Optical Characterization of GO/Mg:ZnO Heterostructure ZnO Nanorod

Electrical Measurement
To obtain better sensitivity, sample B (i.e., the rgO/Mg:ZnO system) is annealed at 700 and 800 °C. It is found that treatment under annealing temperature 800 °C

Fig. 16 For different
concentrations transient
response of CH$_4$ sensor (S4)
(© 2017 Elsevier Ltd.) [17]

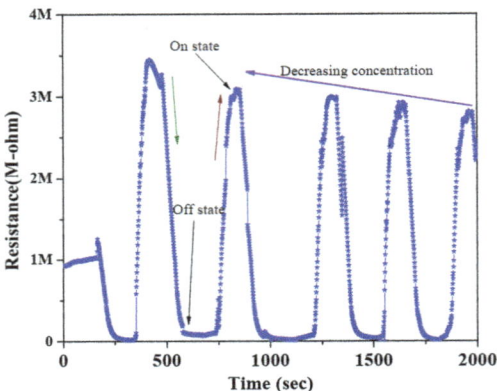

better sensitive material than the annealing temperature 700 °C, which is imputed
to the quality development of the film. Nevertheless, higher annealing temperatures
may turn into the raise of defect state concentration. It may explicitly result in poor
performance. Figure 14 represents graphene-based sensing platform. The sensor
response with respect to different annealing temperatures is shown in Fig. 17.

Optical Measurement

After the spin coating is done under prevailing conditions, the morphology and
arrangement of graphene oxide are characterized through SEM. Several ages are
found in the wrinkled structured flakes which is shown in Fig. 18a. Figure 18b indicates the graphene sheet with monolayer or hardly a few layers. Different magnitudes
of rotation per minute (rpm) are set for depositing graphene over the substrate. SEM
micrographs for 2500 rpm and 4000 are taken and shown in Fig. 18. It is seen that
due to increase of spin rotation, almost monolayer can be achieved.

Fig. 17 Device response
with respect to sample (A)
and sample (B) annealed at
temperatures [18]

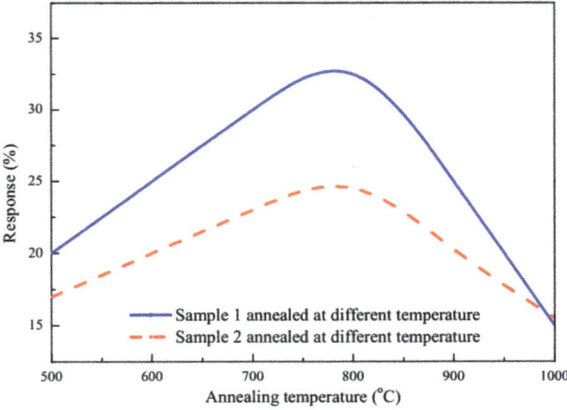

Figure 19a, b compares the distribution of GO and modified GO flakes over effective surface areas; uniform distribution significantly influences on sensitivity [32, 33]. Figure 19b proves the formation of monolayer and uniform distribution.

X-ray diffraction plots of synthesized nanoparticles are shown in Fig. 20. Indexing of the patterns is done in accordance with the JCPDS card no. 41-1487. A sharp peak is found in the XRD spectra at 26.598 when sonication is done to shed from graphite oxide in layers or scales. It is anticipated that an intense XRD peak will be formed and depending on the gap between the scales of graphite after oxidation, 'd' value will change proportionately. The result is the right shift (toward 9–11°). A complete dissection of the peak (002) gives interplanar spacing ($d_{h;k;l}$) of 00.115 nm and a full width half maximum (FWHM) of 00.051. The crystal size is found to be ~209.540 nm (by applying Debye–Scherrer equation) [34].

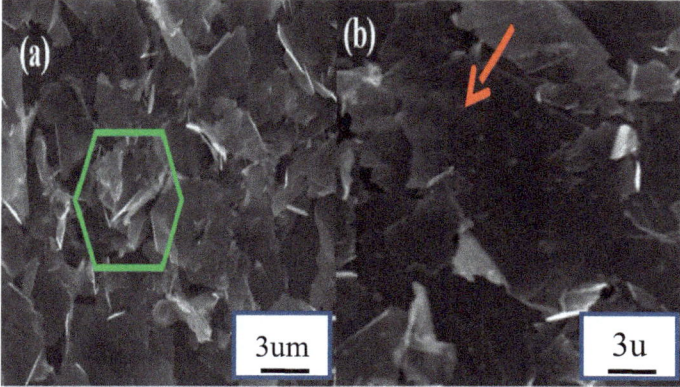

Fig. 18 SEM microimages: **a** multilayer graphene (2500 rpm spin) and **b** monolayer graphene (4000 rpm spin)

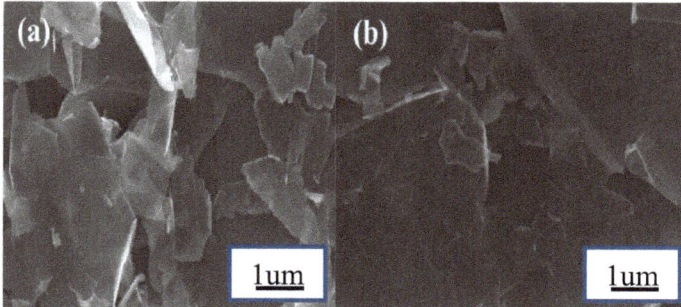

Fig. 19 SEM image for **a** GO with poor uniform size distributions **b** and modified GO improved uniform size distributions [18]

Fig. 20 XRD spectrum for annealed and non-annealed GO sheet (sample B) [18]

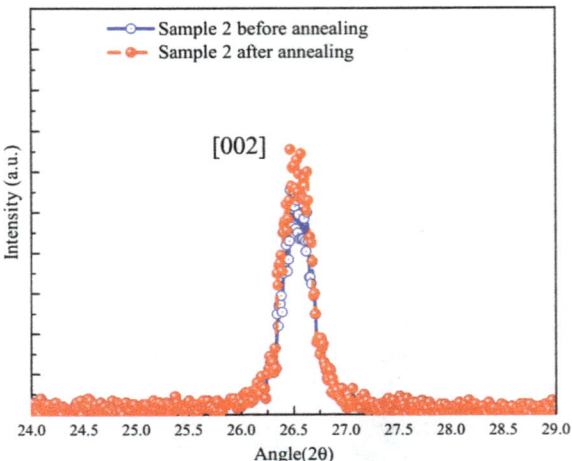

The characterizations of graphene and rGO have also been done by Raman spectra analysis, which are represented in Fig. 21. In case of graphene, double bonds between carbon atoms point to the highly intense peaks in Raman spectrum. Occurrence of G band is observed at ~1605 cm^{-1} for rGO, which resembles with the E2g phonon of the sp^2 C atom. The D band is seen at 1353 cm^{-1}, and it resembles with k point phonons. Here D band is the evidence of disorganization, which in turn comes from some definite defects (like vacancies and grain boundaries). The band intensity proportion, ID/IG, reduces while reducing graphene to rGO owing to the rectification of defects by retrieving the aromatic structures. It is an indication of presence of less defects in annealed reduced graphene oxide (Table 2).

X-ray spectroscopy spectra reveal the formation several peaks which are related to C1s and O1s bands. Due to contamination in the C1s spectrum, C–C components

Fig. 21 Raman spectrum of annealed rGO and as grown sample [18]

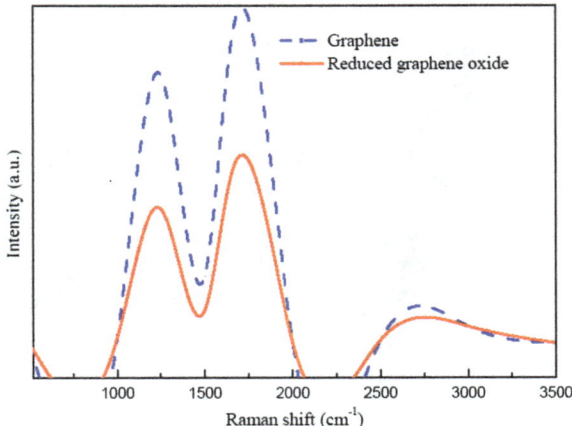

Table 2 X-ray spectroscopy for sample A and sample B [18]

	Sample A (%)	Sample B (%)
O1s	51.0	35.2
C1s	49.0	64.8

create more intense peak than C=O, C–O–C, and O–C=O components (low intensity). Figure 22a–c indicates the C1s spectrum for two samples. In Fig. 22a, the peak is grown at 284 eV which is less intense than that of shown in Fig. 22b. In case of first sample, sp^2 concentration is larger, and the peak is seen at 284 eV. In Fig. 22c, shape of the C1s peak is much more symmetrical and has moved to 284.8 eV. The existence sp^3 bonded carbon is the main reason of the symmetry [18, 19]. Table 1 provides XPS spectra of samples A and B.

6 Enhancement of Gas-Sensing Performance

Graphene, GO and rGO (derivatives of graphene) are relevant materials for sensing mechanism with enhanced sensitivities, and consequently, they are likely to apply in the advances in next generation sensor technologies [35–40]. In addition, rGO has the important properties such as high carrier mobility and minimized $1/f$ noise which could benefit the development of advance system of precision devices [41]. The extraordinary performance of graphene toward methane is the predominant factor for its extensive use in methane gas sensors [42, 43]. When CH_4 gas is applied to the top of the graphene layer, adsorption starts occurring, and it behaves like an electron donor. It results in significant variation in electrical properties (here resistance) [34]. By this way, electrons are rapidly transferred to the surface layer of the sensing material. Free electrons are acquired and ionize the chemisorbed atmospheric oxygen species.

$$2O_2^-(\text{ads}) + CH_4 \rightarrow CO_2 + 2H_2O + 4e^- \tag{6}$$

$$2O_2^-(\text{ads}) + CH_4 \rightarrow CO_2 + 2H_2O + 2e^- \tag{7}$$

Ion concentrations of oxygen are diminished for these reactions, and the electrons are retrieved. And these high mobility electrons are returned back to the delocalized energy band of the graphene oxide. Figure 23a depicts the positioning of electrodes above the reduced graphene oxide sensing layer, and Fig. 23b is the representation of methane gas exposed on graphene layer. The use of center electrode is mainly due to reduction in the electrical path length, so that the device electrical performance is increased. According to Fig. 24, proper set of analyzing the target gas is used, and little amount of methane (0.01–1.0%) is applied as target analyte on the surface of the sensing platform. Sensing responses are noted for the samples A and B and presented in Fig. 25. In this context, the methodology of preparing samples (A and B) and the

Fig. 22 Material characterization: XPS spectra for **a** sample (A) and **b** for sample (B), and **c** spectra for C=C bonding of those samples [18]

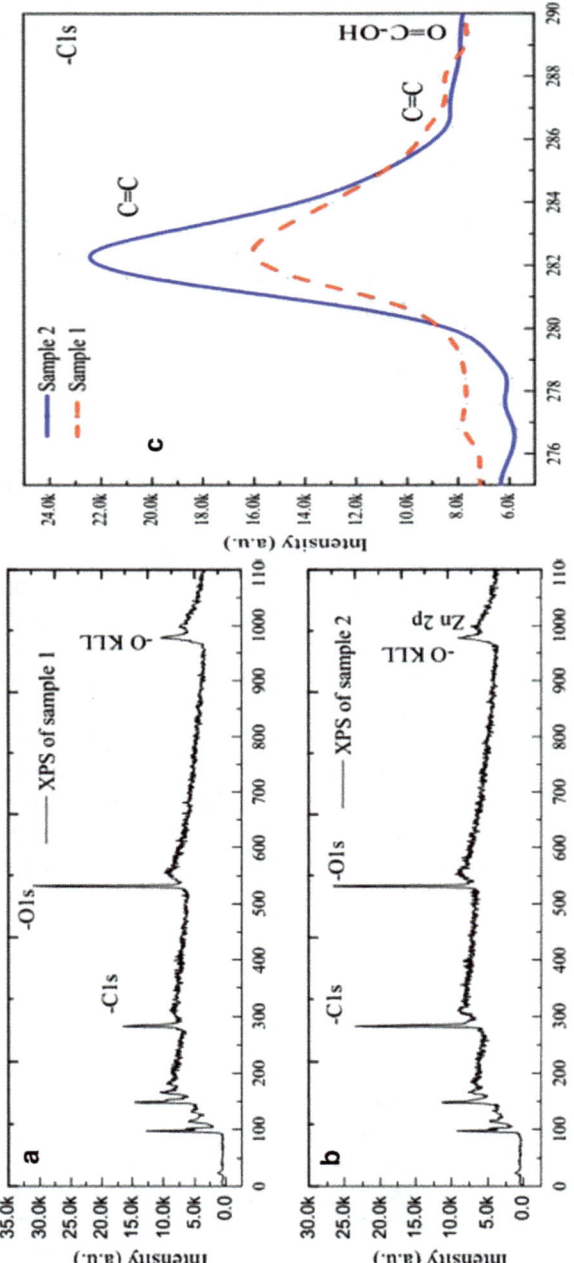

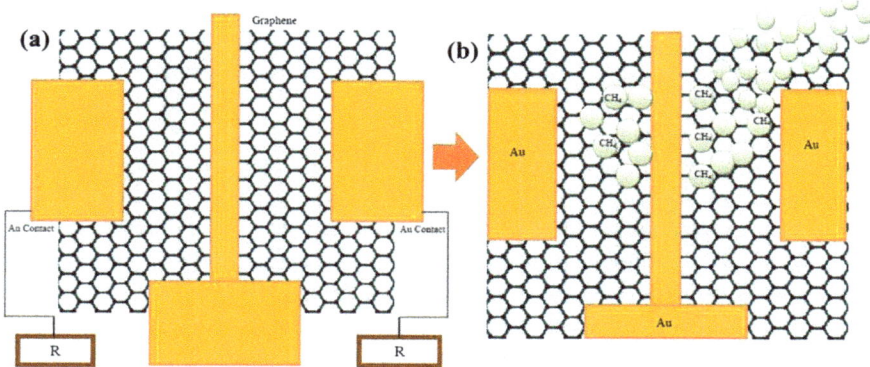

Fig. 23 Schematic representation of sensing mechanism for doped rGO: **a** graphene sensing layer in absence of CH_4 gas and **b** on exposure to CH_4 gas [18]

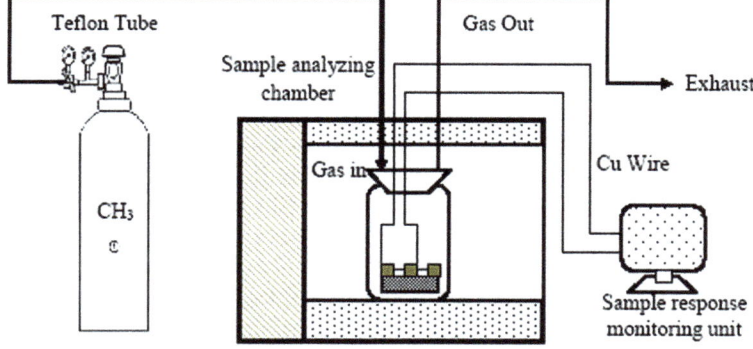

Fig. 24 Schematic diagram of gas-analyzing setup: substrate is at sample-analyzing chamber [15]

Fig. 25 Device transient response: better electrical performance of rGO for CH_4 sensor [18]

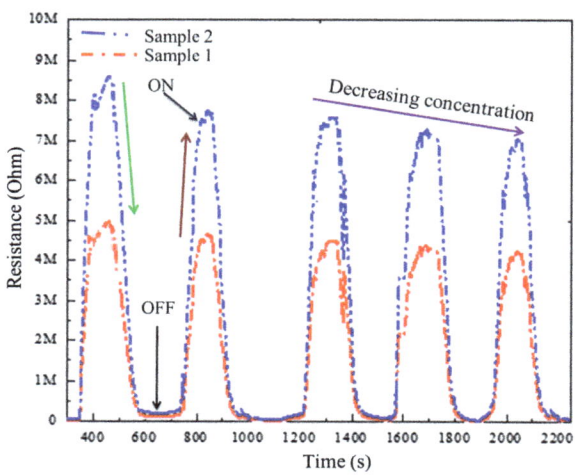

chemo-thermal treatment on samples A and B have been reported in the previous chapter. The properties of graphene oxide like high signal-to-noise ratio make it possible to induce significant changes in resistance by creating a small change in the concentration of the carriers. And it is only because of the process of chemisorption of target gas molecules [35].

Gas-sensing performance is noted for those samples (sample A and sample B). As it is a resistive sensor, there is a reduction of resistance while the gas is in contact with sensing layer. And this change in resistance is due to the depletion of carriers. Figure 24 gives the comparative graph of resistance change for sample A and sample B. The response time of transient and recovery is found to be satisfactory. From the comparative analysis shown in Fig. 24, it is clear that the rGO/Mg:ZnO sensor (i.e., annealed sample B-based setup) shows higher sensitivity (R_{air}/R_{gas}) than the simple rGO-based device (sample A-based setup). Noticeably, at room temperature, Mg:ZnO is having a less excitation energy of 180 k/cm^2.

7   Conclusion

This chapter details about the hydrothermal growth process which is followed to synthesize the hexagonal-structured high sensitive zinc oxide (ZnO) with a large surface area and volume ratio by optimizing defects state for getting better sensing performance by doing chemical treatment. An easy effective way to synthesize reduced graphene oxide (rGO) and zinc oxide (ZnO)/zinc magnesium oxide (ZnMgO)-doped rGO is also discussed. The results of the work are also discussed, and zinc magnesium oxide (ZnMgO)-doped rGO is found to be comparatively more sensitive. The major highlights of the chapter are summarized below:

- The responsivity of gas sensor can be optimized through the structural improvement of ZnO nanorods and reduction of defect state.
- Highly sensitive hexagonal ZnO nanorod may provide large surface area and volume ratio. It is uniform in shape, and distribution is an important phenomenon for sensing application.
- The study of the positioning of IDE below/above the sensing material with the conclusion that the IDE poisoned below the sensing layer shows better sensitivity.
- Temperature-dependent study is on sensing material, device output.
- Graphene oxide/metal oxide heterojunction results enhanced sensitivity as a sensing layer.
- Sensitivity is enhanced due to the existence of magnesium (Mg) in the ZnO/rGO lattice structures.

References

1. Lim HJ, Lee DY, Oh YJ et al (2006) Gas sensing properties of ZnO thin films prepared by microcontact printing. Sens Actuators A Phys 125:405–410
2. Hsueh TJ, Hsu CL, Chang SJ, Chen IC (2007) Laterally grown ZnO nanowire ethanol gas sensors. Sens Actuators B Chem 126:473–477
3. Dikovska AO, Atanasov PA, Tonchev S, Ferreira J, Escoubas L (2007) Periodically structured ZnO thin films for optical gas sensor application. Sens Actuators A Phys 140:19–23
4. Németh Á, Horváth E, Lábadi Z, Fedák L, Bársony I (2007) Single step deposition of different morphology ZnO gas sensing films. Sens Actuators B Chem 127:157–160
5. Baratto C, Sberveglieri G, Onischuk A, Caruso B, Di Stasio S (2004) Low temperature selective NO_2 sensors by nanostructured fibres of ZnO. Sens Actuators B Chem 100:261–265
6. Choi D, Choi GM (2000) Electrical and CO gas sensing properties of layered ZnO–CuO sensor. Sens Actuators B Chem 69:120–126
7. Comini E (2006) Metal oxide nano-crystals for gas sensing. Anal Chim Acta 568:28–40
8. Wang M, Ye CH, Zhang Y, Hua GM, Wang HX, Kong MG, Zhang LD (2006) Synthesis of well-aligned ZnO nanorod arrays with high optical property via a low-temperature solution method. Cryst Growth 291:334–339
9. Azad AM, Akbar SA, Mhaisalkar SG, Birkefeld LD, Goto KS (1992) Solid state gas sensors: A review. J Electrochem Soc 139:3690
10. Arafat MM, Dinan B, Akbar SA, Haseeb ASMA (2012) Gas sensors based on one dimensional nanostructured metal-oxides: a review. Sensors 12:7207–7258
11. Tan JY, Avsar A, Balakrishnan J, Koon G KW, Taychatanapat et al (2014) Electronic transport in graphene-based heterostructures. Appl Phys Lett 104:183504
12. Novoselov KS, Fal VI, Colombo L, Gellert PR, Schwab MG, Kim K (2012) A roadmap for graphene. Nature 490:192–200
13. Craciun MF, Russo S, Yamamoto M, Tarucha S (2011) Tuneable electronic properties in graphene. Nano Today 6:42–60
14. Schwierz F (2010) Graphene transistors. Nat Nanotechnol 5:487
15. Fattah A, Khatami A (2014) IEEE Sens J 14:4104
16. Sharma S, Madou M (2012) A new approach to gas sensing with nanotechnology. Phil Trans R Soc A 370:2448–2473
17. Sarkar A, Maity S, Bhunia CT, Sahu PP (2017) Responsivity optimization of methane gas sensor through the modification of hexagonal nanorod and reduction of defect states. Superlatt Microst 102:459–469
18. Sarkar A, Maity S, Joseph AM, Chakraborty SK, Thomas T (2017) Methane-sensing performance enhancement in graphene oxide/Mg:ZnO heterostructure devices. J Electron Mater 46:5485–5491
19. Öztürk S, Kılınç N, Öztürk ZZ (2013) Fabrication of ZnO nanorods for NO_2 sensor applications: effect of dimensions and electrode position. J Alloy and Compd 581:196–201
20. Saleem M, Fang L, Wu F, Huang QL, Xu CL et al (2012) Effect of zinc acetate concentration on the structural and optical properties of ZnO thin films deposited by sol-gel method. Intl J Phy Sci 7:2971–2979
21. Morkoç H, Özgür Ü (2008) Zinc oxide: fundamentals, materials and device technology. Wiley
22. Bindu P, Thomas S (2014) Estimation of lattice strain in ZnO nanoparticles: X-ray peak profile analysis. J Theor Appl Phys 8:123–134
23. Zhang T, Zeng Y, Fan HT, Wang LJ et al. Synthesis, optical and gas sensitive properties of large-scale aggregative flowerlike ZnO nanostructures via simple route hydrothermal process. J Phys D Appl Phys 42:045103
24. Zhang L, Yin Y (2013) Large-scale synthesis of flower-like ZnO nanorods via a wetchemical route and the defect-enhanced ethanol-sensing properties. Sens Actuators B Chem 183:110–116
25. Han N, Hu P, Zuo A, Zhang D et al (2010) Photoluminescence investigation on the gas sensing property of ZnO nanorods prepared by plasma-enhanced CVD method. Sens Actuators B Chem 145:114–119

26. Waclawik ER, Chang J, Ponzoni A, Concina I, Zappa D, Comini E et al (2012) Functionalised zinc oxide nanowire gas sensors: enhanced NO_2 gas sensor response by chemical modification of nanowire surfaces. Beilstein J Nanotechnol 3:368–377

27. Taunk PB, Das R, Bisen DP, Tamrakar RK (2015) Structural characterization and photoluminescence properties of zinc oxide nano particles synthesized by chemical route method. J Radiat Res Appl Sci 8:433–438

28. Schierbaum KD, Kirner UK, Geiger JF, Göpel W (1991) Schottky-barrier and conductivity gas sensors based upon Pd/SnO_2 and Pt/TiO_2. Sens Actuators B Chem 4:87–94

29. Rai P, Kim Y, Song H, Song M, Yu Y (2012) The role of gold catalyst on the sensing behavior of ZnO nanorods for CO and NO_2 gases. Sens Actuators B Chem 65:133–142

30. Zhang L, Yin Y (2013) Large-scale synthesis of flower-like ZnO nanorods via a wet-chemical route and the defect-enhanced ethanol-sensing properties. Sens Actuators B Chem 183:110–116

31. Dilonardo E, Penza M, Alvisi M, Di Franco C, Palmisano F, Torsi L, Cioffi N (2016) Evaluation of gas-sensing properties of ZnO nanostructures electrochemically doped with Au nanophases. Beilstein J Nanotechnol 7:22–31

32. Bhattacharya S, Dhar P, Das SK, Ganguly R, Webster TJ, Nayar S (2014) Colloidal graphite/graphene nanostructures using collagen showing enhanced thermal conductivity. Int J Nanomed 9:1287

33. Salih E, Mekawy M, Hassan RY, El-Sherbiny IM (2016) Synthesis, characterization and electrochemical-sensor applications of zinc oxide/graphene oxide nanocomposite. J Nanostruct. Chem. 6:137

34. Zhang D, Yin N, Xia B (2015) Facile fabrication of ZnO nanocrystalline-modified graphene hybrid nanocomposite toward methane gas sensing application. J Mater SCI Mater Electron 26:5937

35. Chen L, Tang Y, Wang K, Liu C, Luo S (2011) Direct electrodeposition of reduced graphene oxide on glassy carbon electrode and its electrochemical application. Electrochem Commun 13:133

36. Schedin F, Geim AK, Morozov SV, Hill EW, Blake P et al (2007) Detection of individual gas molecules adsorbed on graphene. Nat Mater 6:652

37. Chen CW, Hung SC, Yang MD, Yeh CW, Wu CH, Chi GC et al (2011) Oxygen sensors made by monolayer graphene under room temperature. Appl Phys Lett 99:243502

38. Kumar S, Kaushik S, Pratap R, Raghavan S (2015) Graphene on paper: a simple, low-cost chemical sensing platform. ACS Appl Mater Interface 7:2189

39. Nemade KR, Waghuley SA (2013) Chemiresistive gas sensing by few-layered graphene. J Electron Mater 42(10):2857

40. Peng Y, Li J (2013) Ammonia adsorption on graphene and graphene oxide: a first-principles study. Front Environ Sci Eng 7:403

41. Varghese SS, Varghese SH, Swaminathan S, Singh KK, Mittal V (2015) Two-dimensional materials for sensing: graphene and beyond. Electronics 4:651

42. Wang DH, Hu Y, Zhao JJ, Zeng LL, Tao XM, Chen W (2014) Holey reduced graphene oxide nanosheets for high performance room temperature gas sensing. J Mater Chem A 2:17415

43. Leenaerts O, Partoens B, Peeters FM (2009) Adsorption of small molecules on graphene. Microelectron J 40:860

Significance of Optimal Positioning of the Reference Electrode for an ISFET

Santanu Sharma, Chinmayee Hazarika, and Sujan Neroula

Abstract Implantable sensors capable of providing precise in-vivo measurements are the need of the hour. Many challenges still prevail in the path of these sensors to become compact and minimally invasive. For the application of ISFET based biosensors, especially in the in-vivo operations, optimization of the device dimension becomes crucial. As the reference electrode is an integral part of an ISFET based sensing system, the optimal positioning of the former with respect to the sensing layer of the ISFET can be a decisive factor in the overall dimension of the system. A comprehensive physics-based model for determination of appropriate separation between the silicon nitride (Si_3N_4) sensing layer and reference electrode has been developed and validated by an experimental setup. For a respective pH value of the electrolyte, it was found that the optimal reference electrode positioning should be at three times the Debye length. An electro-mechanical system has been developed which can place the reference electrode in the desired position with an accuracy of 0.126 μm. The chapter presents the experimental results to support the theory associated with the optimal positioning of the reference electrode. Due to limitation of the system, the experiments were carried out only for three different highly alkaline electrolytes.

Keywords ISFET · Sensors · MOSFET · Reference electrode · Threshold voltage

1 Introduction

The model of ISFET can be explained well by drawing an analogy with the metal oxide semiconductor field effect transistor (MOSFET). The theoretical concept of a MOSFET can be used to explain the operation mechanism of an ISFET. Unlike the

S. Sharma (✉) · S. Neroula
Department of Electronics and Communication Engineering, Tezpur University, Tezpur, India

C. Hazarika
Assam Energy Institute Sivasagar, Centre of Rajiv Gandhi Institute of Petroleum Technology, Jais, Amethi, India

© The Author(s), under exclusive license to Springer Nature Singapore Pte Ltd. 2022
R. Goswami and R. Saha (eds.), *Contemporary Trends in Semiconductor Devices*,
Lecture Notes in Electrical Engineering 850,
https://doi.org/10.1007/978-981-16-9124-9_10

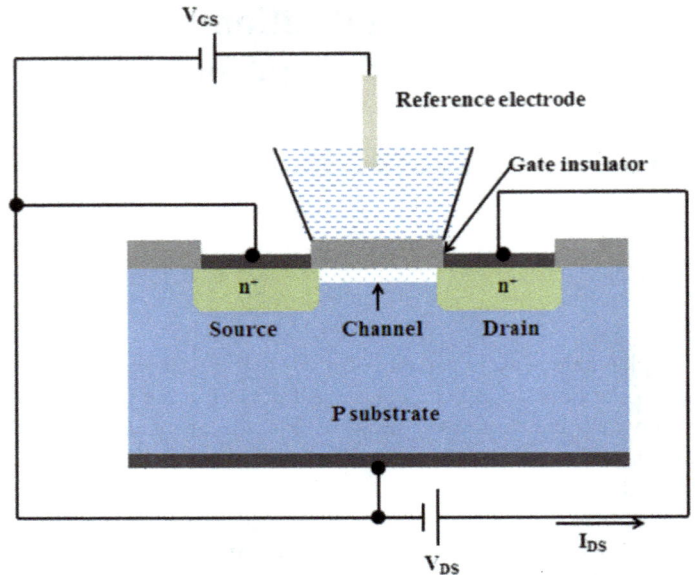

Fig. 1 Depicting the schematic of the ISFET device

MOSFETs, ISFETs do not have the gate layer (metal layer). Therefore, the ISFET device when introduced in the electrolyte, the sensing layer (Si_3N_4 layer) comes directly in contact with the solution. Figure 1 illustrates an ISFET device, wherein the presence of a reference electrode immersed in the solution completes the gate circuit [1].

The ion of interest of the electrolyte solution is buffered by the sensing membrane of the ISFET device. For a MOSFET, the threshold voltage signifies the onset of drain current, and this comprises of voltages due to work function difference between the metal and the semiconductor, the surface potential, and the charges trapped inside the oxide layer [1, 2].

The threshold voltage of an ISFET ($V_{th(ISFET)}$) can further be written in terms of the threshold voltage of the MOSFET ($V_{th(MOSFET)}$) can be expressed as[2]

$$V_{th(ISFET)} = V_{th(MOSFET)} + E_{ref} + \phi_{lj} + \chi_{sol} - \phi_{eo} - \phi_m \qquad (1)$$

Here

E_{ref} is a constant termed as the potential of the reference electrode.

Φ_{eo} is the interfacial electrolyte-insulator potential (function of the pH of the solution).

χ_{sol} is a constant termed as surface dipole potential of the solvent.

Φ_{lj} is the potential difference at the junction of reference solution and the electrolyte, and.

Φ_m is the work function of the metal.

The exchange of hydrogen ions between the electrolyte and surface groups changes the charge at the electrolyte-insulator interface. The active sites are formed by the interaction between the protons and the surface groups. The redistribution of charge changes the potential at the interface. This leads to the formation of the electrical double layer.

There are three theoretical approaches to the double layer on the interface;

(i)   Helmholtz Plane Theory,
(ii)   Gouy-Chapman Theory, and
(iii) Gouy-Chapman Stern Model.

The Helmholtz Double Layer theory proposed in 1879, is a lucid approximation, wherein the neutralization of surface charges are done by the counter ions of opposite polarity [3]. The charges present in the solution are located at the Outer Helmholtz plane (OHP) and the surface is analogous to a parallel plate capacitor. However, this model considers rigid layers of opposite charges, which is not possible in actual operation. Therefore, the Gouy-Chapman double layer theory was an improvisation of the Helmholtz Double Layer theory [4, 5]. In this improvised theory, the counter ions are not considered rigidly held, and they tend to diffuse into the liquid phase unless the counter potential restricts this tendency. The Boltzmann equation gives the ion concentration distribution in the diffusion layer [6]. The assumption was made that only oppositely charged ions diffuse into the bulk solution which contradicts the thickness of the double layer found experimentally and mathematically. This is because both the oppositely charged ions and the same sign charges are found within the double layer. Therefore, the third model (inclusion of a Stern layer) came into being. Thus, the electrical double layer has been explained well by Gouy-Chapman Stern theory as illustrated in Fig. 2 [7–10].

Precise in-vivo measurements are the primary requirement for implantable sensors. Biocompatibility, cost-effectiveness, compactness, minimally invasiveness, etc. are the challenges in the path of development of such sensors. ISFET based sensors can be employed for both *in-vivo* and in-vitro measurements. The reference electrode completes the gate to source circuit and hence it is an inseparable part of ISFET. The positioning of the reference electrode is a crucial factor, as the proper knowledge of its position can optimize the device dimensions. This is relatable to all ISFET based sensors. A theoretical model that was proposed in 2009, suggested that the optimal positioning of the reference electrode should be at three or more than three times the Debye length for the respective pH values [11]. The model was proposed for silicon dioxide. In this chapter, this theory is extended for silicon nitride sensing surface and the validation of the theoretical data has been done experimentally [12]. This validation of this theory can prove to be a milestone in designing biomedical sensors and also can be put to use for the reference electrode positioning for all ISFET based biosensors.

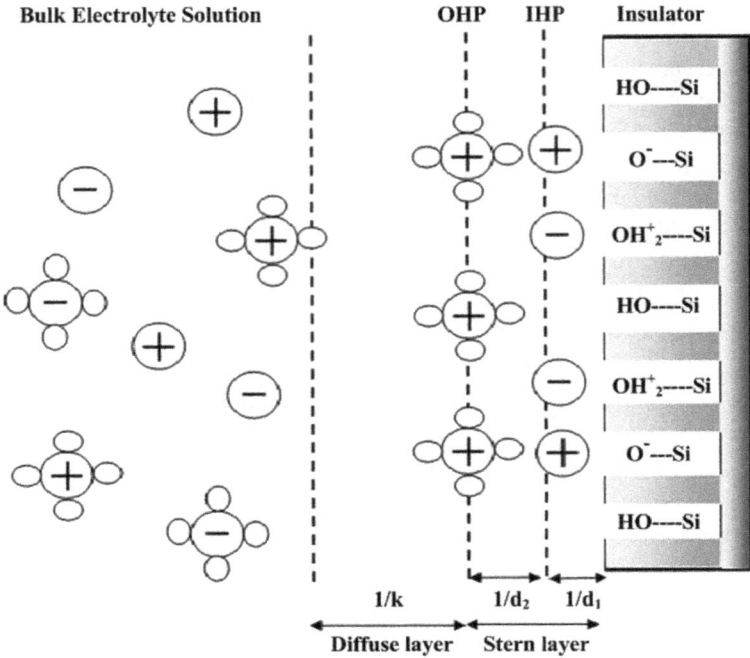

Fig. 2 Illustration of Stern and Gouy–Chapman layers in an electrical double layer

2 Theory

An ISFET device is devoid of the gate over layer hence it requires a reference electrode to complete the gate to source circuit [1]. The reference electrode should be non-polarizable. The conductance of the device is controlled by the voltage applied to the reference electrode. The applied voltage to the reference electrode decides the state of the device, i.e., active or saturation. Change in the surface interfacial potential is detected by the change in drain current. The reference electrode positioning can affect this interfacial potential. When the electrode is moved to the bulk of the solution the potential gradually decreases and attains a steady value [2]. When the electrode is placed in the close vicinity of the sensing layer, the effective electrolyte-insulator potential is not complete. Therefore, to get the complete response of the ISFET, the interfacial electrolyte-insulator potential has to be expressed as a function of the position of the reference electrode.

The reference electrode, when placed in the bulk of the solution the potential profile of the electrolyte-insulator semiconductor (EIS) structure for a planar surface varies as depicted in Fig. 3. Here, a pH value greater than point of zero charge (pH_{PZC}) has been considered. The potential decreases exponentially and attains a steady-state in the bulk of the solution. The surface potential in the semiconductor is denoted as ϕ_s. The potential variation across the diffused layer has been explained using

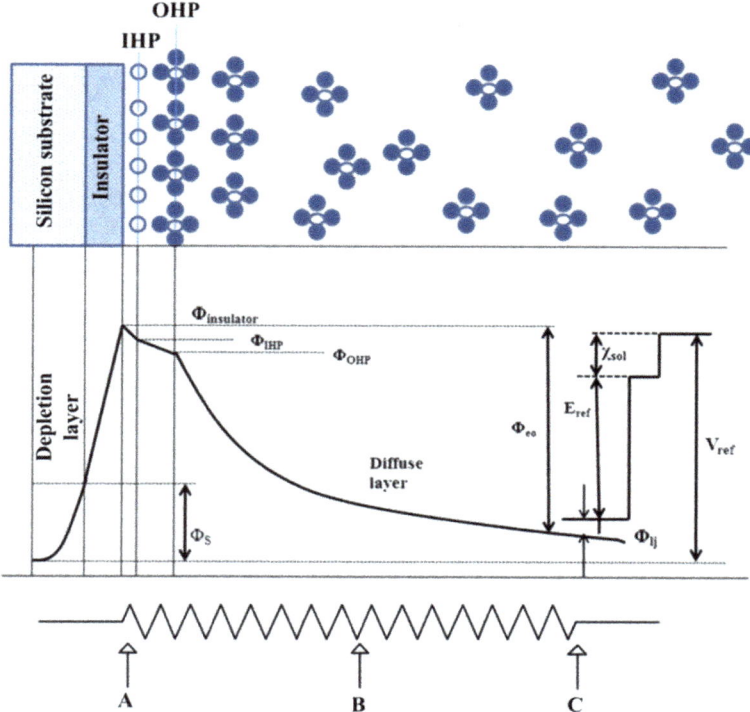

Fig. 3 Potential profile of the EIS structure for a planar surface (when the reference electrode is placed in the bulk of the solution)

an analogy of a variable resistor as illustrated in Fig. 3. Considering the reference electrode to be the slider 'B' moving from 'A' to 'C' will increase the resistance between points A and B. Consequently, this decreases the resistance between points B and C. Here point 'A' is analogous to the sensing layer and the point 'C' describes three times of the Debye length as proposed in the theoretical model [11, 12]. In Fig. 3, the ϕ_{eo} when considered as the potential across the considered resistance, placing of the reference electrode at point 'C' can provide with the full range of ϕ_{eo}.

Further, from Fig. 4 it can be observed that the full range of ϕ_{eo} cannot be achieved when the reference electrode is positioned within the diffuse layer, i.e., before point 'C'. Here, ϕ_{s1} is the semiconductor surface potential. Both Figs. 3 and 4 suggest that ϕ_{s1} is less than ϕ_S. Lesser charge at the inversion layer leads to a decrease in semiconductor potential causing a subsequent drop in the value of the drain current. Hence, as depicted in Fig. 4, the potential profile obtained in this case is incomplete.

In the prior proposed theoretical model, silicon dioxide was considered as a sensing layer. Here in this chapter, the model is further extended to silicon nitride and

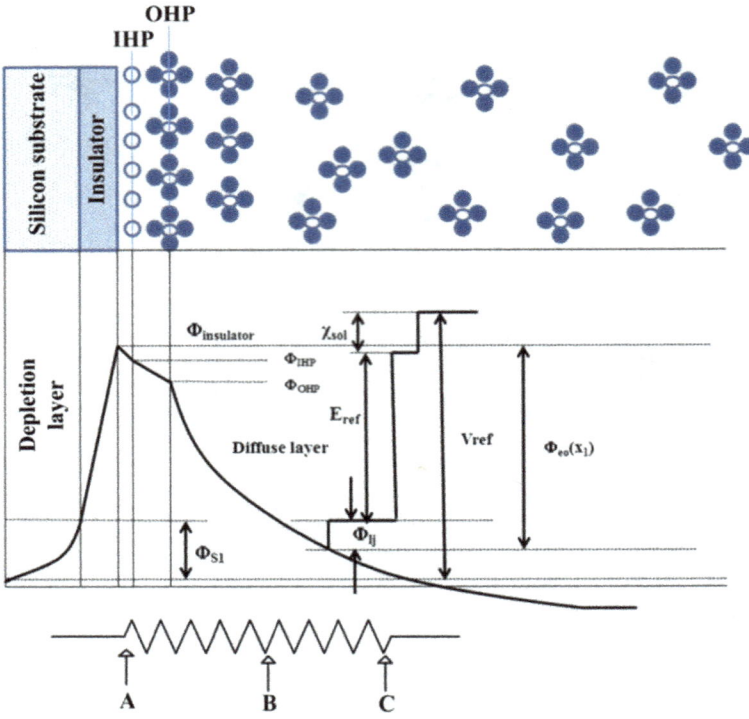

Fig. 4 Potential profile of the EIS structure for a planar surface when the reference electrode is placed in the diffused layer [1]

mathematical modeling has been developed. In the following sections, the mathematical modeling for silicon nitride is presented. This is followed by the methodology involved in the validation of this theoretical model.

3 Mathematical Modeling

In Si_3N_4 gate pH ISFET, the insulating layer consists of two different types of active sites, i.e., silanol and amine groups as depicted in Fig. 5. The exchange of hydrogen ion takes place between the active sites and the ions present in the electrolyte [1]. This results in the redistribution of charges in the electrical double layer (EDL) formed in the vicinity of the insulator layer. The redistribution of charges in this EDL affects the flat-band voltage which further modifies the threshold voltage of the ISFET device. The threshold voltage stated in Eq. 1 (excluding the trapped charges, interface charges, and the fixed immobile charges) indicates that all the terms are constant except the interfacial electrolyte-insulator potential (ϕ_{eo}). Therefore, the

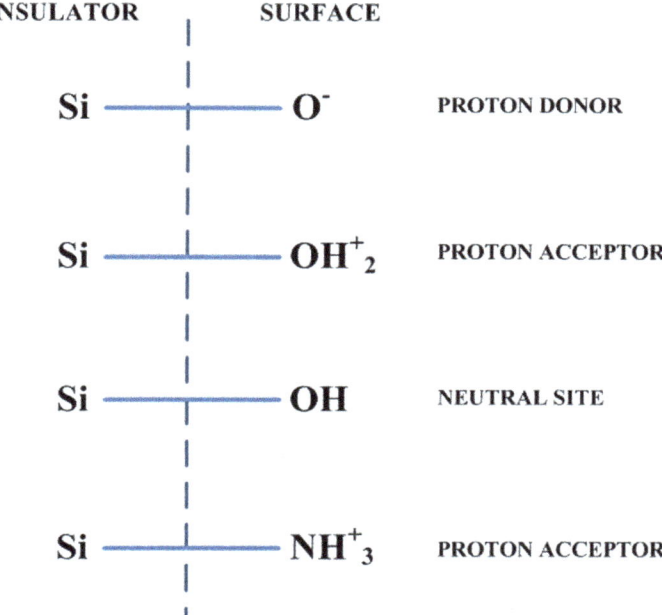

Fig. 5 Site binding model of the silicon nitride surface

interfacial electrolyte-insulator potential influences the response of the ISFET. Here for the Si_3N_4 surface, a mathematical relation between the interfacial electrolyte-insulator potential (ϕ_{eo}) and the distance (x) of the reference electrode from the OHP has been established.

3.1 Model Description

The insulating surface of silicon nitride when in contact with water, yields the silanol SiOH and amine $SiNH_2$ sites according to the schematic reactions below [13–15].

$$Si_3N(s) + H_2O(l) \leftrightarrow Si_2NH(s) + SiOH(s) \tag{2}$$

$$Si_2NH(s) + H_2O(l) \leftrightarrow SiNH_2(s) + SiOH(s) \tag{3}$$

Here, s and l represents solid and liquid phase, respectively. Here in the model, only those surface sites which are involved in the absorption or desorption of hydrogen ion are considered.

In an acidic environment (pH $\leq$ 7), silanol and amine sites adsorb and desorb protons according to the following reactions [14–16]

$$\text{SiNH}_2(s) + \text{H}^+(l) \leftrightarrow \text{SiNH}_3^+(s) \tag{4}$$

$$\text{SiOH}(s) + \text{H}^+(l) \leftrightarrow \text{SiOH}_2^+(s) \tag{5}$$

$$\text{SiOH}(s) \leftrightarrow \text{SiO}^-(s) + \text{H}^+(l) \tag{6}$$

At thermodynamic equilibrium,

$$K_N[\text{SiNH}_2][\text{H}^+] = [\text{SiNH}_3^+] \tag{7}$$

$$K_S^+[\text{SiOH}][\text{H}^+] = [\text{SiOH}_2^+] \tag{8}$$

Here [H$^+$] denotes the concentration of protons. K_N, $K^+{}_S$, and $K^-{}_S$ are the relative equilibrium constants, which are found to be [17–20] $K_N \approx 7.7 \times 10^9$, $K^+{}_S \approx 6.3 \times 10$, $K^-{}_S \approx 63.1 \times 10^{-9}$, $N_S \approx 5 \times 10^{14}$ cm^{-2}.

3.2   Model Formulation

The response of the reference electrode is related to the interfacial electrode potential that has been already mentioned above. Further, the surface potential is related to the pH value of the electrolyte. Therefore, when the pH < pH$_{PZC}$ the surface is positively charged. However, when pH > pH$_{PZC}$ the surface gets negatively charged. The potential of the diffuse layer can be stated using the Poisson Boltzmann equation for a planar surface along with the following boundary conditions

$$\frac{d^2\phi}{dx^2} = -\frac{\sum_{i=1}^{N} C_i^{\text{bulk}} z_i q e^{-\frac{z_i q \phi}{K_B T}}}{\varepsilon_o \varepsilon_r} \tag{9}$$

The boundary conditions used are [21]

$$\left.\frac{d\phi}{dx}\right|_{x=0} = -\frac{\rho}{\varepsilon_o \varepsilon_r} > 0 \quad \text{when pH} < \text{pH}_{PZC}$$

$$\left.\frac{d\phi}{dx}\right|_{x=0} = -\frac{\rho}{\varepsilon_o \varepsilon_r} < 0 \quad \text{when pH} > pH_{PZC}$$

With the following transformation for Eq. (9)

$$\frac{d^2\phi}{dx^2} = \frac{1}{2}\frac{d}{d\phi}\left(\frac{d\phi}{dx}\right)^2 \tag{10}$$

Therefore, Eq. (9) can be written as-

$$\frac{1}{2}\frac{d}{d\phi}\left(\frac{d\phi}{dx}\right)^2 = -\frac{\sum_{i=1}^{N} C_i^{\text{bulk}} z_i q e^{-\frac{z_i q \phi}{K_B T}}}{\varepsilon_0 \varepsilon_r} \tag{11}$$

Integrating Eq. (11)

$$\int \frac{1}{2}\frac{d}{d\phi}\left(\frac{d\phi}{dx}\right)^2 = -\int \frac{\sum_{i=1}^{N} C_i^{\text{bulk}} z_i q e^{-\frac{z_i q \phi}{K_B T}}}{\varepsilon_0 \varepsilon_r}$$

$$\frac{1}{2}\left(\frac{d\phi}{dx}\right)^2 = \frac{K_B T}{\varepsilon_0 \varepsilon_r} \sum_{i=1}^{N} C_i^{\text{bulk}} e^{-\frac{z_i q \phi}{K_B T}} + C_1$$

$$\left(\frac{d\phi}{dx}\right)^2 = \frac{2 K_B T}{\varepsilon_0 \varepsilon_r} \sum_{i=1}^{N} C_i^{\text{bulk}} e^{-\frac{z_i q \phi}{K_B T}} + C_1 \tag{12}$$

Using the boundary conditions.
$x = 0, \phi = \phi_{eo}, x = \infty, \phi = 0, d\phi/dx = 0.$
Applying the above conditions, the constant is obtained.

$$C_1 = -\frac{2 K_B T}{\varepsilon_0 \varepsilon_r} \sum_{i=1}^{N} C_i^{\text{bulk}} \tag{13}$$

From Eq. (12) and (13) we can have the following form of the equation

$$\left(\frac{d\phi}{dx}\right)^2 = \frac{2 K_B T}{\varepsilon_0 \varepsilon_r} \sum_{i=1}^{N} C_i^{\text{bulk}} \left(e^{-\frac{z_i q \phi}{K_B T}} - 1\right) \tag{14}$$

Since 1:1 electrolyte has been considered.
$|z_+|=|z_-|=z = 1.$
Where z is the valency

$$C_+^{\text{bulk}} = C_-^{\text{bulk}} = C^{\text{bulk}}$$

Hence $N = 3$ (two for cationic and one for anionic)

$$\left(\frac{d\phi}{dx}\right)^2 = \frac{2 K_B T}{\varepsilon_0 \varepsilon_r} C^{\text{bulk}} \left(e^{\frac{zq\phi}{K_B T}} - 1 + e^{-\frac{zq\phi}{K_B T}} - 1 + e^{-\frac{zq\phi}{K_B T}} - 1\right)$$

$$\left(\frac{d\phi}{dx}\right)^2 = \frac{2 K_B T}{\varepsilon_0 \varepsilon_r} C^{\text{bulk}} \left(2e^{-\frac{zq\phi}{K_B T}} + e^{\frac{zq\phi}{K_B T}} - 3\right)$$

$$\left(\frac{d\phi}{dx}\right)^2 = \frac{2K_BT}{\varepsilon_o\varepsilon_r}C^{\text{bulk}}\left(-\sinh\left(\frac{zq\phi}{K_BT}\right) + 3\cosh\left(\frac{zq\phi}{K_BT}\right) - 3\right)$$

For simplicity –

$$\cosh\left(\frac{zq\phi}{K_BT}\right) \approx 1 + \frac{1}{2}\left(\frac{zq\phi}{K_BT}\right)^2$$

$$\sinh\left(\frac{zq\phi}{K_BT}\right) \approx \frac{zq\phi}{K_BT}$$

$$\left(\frac{d\phi}{dx}\right)^2 = \frac{2K_BT}{\varepsilon_o\varepsilon_r}C^{\text{bulk}}\left\{\frac{3}{2}\left(\frac{zq\phi}{K_BT}\right)^2 - \frac{zq\phi}{K_BT}\right\} \tag{15}$$

$$\frac{d\phi}{dx} = \pm\left[\left(\frac{2K_BTC^{\text{bulk}}z^2q^2}{\varepsilon_o\varepsilon_r K_BT}\right)\left(\frac{3}{2}\phi^2 - \frac{K_BT}{zq}\phi\right)\right]^{\frac{1}{2}} \tag{16}$$

Since the electric field is a negative gradient of potential here, so we consider only the negative term. Therefore Eq. (16) becomes

$$\frac{d\phi}{dx} = -\left[\left(\frac{2C^{\text{bulk}}z^2q^2}{\varepsilon_o\varepsilon_r K_BT}\right)\left(\frac{3}{2}\phi^2 - \frac{K_BT}{zq}\phi\right)\right]^{\frac{1}{2}}$$

$$\frac{d\phi}{dx} = -\frac{1}{L_D}\left[\left(\frac{3}{2}\phi^2 - \frac{K_BT}{zq}\phi\right)\right]^{\frac{1}{2}}$$

$$\frac{d\phi}{dx} = -\frac{1}{L_D}\left(\frac{3}{2}\right)^{\frac{1}{2}}\left(\phi - \frac{K_BT}{3zq}\right) \tag{17}$$

Further, it can be assumed that

$$\phi - \frac{K_BT}{3zq} \approx \phi \tag{18}$$

Therefore Eq. (17) becomes

$$\frac{d\phi}{dx} = -\frac{1}{L_D}\left(\frac{3}{2}\right)^{\frac{1}{2}}\phi$$

$$\int\frac{d\phi}{\phi} = -\left(\frac{3}{2}\right)^{\frac{1}{2}}\frac{1}{L_D}\int dx$$

$$\ln(\phi) = -\left(\frac{3}{2}\right)^{\frac{1}{2}}\frac{1}{L_D}x + C_2$$

$$\phi = \exp\left[-\left(\frac{3}{2}\right)^{\frac{1}{2}}\frac{x}{L_D}\right].C_2 \tag{19}$$

Now following the boundary condition $x = 0$, $\varphi = \varphi_{eo}$, and $x = \infty$, $\varphi = 0$

$$\phi_{eo} = C_2$$

Therefore, we obtain

$$\phi = \phi_{eo}\left[\exp\left(-1.225\frac{x}{L_D}\right)\right] \tag{20}$$

The Eq. (20) gives the potential profile of a planar surface silicon nitride gate pH ISFET.

It has been found that the effective electrolyte oxide interface potential appearing in Eq. (20) should not be ϕ_{eo}. Rather, it should be an effective potential which has been found as

$$\phi_{eo\_eff}(x) = \phi_{eo} - \phi(x)$$

Therefore, Eq. (20) can be written as

$$\phi_{eo\_eff}(x) = \phi_{eo}\left[1 - \exp\left(-1.225\frac{x}{L_D}\right)\right] \tag{21}$$

The mathematically obtained ϕ_{eo_eff} is further compared with the experimentally obtained effective interfacial electrolyte-insulator potential which is obtained by precise positioning of the reference electrode and the acquired data using a system described in the following section.

4 Experimental Setup

The experimental setup has been divided into two sections, first being the ISFET biasing circuitry including the data acquisition system, and second is the mechanical system for the precise movement and positioning of the reference electrode. Figure 6 depicts the block diagram of the setup for the movement of the reference electrode along with the necessary data acquisition circuitry. Figure 7 illustrates the actual setup along with the data acquisition system. The drain to source voltage to the ISFET was applied using a Keithley Electrometer which is of two electrodes systems (Fig. 8).

A high resolution 10-bit digital to analog converter (DAC) was used to generate a stable voltage at the gate terminal. The DAC was controlled using a microcontroller. The DAC with a resolution of 4.8 mV was used to generate a pulse-width modulated

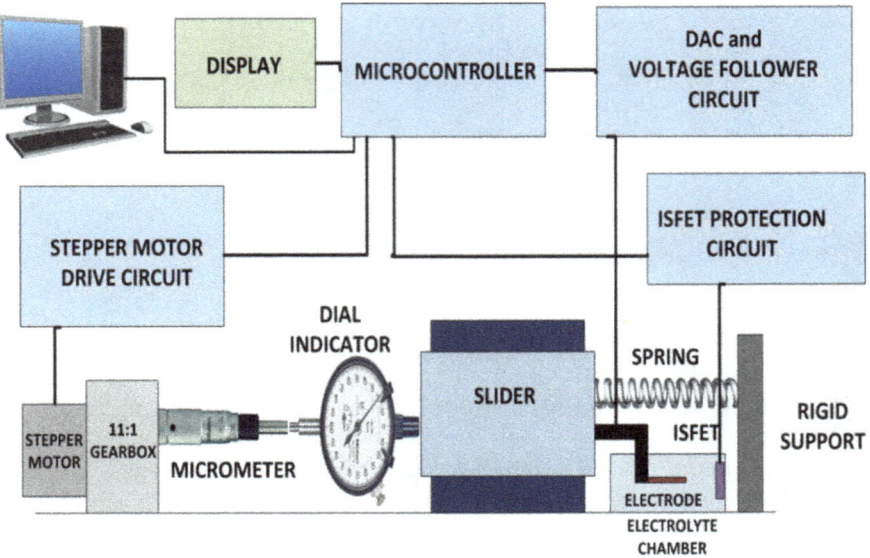

Fig. 6 Block diagram of the experimental setup

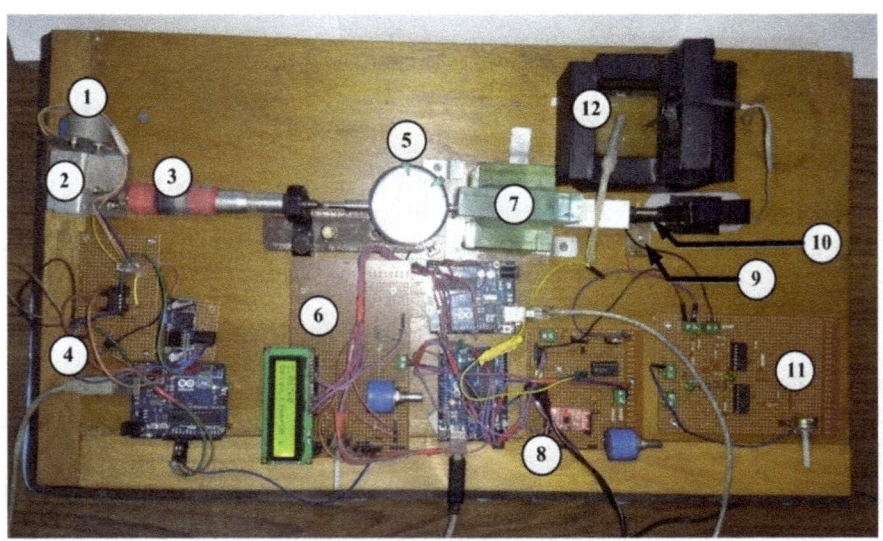

1. Stepper motor
2. Reduction gear box
3. Micrometer
4. Stepper motor driver circuitry
5. Dial indicator
6. LCD along with its circuitry
7. Glass stabilizer
8. Stable voltage generation circuitry
9. Virtual contacts
10. Spring mechanism
11. Protection circuitry
12. Reference electrode along with ISFET immersed in measurand

Fig. 7 The complete setup with the data acquisition system

(PWM) signal with a highly stable amplitude and was later followed by a voltage follower circuit made with an operational amplifier LM324 (Figs. 7 and 8).

The presence of the voltage follower circuit prevents any loading effect on the DAC. Part no 8 in Fig. 7 illustrates the stable gate voltage generation circuitry. The voltage stabilized by the voltage follower was then applied to the reference electrode that acts as the gate terminal for the ISFET.

Data acquisition was carried out during the ON time of the gate voltage signal and movement of the reference electrode was accomplished during the OFF time of the PWM signal as indicated in Fig. 9.

Fig. 8 Keithley Electrometer 6517B

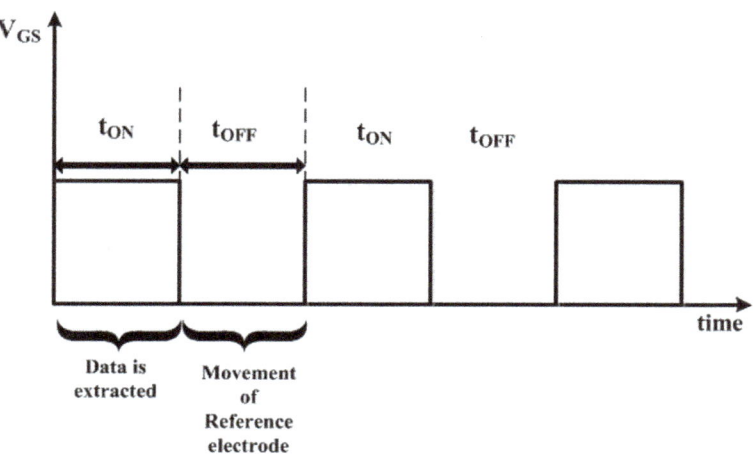

Fig. 9 The gate voltage signal generated by the DAC

The movement of the reference electrode is completed during the OFF time of the gate voltage signal to carry out a stable data acquisition process. As the discharging time of the gate capacitance varies with pH values, hence the gate signal was programmed to have a variable duty cycle and frequency. The duty cycle of the gate signal was programmed to be varied in real-time using a reference voltage provided by a stable 10 kΩ multi-turn wire-wound potentiometer. During the off period, the output of the DAC, i.e., the gate voltage was maintained at zero volts and hence a zero drain current was observed. For convenience, a 16 $\times$ 2 LCD was programmed to display the real-time gate voltage generated by the DAC along with the off-time of the PWM in seconds (part no 6 in Fig. 7).

The reference electrode was needed to be moved to the desired position while the gate voltage is zero. To accomplish this, the mechanical system was used in synchronization with the data acquisition system. The Debye length varies from 300 to 3000 μm for pH 12 to 14. Henceforth the reference electrode must be accurately positioned at various distances from the sensing layer. A micrometer with a least count of 0.2 mm was used for the movement of the reference electrode. With one complete turn of the micrometer's thimble, the spindle (of the micrometer) along with the reference electrode moves forward by a distance of 0.5 mm (500 μm). The thimble of the micrometer was turned using a stepper motor with necessary gear reduction. With a reduction gearbox (Fig. 10) between the stepper motor and the micrometer, a one-degree turn of the stepper motor shaft makes the thimble of the micrometer to turn 0.090 degrees. Therefore with a one-degree turn of the stepper motor, the spindle and reference electrode move forward by 0.126 μm which is illustrated in Fig. 11. The actual picture of the micrometer and the dial indicator is shown in Fig. 12.

1. Stepper motor
2. Worm gear with 1:11 reduction

Fig. 10   Stepper motor with 1:11 worm gear reduction

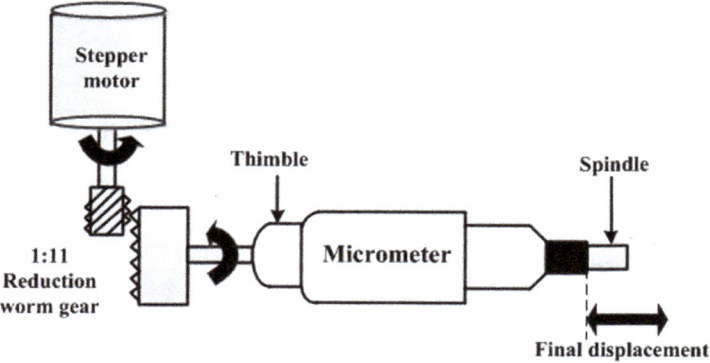

360° turn of the thimble = 500 μm final displacement

1°(movement of thimble of the micrometer) =1.389 μm (final displacement)

1° step of stepper motor = 1.389/11 μm displacement of the spindle

= 0.126 μm displacement of the spindle of the micrometer

Fig. 11 The translation mechanism

1. Stepper motor   2. Worm gear with 1:11 reduction
3. Thimble of micrometer   4. Spindle of micrometer
5. Dial indicator

Fig. 12 The micrometer and the dial indicator

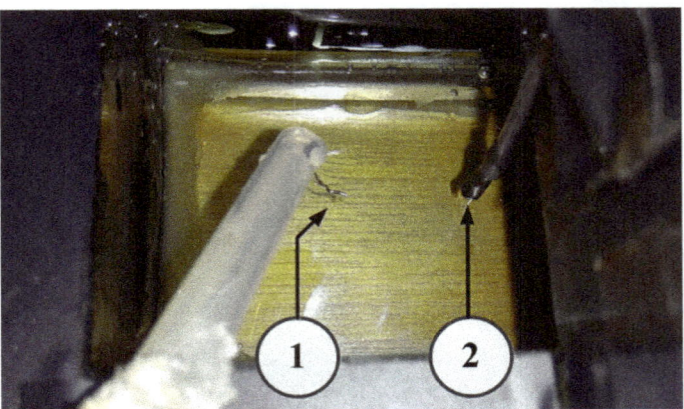

1. Insulated platinum reference electrode with exposed tip
2. Sensing layer of Si₃N₄ ISFET immersed in measurand

Fig. 13 Reference electrode and the ISFET immersed in the measurand

The precise forward and retrieval movement of the reference electrode was accomplished using a stepper motor (controlled by the microcontroller), driven using an H-Bridge IC (L293D). Proper coordination between the two microcontrollers used for the data acquisition and movement of the reference electrode was maintained. The ISFET along with the reference electrode was kept immersed in an electrolyte placed in a glass chamber as indicated in Fig. 13.

Initially, the reference electrode was positioned close to the silicon nitrite sensing layer of the ISFET (WINSENSE S01010) and then the reference electrode was moved away using the stepper motor and micrometer assembly discussed above.

The reference electrode, if touches the sensing layer of the ISFET can cause permanent damage to the later. To counter this concern, an additional safety measure was employed using mechanical contacts that move in a parallel plane along with the reference electrode. For the lateral stability of the reference electrode, a sliding mechanism using glass was designed to ensure the reference electrode movement was confined in one dimension (part no 7 in Fig. 7). A dial indicator (Mitutoyo 2046S) was placed between the micrometer spindle and the slider (part no 5 in Fig. 7). The micrometer's spindle pushes the sensor button of the dial indicator and the dial indicator's spindle extended to the diagonally opposite side pushes the slider along with the reference electrode.

To prevent any backlash and undesired vibrations in the system a spring (part no 10 in Fig. 7) was compressed during the forward movement of the slider which later helps in the smooth retrieval of the slider along with the reference electrode. Therefore, the complete setup for the movement of the reference electrode and data acquisition system is divided into various sub-sections as discussed below. The first and very important component of the setup is the circuitry for the generation of an accurate gate voltage which is applied to the reference electrode. The next section is the circuit for displaying gate voltage using LCD. The next important section illustrates the circuit for the protection of the sensing layer, as any impact by the reference electrode on the sensing layer may cause permanent damage to the latter. This is followed by the circuit for driving the stepper motor and its synchronization with the data acquisition system. Finally, the mechanical setup for the precise movement of the reference electrode has been discussed.

4.1   Generation of Gate Voltage

As already indicated earlier, part no 8 of Fig. 7 illustrates the setup for the generation of gate voltage which is applied to the reference electrode. The block diagram and the circuit diagram are illustrated in Figs. 14 and 15, respectively.

A 10-bit serial DAC module TLC5615 with a resolution of 4.8 mV was controlled using a microcontroller (Arduino UNO R3) to generate a PWM signal with stable amplitude. The generated voltage from DAC was applied to the voltage follower using an operational amplifier IC LM324 with an input bias current of 20nA. The PWM signal with a variable duty cy-le was required, as the time required for the discharging of built gate charges may differ for different pH values. The movement of the reference electrode is required to be carried out only during the off-time of the PWM as already stated. The off-time of the PWM can be varied using a reference voltage from a potentiometer. A LED was used as a visual indicator that provides information regarding the state of the gate voltage if it was low or high. In addition to this, a 16×2 LCD was used to display the gate voltage along with the time for which it is zero.

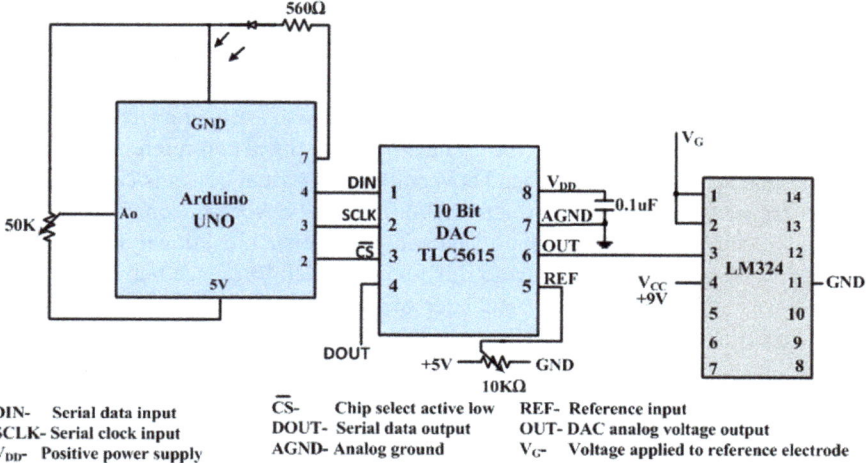

DIN-   Serial data input
SCLK- Serial clock input
V_{DD}- Positive power supply

$\overline{CS}$-   Chip select active low
DOUT- Serial data output
AGND- Analog ground

REF- Reference input
OUT- DAC analog voltage output
V_G-   Voltage applied to reference electrode

Fig. 14 The block diagram for the generation of gate voltage

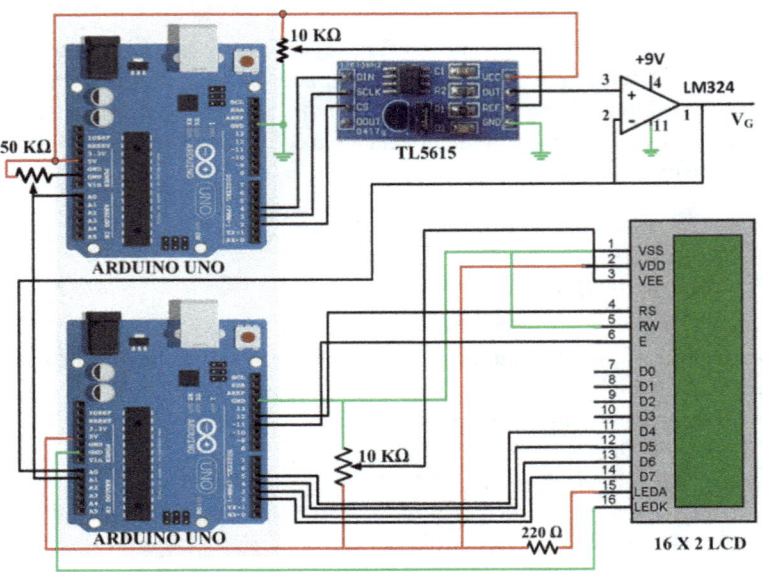

Fig. 15 Circuit diagram for the generation of the gate voltage and its display in 16 × 2 LCD

The reference electrode when moved toward the sensing layer, the sensing layer was subjected to a harsh impact that could result in permanent damage to the sensing layer of the ISFET. Henceforth to avoid such a condition a protection system was designed that could avoid any situation that can lead to damage of the sensing layer as depicted in Fig. 16. As the reference electrode was placed inside the electrolyte hence the microscopic separation between the sensing layer and the reference electrode was inconvenient to be inspected. To counter this, separate contact was made to move in parallel with the reference electrode on the same plane. The two contacts acted as the virtual reference electrode (point A, moving along the reference electrode) and sensing layer (point B, stationary) as indicated in Fig. 17. Any connection between these external contacts indicated the contact between the actual reference electrode and the sensing layer. The distance between the two virtual contacts can be as low as in micrometers hence a higher potential difference between the contacts can lead to a high electric field between them and cause dielectric breakdown of the air and can lead to false indication. To avoid this, the potential difference between the two contacts was maintained at 5 mV and later amplified to 5 V using an op-amp with a gain of 1000. The 5 V output was used as the signal to control the movement of the stepper motor. Any contact between the virtual reference electrode and the sensing layer disabled the motor driver IC (L293D) and aborted any further movement of the reference electrode toward the sensing layer (Fig. 16).

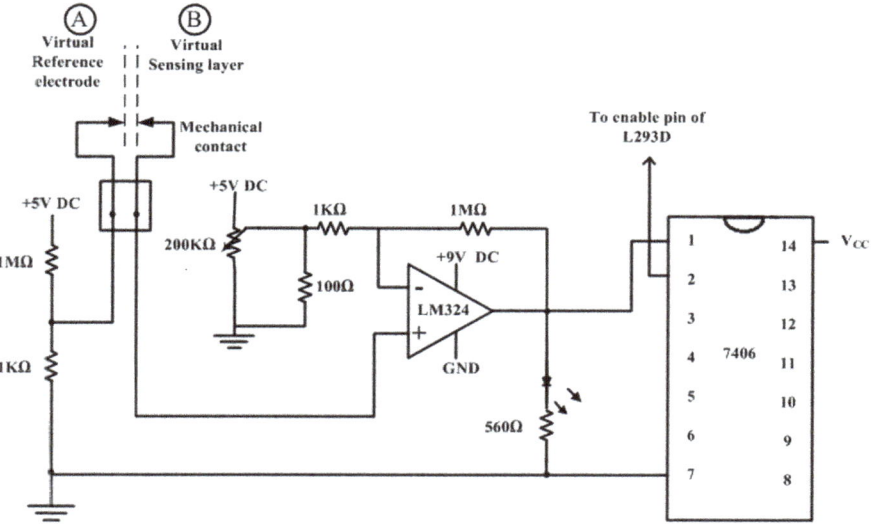

Fig. 16 Protection circuit

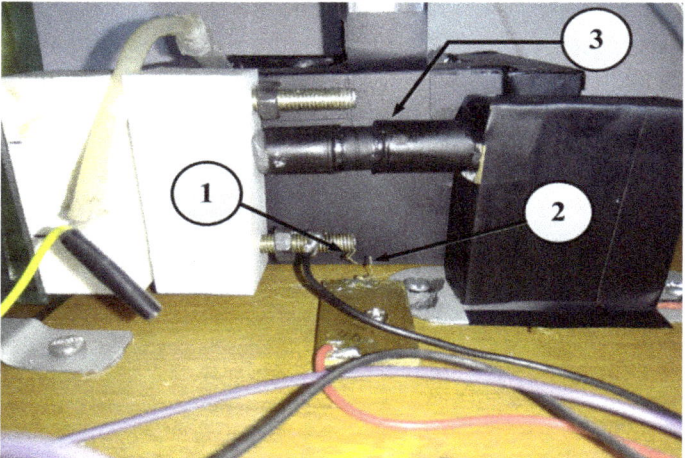

1. Virtual reference electrode (contact "A")
2. Virtual sensing layer (contact "B")
3. Spring mechanism

Fig. 17 Virtual contact points acting as reference electrode and sensing layer along with the spring mechanism

5 Results and Discussion

The Debye length along with the pH values for a silicon nitride sensing surface is tabulated in Table 1.

Further, when the reference electrode is placed close to the OHP, it has been observed that the potential possesses a higher value. As the electrode is moved far from the OHP into the bulk of the solution, this potential drops gradually. As the pH value increases the charge density attains a negative value which results in negative potential in the vicinity of the sensing layer. It has been observed that the potential has a higher value in the acidic range and it decreases with an increase in pH value. This can be explained well by the following plots in Figs. 18, 19, and 20. Figure 18 is for pH 4, where the charge density is positive. Further, for pH 7, the charge density is negative (Fig. 19). So is the case for pH 10, where the potential is more negative than in pH 7 as depicted in Fig. 20. The parameters for the simulation are considered as close as the actual device parameters. The simulation parameters are $W = 1400\,\mu$m, $L = 3550\,\mu$m and the V_{th} values for pH 12, 13 and 14 are 1.56 V, 1.61 V and 1.66 V, respectively, $\varepsilon_0 = 8.854 \times 10^{-12}$ F/m. For the experimentation, the initial ionic concentration of the pH solution was adjusted to be 0.1 M using a digital pH meter for pH 14. Further, for the consecutive pH solution, the initial solution was diluted using Milli Q water.

The system which is developed here has its limitations and can validate the theoretical model only for higher pH values. This limitation of the translational mechanism along with the calculations involved is already indicated in Fig. 11. The effective insulator interfacial potential versus position (in μm) of the reference electrode with respect to the sensing layer for pH 14 is illustrated in Fig. 21. In the experiment, a

	pH value of the measurand	Debye length (in m)
Table 1 Table for Debye length values for the corresponding pH of the measurand for a Si_3N_4 surface	1	9.6484×10^{-10}
	2	3.0511×10^{-9}
	3	9.6484×10^{-9}
	4	3.0511×10^{-8}
	5	9.6484×10^{-8}
	6	3.0511×10^{-7}
	7	9.6484×10^{-7}
	8	3.0511×10^{-6}
	9	9.6484×10^{-6}
	10	3.0511×10^{-5}
	11	9.6484×10^{-5}
	12	3.0511×10^{-4}
	13	9.6484×10^{-4}
	14	3.0511×10^{-3}

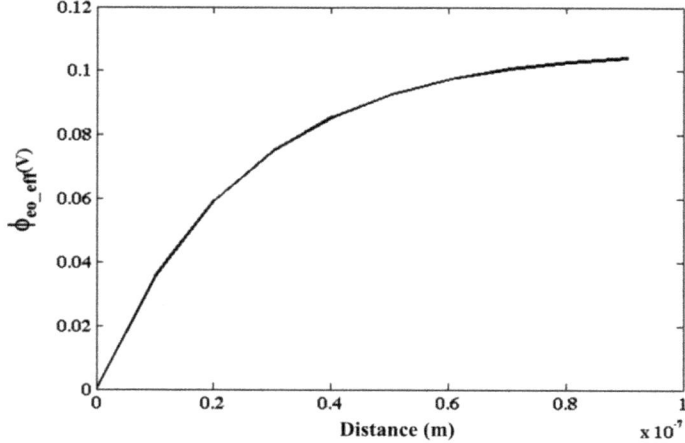

Fig. 18 Variation of interfacial effective electrolyte-insulator potential vs distance for pH 4

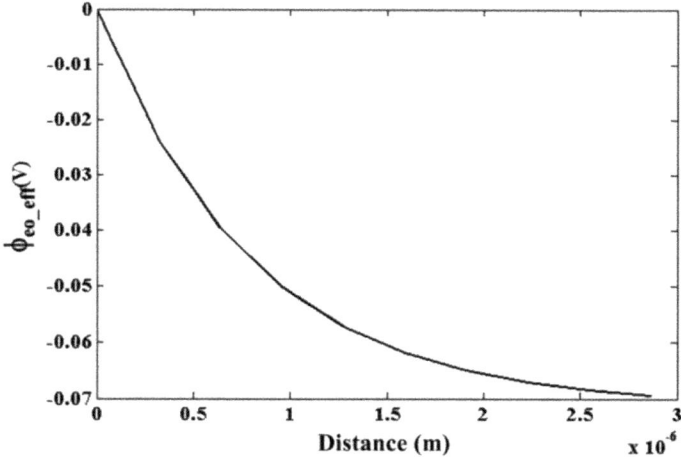

Fig. 19 Variation of interfacial effective electrolyte-insulator potential vs distance for pH 7

complex mechanical setup was used to position the reference electrode at various distances from the sensing layer. For all pH values, the measurements of drain current were carried out with a constant gate voltage of 2 V and a constant drain to source voltage of 0.27 V.

Here the plot associated with pH values of 14, 13, and 12 are discussed. The setup is capable of a minimum movement of the reference electrode by 0.126 μm. This limitation as already stated, makes it difficult to carry out this experiment for values below pH 12, as the Debye length of pH values below 12 is beyond the operating range of the design. Three times of Debye length is 9 mm for pH 14, henceforth for

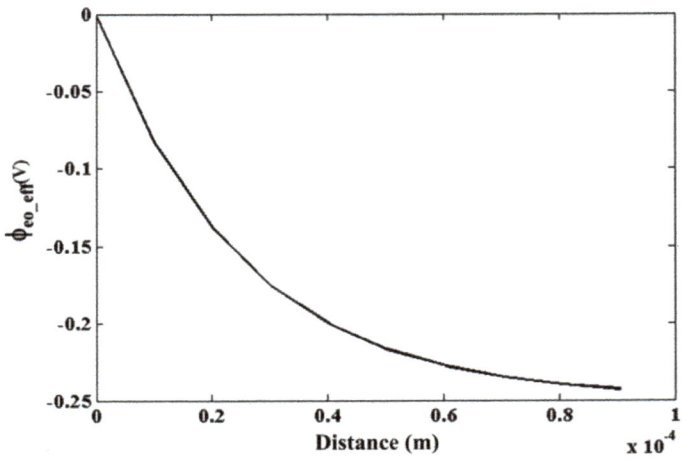

Fig. 20 Variation of interfacial effective electrolyte-insulator potential vs distance for pH 10

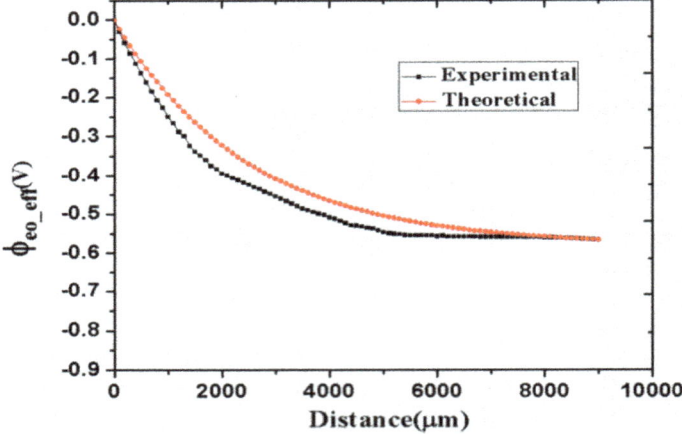

Fig. 21 Variation of effective electrolyte-insulator interface potential with an increase in distance of the reference electrode from the sensing layer ranging from 0 mm (i.e., the sensing layer) to 9 mm (i.e., three times of Debye length for pH 14)

measurements, 90 equidistant points with a spacing of 100 μm from the sensing layer were selected as illustrated in Fig. 22. Initially, the reference electrode was positioned in contact with the sensing layer of the ISFET ensuring no damage was done to the latter. The reference electrode was then retrieved back into the bulk of the electrolyte away from the sensing layer. After the steady positioning of the reference electrode, the drain current was measured at any desired point by applying a gate voltage which was kept at zero when the reference electrode is in motion. ϕ_{eo_eff} can be conveniently

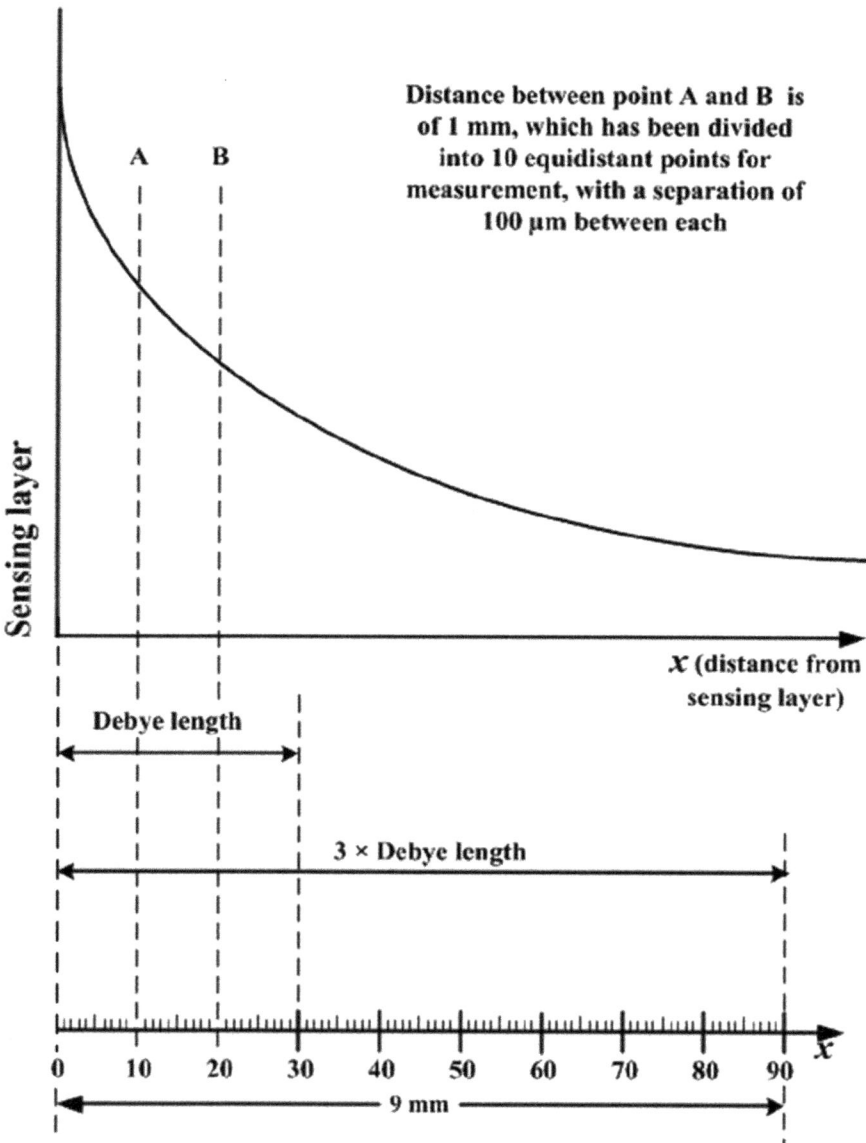

Fig. 22 Figure showing 90 different equidistant points between the sensing layer and 3 times of Debye length, chosen for the data extraction for pH 14

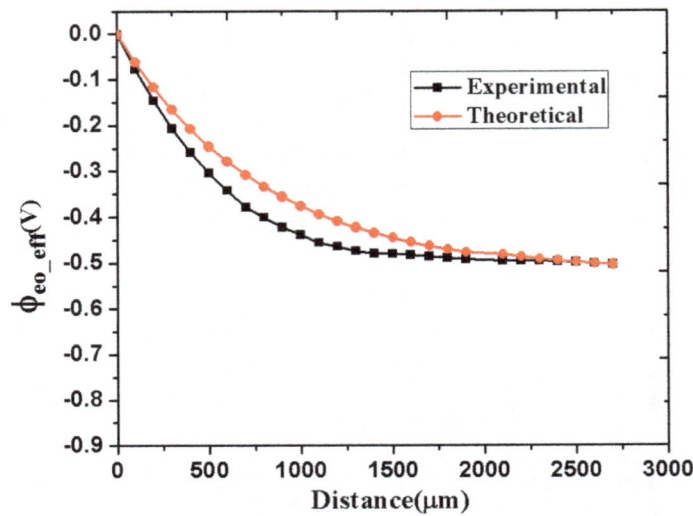

Fig. 23 Variation of effective electrolyte-insulator interface potential with an increase in distance of the reference electrode from the sensing layer ranging from 0 mm (i.e., the sensing layer) to 2.7 mm (i.e., three times of Debye length for pH 13)

derived for all value of the drain current using the threshold voltage equation and the equation defining the drain current for an ISFET device [1, 2]. Thus, ϕ_{eo_eff} was experimentally obtained for each position of the reference electrode. Further, it was observed that the theoretically and experimentally obtained values of ϕ_{eo_eff} almost resemble each other. Using the previous equations for pH 14, for pH 13 and 12 the theoretical values are plotted alongside the experimentally obtained value of ϕ_{eo_eff}, for different distances from the sensing layer as illustrated in Figs. 23 and 24, respectively. It is evident from the figures that the practically acquired value of ϕ_{eo_eff} at any point follows a resembling trend as that of the theoretically derived value of ϕ_{eo_eff} for that point.

It can be observed from Fig. 18 that as the reference electrode touches the sensing surface, i.e., x is zero the effective interfacial electrolyte-insulator potential becomes zero. The effective potential decreases exponentially when the reference electrode is retracted back from the sensing surface.

For pH = 14, the ϕ_{eo} is negative (as $pH_{PZC} = 5.8$ [22]). ϕ_{eo_eff} further decreases exponentially with the increase in distance and attains a stable value of $- 0.567$ V at three times the Debye length, i.e., 9 mm for pH 14. From Fig. 18, it can be observed that at a distance of approximately 9 mm from the sensing layer both the experimental and the theoretical plots are in close agreement with each other and attain a stable value, thus clearly indicating that it is redundant to place the reference electrode beyond this distance. Plots associated with pH values 13 and 12 also suggest the same as can be witnessed in Figs. 23 and 24, respectively. It was observed in our experiments that immersion of the ISFET device in the measurand for prolonged

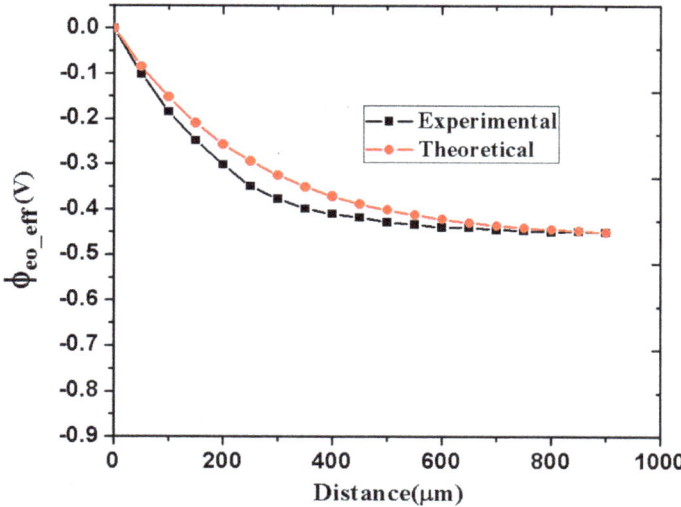

Fig. 24 Variation of effective electrolyte-insulator interface potential with an increase in distance of the reference electrode from the sensing layer ranging from 0 mm (i.e., the sensing layer) to 0.9 mm (i.e., three times of Debye length for pH 12)

hours causes change in the threshold voltage and the characteristics of ISFET due to an effect called drift [23].

Ambiguity in measurement can be the consequence of positioning of the reference electrode within three times the Debye length which can be observed in Fig. 25. Here in this figure, the reference electrode is considered to be positioned at 100 μm from the sensing layer and the variation of electrolyte-insulator interface potential with pH has been discussed.

For pH = 11, the three times of Debye length is 2.88 μm and hence lies well within 100 μm whereas for pH values above 11 this 100 μm separation is not sufficient to get the complete potential profile for the particular pH value.

From Fig. 25, it is evident that the theoretical curve becomes nonlinear beyond pH 11 and deviates from the ideal case. Also, it can be noticed that one value of ϕ_{eo_eff} corresponds to two different values of pH for the curves beyond pH 11, thus leading to ambiguity in measurement. For example, $\phi_{eo\_eff} = -0.0812$ V corresponds to pH 7.9 and pH 12.08 of the theoretical curve [24]. Therefore, it can be clearly stated that for accuracy in measurement the position of the reference electrode should be at or beyond the three times of Debye length of each pH values of the measurand.

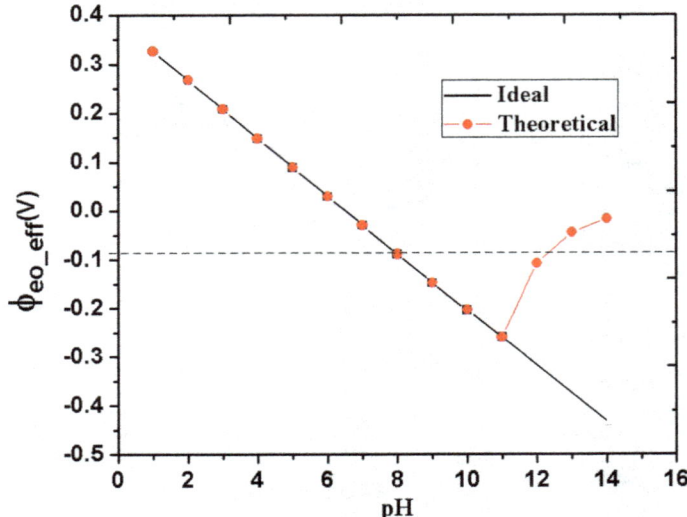

Fig. 25 Effective electrolyte-insulator interface potential (ϕ_{eo_eff}) vs. pH with the reference electrode positioned at 100 μm from the OHP

Acknowledgements We are grateful to Dr. Panchanan Puzari, Department of Chemical Sciences, Tezpur University, and Late Dr. Tapas Medhi, Department of Molecular Biology and Biotechnology for their help and valuable suggestions in this research work.

Bibliography

1. Bergveld P (2003) Thirty years of ISFETOLOGY. Sens Actuators, B Chem 88:1–20. https://doi.org/10.1016/s0925-4005(02)00301-5
2. Massimo G, Massobrio G (1998) Bioelectronics handbook: Massobrio G. Bioelectronics handbook: MOSFETs, biosensors, and neurons. McGraw-Hill
3. Yates DE, Levine S, Healy TW (1974) Site-binding model of the electrical double layer at the oxide/water interface. J Chem Soc Faraday Trans 1 Phys Chem Condens Phases 70:1807. https://doi.org/10.1039/f19747001807
4. Fung C, Cheung P, Ko W (1986) A generalized theory of an electrolyte-insulator-semiconductor field-effect transistor. IEEE Trans Electron Devices 33:8–18. https://doi.org/10.1109/t-ed.1986.22429
5. Siu MW, Cobbold RSC (1979) Basic properties of the electrolyte—SiO$_2$—Si system: physical and theoretical aspects. IEEE Trans Electron Devices 26:1805–1815
6. Gouy M (1910) Sur la constitution de la charge électrique à la surface d'un électrolyte. J. Phys. Theor. Appl. 9(1): 457–468. https://doi.org/10.1051/jphystap:019100090045700. jpa-00241565
7. Chapman DL (1913) LI. A contribution to the theory of electrocapillarity. The London, Edinburgh, and Dublin philosophical magazine and journal of science 25:475–481
8. Stern O (1924) The theory of the electrolytic double-layer. Z Elektrochem 30:1014–1020

9. Grahame DC (1947) The electrical double layer and the theory of electrocapillarity. Chem Rev 41:441–501. https://doi.org/10.1021/cr60130a002
10. Massobrio G, Martinoia S, Grattarola M (1994) Use of SPICE for modeling silicon-based chemical sensors. Sens Mater 6:101–101
11. Sharma S (2009) Modeling and Simulation of nanobioelectronic device the cylindrical ion sensitive field effect transistor. Dissertation, PhD Thesis, Tezpur University
12. Hazarika C (2018) Modeling and positioning of reference electrode in ion sensitive field effect transistor (ISFET) and development of a Schottky enzyme field effect transis-tor (ENFET) for hydrocarbon detection. Dissertation, PhD Thesis, Tezpur University
13. Bergström L, Pugh RJ (1989) Interfacial characterization of silicon nitride powders. J Am Ceram Soc 72:103–109
14. Harame DL (1985) Integrated circuit chemical sensors, Dissertation, PhD Thesis, Stanford University
15. Grattarola M, Massobrio G, Martinoia S (1992) Modeling H/sup +/-sensitive FETs with SPICE. IEEE Trans Electron Devices 39:813–819. https://doi.org/10.1109/16.127470
16. Beruto D, Mezzasalma S, Baldovino D (1995) Theory and experiments for evaluating the number and the dimensions of solid particles dispersed in a liquid medium. Application to the system Si_3N_4/H_2O (I). J Chem Soc Faraday Trans 91:323–328
17. Peri J (1966) Infrared study of OH and NH2 groups on the surface of a dry silica aerogel. J Phys Chem 70:2937–2945
18. Armistead CG, Tyler AJ, Hambleton FH, Mitchell SA, Hockey JA (1969) Surface hydroxylation of silica. J Phys Chem 73:3947–3953
19. Laidler KJ (1965) Analysis of kinetic results. Chemical Kinetics. McGraw-Hill, New York, pp 19–21
20. Mezzasalma S, Baldovino D (1996) Characterization of silicon nitride surface in water and acid environment: a general approach to the colloidal suspensions. J Colloid Interface Sci 180:413–420. https://doi.org/10.1006/jcis.1996.0320
21. Andelman D (2004) Introduction to electronics in soft and biological matter. http://citeseerx.ist.psu.edu/viewdoc. Accessed 18 Sept 2020
22. Niu MN, Ding XF, Tong QY (1996) Effect of two types of surface sites on the characteristics of Si3N4-gate pH-ISFETs. Sens Actuators, B Chem 37:13–17
23. Hazarika C, Neroula S, Sharma S (2019) Long term drift observed in ISFET due to the pene-tration of H+ ions into the oxide layer. In: International conference on pattern recognition and machine intelligence, Springer, Cham, pp 543–553
24. Hazarika C, Dutta A, Sharma S. (2017) Modelling of reference electrode for a $S\grave{i}_3N_4$ gate pH ISFET. In: International conference on innovations in electronics, signal processing and communication (IESC), IEEE, pp 149–154

Applications

High Speed Interconnects Made of Composite Materials for VLSI Application

Souradeep De, Bhabana Baruah, and Santanu Maity

Abstract The role of interconnects has become remarkably significant with the downscaling of VLSI technology, affecting the performance of integrated circuits to a great extent. Therefore, researchers have been investigating the use of different materials other than traditional materials to fabricate superior high-speed VLSI interconnects. In this review book chapter, a comparative study is carried out on the different interconnect materials. Analysis of recent progress in materials with emphasis on composite materials for high-speed VLSI interconnects has been done. Book chapters published in the last 10 years have been reviewed and some earlier book chapters that played an important role in shaping the field of interconnects are included.

Keywords Copper (Cu) · Aluminum(Al) · Carbon nanotubes (CNTs) · Graphene nanoribbons (GNRs) · Mixed carbon nanotube bundle (MCB) · Through silicon vias (TSVs) · Mean free path (MFP) · Interconnect · Crosstalk · Delay · CNT composite material

1 Introduction

In recent years, there have been consistent improvements in the performance of integrated circuits due to the continuous downscaling of feature sizes. However, as technologies have been scaled down, the role of interconnects has become increasingly important. Although scaling improves the performance, power consumption, and speed of active circuits, the same does not apply to interconnect as shown by Davis et al. [1], Coulibaly [2]. There are three classes of interconnects, namely,

S. De · S. Maity (✉)
School of Advanced Materials Green Energy and Sensor Systems, Indian Institute of Engineering Science and Technology, Shibpur, Howrah 711103, India
e-mail: smaity.cegess@faculty.iiests.ac.in

B. Baruah
Department of Electronics & Communication Engineering, Tezpur University, Assam 784028, India

© The Author(s), under exclusive license to Springer Nature Singapore Pte Ltd. 2022
R. Goswami and R. Saha (eds.), *Contemporary Trends in Semiconductor Devices*,
Lecture Notes in Electrical Engineering 850,
https://doi.org/10.1007/978-981-16-9124-9_11

local, semi-global, and global interconnects. If the devices are smaller, interconnects which are immediate (local interconnects) will be smaller. The global interconnects (which carry the global signals such as CLOCK, RESET), on the other hand, are scaled not by the transistor size but by the die size so that the global interconnect delay can be the determining factor for the speed of an integrated circuit. We require techniques to mitigate the large interconnect delays. One such solution can be obtained by using new and improved materials for fabricating interconnects. This review has been divided into the following sections. Section 2 (Basic interconnect model) consists of an overview of the basic interconnect model. Section 3 (Progress in the performance of interconnects owing to the usage of diverse materials and procedures) contains the summary of all the work that has been done on interconnects. Section 4 (Composite materials for interconnect fabrication) consists of an overview of the use of composite materials for interconnect fabrication. Section 5 (Mathematical modeling of composite CNT interconnects) comprises of the mathematical modeling of CNT composite interconnects. Section 6 (Mean free path of materials used for fabrication of interconnects) consists of a comparative analysis of the mean free paths of different interconnect materials. Section 7 (Delay in interconnects composed of different materials) consists of a review of the delay of different interconnect materials. An overall summary is presented in Sect. 8 (Conclusion).

2   Basic Interconnect Model

At high frequencies, the parasitic elements of interconnect cannot be neglected as they contribute to the distortion and delay of the waveform propagating from the driver to the receiver. The basic approaches available for on-chip Cu interconnect modeling are as follows:

2.1   Lumped Interconnect Model

Figure 1 shows a CMOS inverter-driven lumped interconnect model. The resistance of each segment of the wire in the lumped RC model is lumped into a single resistor R. Likewise, the total capacitances are combined into a single capacitor C. Thus, a lumped circuit is utilized to approximately represent the interconnect line. The NMOS drives the output to low when the input is high and the PMOS drives the output to high when the input is low. A capacitive load is used to represent the output. For long interconnect wires, this lumped RC model is erroneous and is more effectively represented by a distributed RC model. The overall delay is given by

$$\tau = RC_L \qquad (1)$$

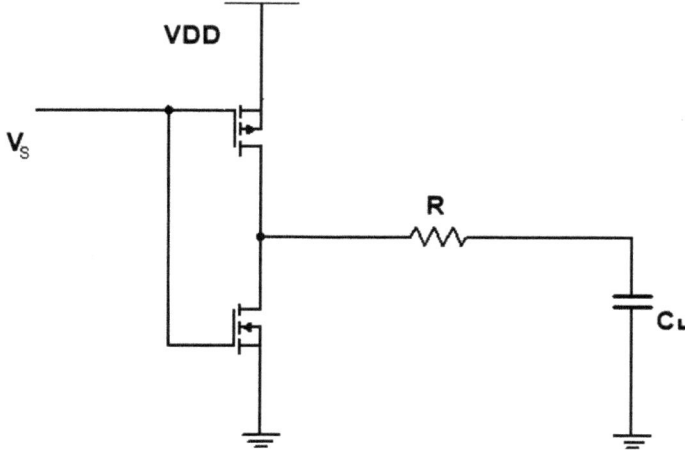

Fig. 1 A CMOS inverter-driven lumped interconnect model

2.2 Distributed Interconnect Model

Figure 2 shows a distributed RLC interconnect model driven by a CMOS inverter which provides better accuracy over a lumped interconnect model. This interconnects with a distributed series of inductors and capacitors are known as a transmission line. Here, r, l, and c are per unit length resistance, inductance, and capacitance, respectively. Distributed RLC interconnects have been modeled by Davis et al. [1], Coulibaly [2], Ullah et al. [3] and the expressions for time delays have been derived and plotted.

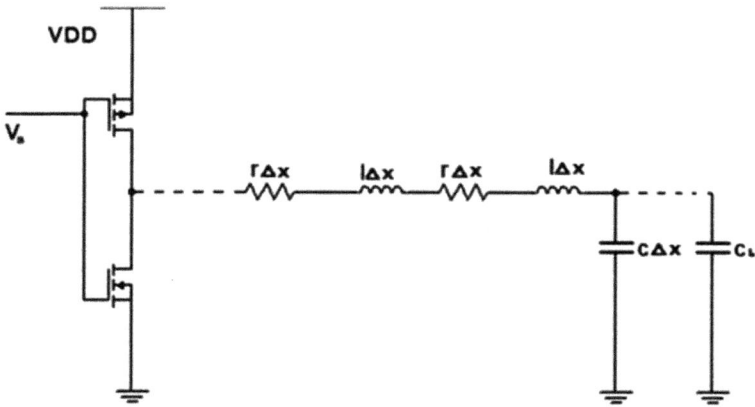

Fig. 2 A distributed RLC interconnect model driven by a CMOS inverter

Fig. 3 Different materials used for VLSI interconnects over the decades

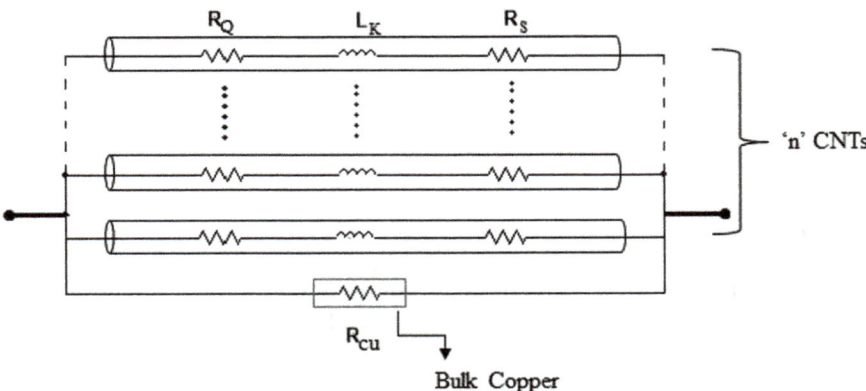

Fig. 4 Equivalent electrical circuit model of Cu/CNT composite interconnect comprising of a bulk Cu conductor connected in parallel to 'n' CNTs. R_{CU} is the resistance of the bulk Cu conductor. R_Q, R_S, and L_K are the quantum resistance, scattering resistance, and kinetic inductance of the CNTs, respectively

3   Progress in the Performance of Interconnects Owing to the Usage of Diverse Materials and Procedures

To start with, Aluminum (Al) was used as a standard interconnect material but with the scaling down of technology, below 130 nm, Cu was found by Bohr (1995) to be a more suitable substitute to Al due to its high electro-migration resistance and low resistivity [4]. Davis (2001) discussed the limits of Cu interconnect in terms of material, circuit, and systems [5]. The effects of reverse scaling of multilevel interconnect have been investigated and the possible usage of 3-D integrated circuits has been demonstrated. Since the 3-D ICs contain multiple active layers as well as separate layers of repeaters, a significant improvement in the performance of interconnects is observed for 50 nm technology node. However, the reverse scaling approach generally leads to higher production costs due to an increase in chip area or levels of interconnects. A dual damascene Cu structure was then reviewed, which has comparatively lower via resistance and was more reliable than a single damascene structure. In the dual damascene structure, metal fill (Cu) is laid over preformed trenches and via. Therefore, only one metal fill was required for each level of interconnect, thus reducing cost. Also, in advanced Cu dual damascene metallization tantalum nitride (TaN) and Co are used over the dielectric as a barrier layer, decreasing the amount of copper used in interconnects, thereby decreasing the problem of electron scattering to some extent.

However, the dual damascene structure has the disadvantages of a higher aspect ratio and misalignment error due to overlap of via with the underlying metal. Eventually, with downward scaling, early failure due to electro-migration was detected in the dual damascene Cu interconnect model shown by Ogawa et al. [6]. When a low-k dielectric was combined with Cu, enhancement in performance was achieved in comparison to Al or Al-Cu metallization as demonstrated by Havemann and Hutchby [7]. Maex et al. (2003) gave an overview of the classifications, properties, and characteristics of different low-k dielectric materials such as organic polymers, amorphous carbon, etc. [8]. The introduction of porosity in the dielectric material further reduced the dielectric constant (Fig. 3).

The porous dielectrics, however, had to be sealed using different processes such as thin-film deposition, plasma surface interaction, surface cross-linking/reconstruction to prevent contamination during subsequent processes. The aim was to obtain a k value of one so that there was an overall reduction in the resistivity and the capacitance between the wires. Yang (2010) fabricated low-k Cu interconnect with tantalum nitride (TaN) diffusion barrier layer and a Ru/TaN liner layer [9]. Ta was used to adhere Cu to the TaN diffusion layer. Ultrathin Cu was deposited over the thin liner layer, which was deposited over different substrates (Ta, Ta/TaN, and Ru/TaN). The ruthenium (Ru) substrates improved the wettability of the ultrathin Cu layer and an improvement in the electro-migration resistance was observed. The limitations of Cu interconnects have been further deliberated by Ceyhan and Naeemi [10]. Sarkar et al. (2011) considered graphene as a probable material for VLSI interconnects [11]. Characteristics of (Arsenic pentafluoride) doped multi-layered graphene ribbons (GRs) were studied and found to surpass Cu and CNT for local interconnects. However, for global interconnects, as the mean free path (MFP) became large and equivalent to skin depth, graphene ribbons were susceptible to anomalous skin effect (ASE) thus degrading their performance for global interconnects. Moreover, Contino et al. (2018) showed the graphene ribbons were prone to edge scattering which limited the conductance to an extent [12]. Graphene nanoribbons (GNRs) were fabricated by Murali et al. (2009) and the 3-D resistivity obtained was compared to that of Cu [13]. Due to the different scattering mechanisms, the average resistivity of GNRs was observed to be considerably higher than that of Cu. The distinctive properties of carbon nanomaterials and their use as interconnect material has been discussed by Li [14]. A comparison between the performance of Cu and CNT interconnects was presented by Naeemi et al. [15]. Several monolayers of single-walled carbon nanotube (SWCNT) were taken in parallel to the ground plane. It was observed that the CNT surpassed the Cu interconnects in performance by having lower latency, provided the mean free path increased with an increase in interconnect length. However, owing to its large contact resistance and characteristic impedance, the performance of a single SWCNT was observed to be lower than a Cu interconnect. Therefore, a model comprising of a bundle of SWCNTs was taken, which outperformed the Cu interconnects in terms of resistance and speed as shown by Naeemi and Meindl [16]. Srivastava (2009) investigated thermal reliability, and power dissipation of single-walled carbon nanotubes as VLSI interconnects [17]. The improvement in latency achieved due to the nanotube bundles as given by Naeemi et al. [15] is

$$\frac{\tau_{copper} - \tau_{bundle}}{\tau_{bundle}} \approx \frac{R_{tr} + 0.5 r_{int} L}{R_{tr} + \frac{0.5 R_0 L}{L_0}} - 1 \tag{2}$$

where L is the length of the interconnect, R_{tr} is the driver resistance, R_0 is the resistance of a bundle of ballistic CNTs, L_0 is the mean free path, and r_{int} is the internal resistance per unit length of the CNT bundle.

The performance of bundled interconnects was observed by Srivastava (2005) to depend on the packing density [18]. Li et al. (2008) investigated multi-walled CNTs (MWCNTs) theoretically and found them to be a better contender for usage as interconnect materials than SWCNTs and Cu for long interconnects (intermediate and global) [19]. However, for short (local) interconnects Cu and SWCNTs can outperform MWCNTs due to decreasing mean free path (MFP) of each shell of MWCNT, leading to a decrease in conductivity. The performance of SWCNT can be equivalent to that of MWCNT provided the SWCNT has a high density and high (100%) metallic fraction. The performance of SWCNT bundles and MWCNT were compared by Majumder [20]. For global interconnects, the improvement in dynamic crosstalk-induced delay was observed to be superior in MWCNT than in SWCNT bundle interconnects. The improvement in the percentage conductance of mixed CNT bundle over Cu was analyzed by Haruehanroengra and Wang [21]. Interconnects having a length equivalent to the MFP had a drastic increase in the improvement of conductance percentage as the length was increased. On the other hand, interconnects having a length greater than the MFP showed a limited increase in improvement of conductance percentage since for longer lengths, the Cu conductance also increased. Therefore, although the conductance of the mixed CNT bundles increased, there was a decrease in the improvement of conductance. Majumder et al. (2014) considered six different models (MCB-I, MCB-II, MCB-III, MCB-IV, MCB-V, MCB-VI) of mixed CNT [22]. MCB-I consisted of randomly arranged CNT (both SWCNT and MWCNT) of different diameters and the rest comprised of specifically arranged CNTs. The propagation delay and dynamic crosstalk of each model were analyzed. It was observed that the MCB-VI model which had SWCNTs located at the center surrounded by MWCNTs at the boundary had the maximum reduction in propagation delay (80%) for global interconnect length in comparison to the MCB-I model.

4 Composite Materials for Interconnect Fabrication

Carbon nanotubes (CNTs) are considered as a potential substitute for Cu owing to their various advantages such as large electro-migration resistance, thermal conductivity, and ampacity. Although theoretically, CNTs show an ideal performance than most traditional materials like copper, practically one has to deal with various criteria. The conductivity of a CNT depends on its chirality. There are three types of chirality for nanotubes—armchair, chiral, and zig-zag. The armchair configured CNTs are always metallic, whereas the ones with other configurations may be either metallic or semiconducting. Thus, armchair CNT has the highest conductivity among the three. However, it is not possible to control the chirality of CNTs during the fabrication process to manufacture ideal metallic CNTs. Statistically, in a nanotube bundle, only one-third of the CNTs is metallic while the remainder is semiconducting as proved by Ilani and Mceuen [23]. Hence, practically their performance is entirely dependent on the fabrication process. Moreover, as per Jarosz et al. (2011), Behabtu et al. (2013), the electrical conductivities of macro-materials composed of nanoscopic CNT were found to be lower than Cu, constraining their physical application [24, 25]. With the scaling down of technology, many attempts have been made in research to study the possibility of the incorporation of composites in interconnect fabrication. Carbon nanotubes (CNT) have been used in a majority of the research work along with metals, ceramics, and polymers to fabricate composite materials, owing to their exceptional properties. Also, the addition of specific materials aids in emphasizing the mechanical properties of copper. A copper-ruthenium-tantalum composite has been synthesized by Sule et al. (2012) using the powder metallurgy technique. Tantalum (Ta) and ruthenium (Ru) were homogeneously incorporated into the Cu matrix [26]. A 94 $\mu\Omega$ cm electrical resistivity was observed in the Cu–Ru (Cu + 1 vol % Ru) composites which increased to 116 $\mu\Omega$ cm with the increase in Ru content. The electrical resistivity of 123 $\mu\Omega$ cm was achieved in the Cu–Ta (Cu + 1 vol % Ta) composites, which increased with an increase in tantalum content. The increase in the resistivity of Cu–Ta and Cu–Ru composites was primarily because of higher bulk resistivities of tantalum (Ta) and ruthenium (Ru). However, the electrical resistivities of the two-phase composite samples of Cu–Ta–Ru ($Cu_{95} + Ta_1 + Ru_4$ and $Cu_{95} + Ta_{2.5} + Ru_{2.5}$) decreased and were evaluated to be 73 $\mu\Omega$ cm and 110 $\mu\Omega$ cm, respectively, which was due to the properties of the constituent phases such as volume fraction, distribution, orientation, electrical conductivity, particle size, and spacing of the system. Further, it was observed that with an increase in the volume of Ta, the density and hardness of Cu decreased while refining the grain size (11.3 to 6 μm). A study on cobalt/copper composite interconnects has been presented by Nogami (2017) for the back end of the line (BEOL) conductor of fine (<7 nm) and wide (>36 nm) dimensions [27]. Two types of composite systems were fabricated- a Co/Cu composite interconnect system with the conventional TaN barrier (4 nm) and a Co/tCoSFB-Cu composite system with a very thin layer of TaN barrier (<1 nm). The composite with the through-cobalt self-formed-barrier (tCoSFB) was found to have significantly lower line resistance for both fine and broad lines due to the decreased TaN barrier thickness which adheres Co to the dielectric meanwhile maximizing the volume fraction of Co.

Electro-deposition method was developed by Wolter et al. (2011) and Sundaram et al. (2017) for depositing Cu onto premade CNT arrays [28, 29]. Electroplating and electrophoresis techniques were combined to fabricate Cu/CNT thin films for interconnect application by Liu et al. [30]. Chai et al. (2008) used blech structure to present a comparative study between the electro-migration properties of pure Cu and Cu-CNT composites [31]. Van der Pau structure and blech structure were fabricated on the same chip and both the test structures were subjected to a current density of 1.2×10^4 A/cm^3. It was observed that there was a significantly high growth rate of voids in the pure Cu segment, whereas in the Cu-CNT structure, gradual voids were formed only after a long duration of electro-migration testing. The resistivity of Cu and Cu-CNT composite were obtained to be 1.9 $\mu\Omega$ cm and 2.2 $\mu\Omega$ cm, respectively. Also, critical length and the average threshold product of the current density of both Cu-CNT and Cu were obtained to be 5400 and 1800, respectively. Since the electro-migration resistance of Cu-CNT composite was greater than that of pure Cu, it was a more reliable and thereby a better substitute to pure Cu as an interconnect material. The maximum stress caused by electro-migration in the samples were calculated by

$$\sigma_{\text{delam}} = \sqrt{\frac{2E_n G_{\text{adh}}}{h}} \tag{3}$$

where G_{adh}, h, and E_n are the adhesion energy, thickness, and modulus of elasticity of the conductor, respectively. Cheng (2017), fabricated a Cu-CNT composite interconnect model by electrodepositing copper over the pores of CNT that has been premade [32]. The length, width, and thickness of interconnect are supposed as 1000 μm, 100 nm, and 200 nm, respectively. A comparison of the performance characteristics of Cu and pure CNT (both SWCNT and MWCNT) such as effective resistivity, Cu-CNT composite propagation time delay was performed using a driver-interconnect-load (DIL) system equivalent circuit model. For SWCNT, the resistivity was observed to decrease with an increase in the CNT filling ratio and the metallic fraction. However, it was significantly greater than that of Cu. The resistivity of Cu-CNT was marginally higher but comparable to Cu. As for the propagation time delay, Cu-CNT was observed to show better performance than that of pure CNT (both SWCNT and MWCNT) due to the high resistivity of the latter. The overall analysis after simulation showed that the Cu-CNT composite structure had comparable conductivity but higher current carrying capacity than that of pure copper. The 50% propagation time delay for the DIL system was calculated by

$$\tau = (1.48\varsigma + e^{-2.9\,\varsigma^{1.35}})\sqrt{Ll(Cl + C_L)} \tag{4}$$

where

$$\varsigma = \frac{Rl(Cl + 2C_L) + 2(R_c + R_d)(Cl + C_L)}{4\sqrt{Ll(Cl + C_L)}} \tag{5}$$

In the above equations, the interconnect length is given by l; R, C, L are the per unit length resistance, capacitance, and inductance, and C_L, R_c, R_d are the load capacitance, contact resistance, and driver resistance, respectively. Some methods of fabricating Cu-CNT composites were powder metallurgy technique developed by Sun et al. [33] and Tu et al. [34]. A Cu/CNT nanocomposite material fabrication process was proposed by Sun (2016) for 3-D interconnects [33]. Vertically aligned CNT bundles were homogenously coated by electroplating Cu to obtain a high aspect ratio of 300:1. CNT grid arrays were grown by chemical vapor deposition (CVD) over a patterned wafer. Ti/Au of nanometer thickness was then sputtered over the CNT grid arrays. The CNT arrays were then aligned and transferred by tape-assisted transfer (thermal release tape) into the through silicon vias (TSVs). The donor wafer was then removed and then Cu was electroplated onto the sample to produce CNT-Cu TSVs. The proposed nanocomposite had a significantly low silicon-order coefficient of thermal expansion (CTE) and low copper-order resistivity. An alternate high yield composite Cu/CNT fabrication process with homogenously distributed CNT was presented by Zheng et al. [35]. Cu/CNT nanocomposite powders were fabricated by co-electrodeposition with the subsequent fabrication of bulk Cu/CNT nanocomposite and Cu by compaction of the previously fabricated powders followed by sintering. The microhardness of Cu in the absence of CNTs was obtained as 500.6 ± 37.2 MPa and that of the Cu/CNT composite as 600.8 ± 63.8 MPa when measured. This two-step fabrication process was significantly advantageous over other fabrication methods as it led to a higher yield of uniformly distributed CNT in the metal matrix in addition to high interfacial bonding between the CNTs and metallic matrix due to the electrostatic force produced throughout the co-electrodeposition process.

A study on the resistivity property of composite systems was presented by Wang et al. [36]. Dense CNT mullite ceramic composite was fabricated by homogeneously dispersing pre-treated MWCNT in a mullite (3Al$_2$O$_3$, 2SiO$_2$) matrix. The inclusion of CNT into the ceramic matrix provided electrical conductivity to the insulating mullite and also improved the mechanical properties. The addition of 1% (by volume) of CNT significantly decreased the electrical resistivity of the overall composite from 2×10^{11} Ω cm to 1.42×10^{11} Ω cm and with an increase in the CNT volume content (10%), the resistivity of the composite was obtained as low as 2Ω cm. However, with a further increase in CNT volume (15%), the resistivity of the composite sample increased compared to the previous samples, because of the lower relative density (<95%) of the composite materials (CNT and mullite). Therefore, the sample having 15% volume of CNT had considerably more pores inside, which subsequently affected the electrical property of the composite CNT, resulting in increasing resistivity. The utilization of conducting polymers (such as polyaniline) in microelectronics has been conjectured by Angelopoulos et al. [37]. Composites that incorporate CNT with conducting polymers have been synthesized by Long et al. [38], Alsheri et al. [39], Seol et al. [40]. In each case, the electrical conductivity was ascertained to increase with an increase in the CNT concentration. However, feasible VLSI interconnects using CNT polymer composites for practical applications are yet to be attained.

5   Mathematical Modeling of Composite CNT Interconnects

The mathematical modeling and analysis of composite SWCNT were performed by Jurn et al. (2016) to evaluate the electromagnetic properties of the composite material [41]. The SWCNT was assumed to be a solid cylinder (Nano-solid tube material) and the composite material was constituted of SWCNT covered with a fine coating of alternative material. The effective conductivity of the composite material was derived as

$$\sigma_{\text{composite}} = \sum_{j=1}^{k} m_j \sigma_{zj} \tag{6}$$

where k is the number of materials constituting the composite ($k = 2$), σ_z is the conductivity of respective material, and m_j is the volume fraction of the material.

$$\sigma_{\text{composite}} == P\sigma_{\text{SWCNT}} + A\sigma_{\text{Coat}} \tag{7}$$

With

$$\sigma_{\text{SWCNT}} = -j \frac{2e^2 V_f}{\pi^2 hr(j - wv)} \tag{8}$$

In the above equations, P is the circumference of the SWCNT ($P = 2\pi r$), A is the average cross-sectional area of the coating layer, e is the charge of an electron, h is the Planck's constant, r is the radius of the SWCNT, $\upsilon = \frac{6\mathrm{T}}{r}$ is the relaxation frequency and $V_f = 9.7 \times 10^5$ ms is the Fermi velocity of CNT. The radius of the SWCNT was calculated using the formula

$$r = \frac{\sqrt{3}b}{2\pi}\sqrt{m^2 + mn + n^2} \tag{9}$$

where m, n are the chiral indices and $b = 0.142$ nm is the graphene interatomic distance. The effective conductivity of the composite material was observed to be impacted by r, t, and σ_{Coat}.

Cheng et al. (2017) presented a Cu-CNT composite interconnect utilizing an equivalent circuit model of a DIL system [32]. The per unit length scattering resistance of the composite interconnect was given by

$$R = \frac{\rho_{\mathrm{eff}}}{wt} \tag{10}$$

where

$$\rho_{\mathrm{eff}} = \mathrm{Re}\left(\frac{1}{\sigma_{\mathrm{eff}}}\right)\frac{1}{wt} \tag{11}$$

$$\sigma_{\mathrm{eff}} = (1 - f_{\mathrm{cnt}})\,\sigma_{\mathrm{cu}} + f_{\mathrm{cnt}}\sigma_{\mathrm{cnt}} \tag{12}$$

In Eq. 11, ρ_{eff} and σ_{eff} are the effective resistivity and conductivity, respectively. In Eq. 12, σ_{cu} is the conductivity of copper, σ_{cnt} is the conductivity of CNT, and f_{cnt} is the CNT filling ratio. Electrical models for Cu/CNT composite were proposed by Feng (2015) and Rao (2017) to investigate the current density distributions [42, 43]. Feng et al. (2015) presented a composite electrical model of uniformly arranged CNT bundles mixed with different ratios of copper (10%, 50%, and 90%) to review the composite interconnect current density distribution [42]. The effects of Joule heating and the dimensions of CNT on electrical properties were evaluated and compared for Cu/SWCNT and Cu/MWCNT composites. In this model, the vias were assumed to be filled with CNTs of varying densities. Cu was used for conduction in the space or voids in between the CNT bundles. The area occupied by the CNT bundles by a single via was assumed to be a square shape and was given by $\sim n \times (D_{\mathrm{CNT}} + 0.34\ \mathrm{nm})^2$. The tube quantity 'n' was calculated by:

$$n = \frac{\pi D_{\mathrm{TSV}}^2}{4(D_{\mathrm{CNT}} + d)^2} \tag{13}$$

where d is the separation between each nanotube, D_{TSV} is the diameter of a TSV, and D_{CNT} is the CNT diameter. The distribution of CNTs in the vias was regulated by the distance between each tube, d. When $d > 0.34$ nm (Vander Waals gap), the CNT density decreases while the interconnect resistance increases. Thus, the resistance between the nanotubes was lowered by packing the voids with Cu. The Cu filling ratio was given by (Fig. 4).

$$\text{Filling ratio} = \left(1 - \frac{4n(D_{CNT} + 0.34 \text{ nm})^2}{\pi D_{TSV}^2}\right) \times 100\% \tag{14}$$

The electrical model was assumed to be a bulk Cu conductor connected in parallel with CNT bundles. Considering Cu was used to fill the void space between CNTs and were not isolated, the contact resistance was neglected. The magnetic inductance of CNT along with the Cu inductance was also excluded. The TSV was assumed to be a Cu/CNT matrix-free of voids, thus neglecting the quantum capacitance and any inter-sheet interactions and coupling with the substrate. The overall impedance of the composite Cu/CNT system in the TSV is given by:

$$Z(w) = \frac{1}{\frac{1}{R_{cu}} + \frac{n}{Z_{CNT}(w)}} \tag{15}$$

where the CNT impedance

$$Z_{CNT}(w) = R_{CNT} + \frac{iwhL_K}{N} \tag{16}$$

The kinetic inductance and CNT resistance are, respectively,

$$L_K = \frac{R_Q}{2v_F}; \quad R_{CNT} = \frac{R_Q + R_S h}{N} \tag{17}$$

The scattering resistance is given by

$$R_S = \frac{R_Q}{D_{CNT}}\left(k_1 + k_2 T + k_3 T^2\right) \tag{18}$$

where $k_1 = 3.005 \times 10^{-3}$, $k_2 = -2.122 \times 10^{-5}$ K^{-1}, $k_3 = 4.701 \times 10^{-8}$ K^{-2}.

Further, it was assumed that the TSV was homogenously filled by the material and the conductivity, in the frequency domain $\sigma > w\varepsilon$. Thus, the distribution of current density at the interconnect cross-section was provided by the solution of the following differential equation

$$\frac{d^2 J}{dr^2} + \frac{1}{r}\frac{dJ}{dr} + k^2 J = 0, \quad \text{where } k = \sqrt{-j\omega\mu\sigma(\omega)} \tag{19}$$

The distribution of the current density of pure Cu, pure SWNT/MWCNT bundles, and Cu/SWNT composite bundles were done. The results implied that through silicon vias (TSVs) filled with Cu/CNT medium exhibited a more uniform current distribution with decreased skin effect compared to the vias filled with only copper. Evaluating the results, SWNTs with a diameter of 2 nm and MWNTs with an outer diameter of 10 nm indicated similar distribution. The current density distribution was examined to be less homogeneous for MWNT with a diameter less than 30 nm compared to an interconnect with a larger diameter. It was also observed that the current density distribution was more homogeneous at high frequencies for higher ratios of CNTs.

A high-frequency electrical model for 3-D interconnects of composite Cu/SWCNT was proposed by Rao [43]. In this model, bundles of SWNT of different diameters having Cu filling ratios of 0%, 10%, 50%, 90%, and 100% were assumed. The tapered TSV was modeled using a lumped RLGC electrical circuit. Rao (2016) developed a model for cylindrical TSV profile and was compared with the previous model [44]. It was observed that a TSV with a cylindrical profile presented a homogenous current density for composite Cu/MWNT interconnect when compared to a tapered TSV as the tapered sidewalls of vias degraded the electrical performance of the interconnects. Using the Elmore delay method, a vertical delay model was derived to calculate the approximate vertical bidirectional signal delay for tapered TSV filled with different ratios of a Cu/MWNT composite. The vertical delays for the top to bottom directed signal and vice-versa were given by the following equations.

$$\text{Delay}_{tb} = \int_0^h \frac{\left(C_{\text{eff}}\frac{h-x}{h} + C_{\text{load}}\right)dx}{\frac{\pi(b_{\text{tsv}} + \frac{(a_{\text{tsv}} - b_{\text{tsv}})x}{h})^2 - \left(a_{\text{tsv}}^2 \pi (1-\text{filling\_ratio})\right)}{\rho} + \frac{h \times n}{R_{\text{MWCNT}}}} \tag{20}$$

$$\text{Delay}_{bt} = \int_0^h \frac{\left(C_{\text{eff}}\frac{h-x}{h} + C_{\text{load}}\right)dx}{\frac{\pi\left(a_{\text{tsv}} + \frac{(b_{\text{tsv}} - a_{\text{tsv}})x}{h}\right)^2 - \left(a_{\text{tsv}}^2 \pi (1-\text{filling\_ratio})\right)}{\rho} + \frac{h \times n}{R_{\text{MWCNT}}}} \tag{21}$$

$$b_{\text{tsv}} = a_{\text{tsv}} + h \times \tan\theta, \tag{22}$$

where b_{tsv} and a_{tsv} are the upper and lower radius, h is via height, and θ is the tapering angle. As compared to the signal flow in the reverse direction, the signal directed from top to bottom of the tapered TSV interconnect was found to be consistently lower.

Table 1 Mean free path of different interconnect materials at room temperature

Mean Free Path at room temperature(nm) [14]					
Copper	SWCNT	MWCNT	Mixed CNT bundle	GNR	CNT composite
40	$> 10^3$	2.5×10^4	$*8.19 \times 10^5$	1×10^3	$\sim 3 \times 10^3$

6   Mean Free Path of Materials Used for Fabrication of Interconnects

Table 1 shows the mean free path (MFP) values of different interconnect materials calculated by Li et al. [14]. The performance of interconnect is significantly dependent on the MFP. MFP is fundamentally the average distance traversed by an electron between collisions with other particles Thus, the conductivity improves with improvement in the MFP.

The mixed CNT bundle's MFP is given by Naeemi [45] as,

$$\lambda = \frac{v_F \mathrm{d}T}{\alpha} \tag{23}$$

where v_F is the Fermi velocity of graphene, α is the total scattering rate, d is the diameter of the tube, and T is the temperature. Mixed CNT bundles with metallic SWCNT and large diameter MWCNT at the center will have a larger MFP. Next, the effective MFP of CNT composite is given by Z. H. Cheng et al. (2017) as

$$\lambda_{\mathrm{eff}} = 1000D \tag{24}$$

D is the CNT diameter (3 nm) [32]. The performance of a pure CNT significantly relies on the MFP and therefore is directly affected by the decrease in MFP. However, the Cu-CNT composites are far less impacted by the degradation in MFP.

7   Delay in Interconnects Composed of Different Materials

With the scaling down of transistor sizes, propagation delay due to interconnects has been a major cause of concern for high-speed VLSI technology. Researchers have been experimenting with different materials to overcome this problem.

7.1   Crosstalk-Induced Delay

Crosstalk is fundamentally the noise that is induced in a non-switching wire (victim) due to switching in adjacent wire (aggressor), leading to increased delay in the

Table 2 Crosstalk induced delay in Cu, SWCNT bundled and MWCNT bundled interconnects

Length of Interconnect (μm)	Delay due to crosstalk (ns)			% improvement in delay between Cu and MWCNT bundle interconnect [51]
	Copper interconnect (22 nm) [51]	SWCNT bundle interconnect (21 nm) [50]	MWCNT bundle interconnect (22 nm) [51]	
200	0.6565	0.046	0.3941	39.97
500	0.8363	0.102	0.3151	62.27
1000	0.8437	0.18	0.2624	68.90
1500	0.9417	0.272	0.2639	71.97
2000	0.11528	0.373	0.0306	73.66

switching wire owing to an increase in load capacitance. An aggressor is a wire in which signal transition occurs and the victim is the wire in which coupling of noise ensues. Crosstalk in VLSI interconnects typically occurs due to capacitive and inductive coupling as analyzed by Mido and Asada [46], Sainarayanan [47], Agarwal et al. [48]. Ismall and Friedman (2003) investigated inductive effects for high-speed VLSI interconnects and was observed to cause large errors in the propagation delay (35% for 250 nm technology) [49].

The effects of crosstalk were investigated in SWCNT for 15 nm and 21 nm technology nodes by Sahoo et al. (2013) using the ABCD parameter approach. Khezeli et al. (2017) presented an analysis of the crosstalk delay of copper and MWCNT bundled interconnects (sub 22 nm technology) for the ternary logic system at the global level without repeater insertion [50, 51].

Table 2 shows the delay induced due to crosstalk in Cu, SWCNT bundled and MWCNT bundled interconnects along with the percentage improvement between the crosstalk induced delay in Cu and MWCNT bundled interconnects for interconnects of different lengths for a sub 22 nm technology node without repeater insertion as reported by Khezeli et al. [51].

For global interconnects, the improvement in dynamic crosstalk-induced delay was observed to be superior in MWCNT than in SWCNT bundle interconnects. The worst-case delay in MWCNT bundled interconnect was observed to be significantly lower than that of Cu interconnect with the increase in interconnect length. Khezeli et al. (2018) presented a comparative analysis between Cu and large diameter MWCNT bundled interconnects (intermediate, repeated global, and unrepeated global) at 14 nm technology nodes for a ternary logic system [52]. MWCNT bundled interconnects were observed to have lower delay than Cu interconnects at the global and semi-global levels (without repeater insertion) due to lower distributed resistance. For repeater inserted global interconnects, the Cu interconnects were observed to have lower delay than the MWCNT bundled interconnects. However, the number of repeaters used for Cu were significantly higher than that used for the MWCNT

Table 3 Comparison between the delay induced due to crosstalk in Cu and GNR interconnects

Interconnect length (μm)	Increase in the rise/fall time (ns)				Decrease in the rise/fall time (ns)			
	Cu		GNR		Cu		GNR	
	$\Delta\tau_{LH}$	$\Delta\tau_{HL}$	$\Delta\tau_{LH}$	$\Delta\tau_{HL}$	$\Delta\tau_{LH}$	$\Delta\tau_{HL}$	$\Delta\tau_{LH}$	$\Delta\tau_{HL}$
1	0.0015	0.0009	0.0006	0.0004	−0.003	−0.0002	−0.0006	−0.0005
5	0.0061	0.0040	0.0024	0.0014	−0.0053	−0.0046	−0.0021	−0.0018
10	0.0103	0.0077	0.0043	0.0026	−0.0086	−0.0077	−0.0036	−0.0032
50	0.0721	0.0664	0.0214	0.0199	−0.0268	−0.0277	−0.0109	−0.0114
100	0.2036	0.2002	0.0614	0.0639	−0.0698	−0.0816	−0.0267	−0.0313
500	3.5591	2.2312	1.1766	1.1849	−1.2465	−0.1436	−0.4538	−0.4681
1000	14.763	13.945	12.230	12.033	−4.7872	−3.3192	−1.6436	−1.1974

bundled interconnects, thus subsequently consuming more routing area. Next, Bhattacharya et al. (2014) presented a comparison between the delay induced due to crosstalk in Cu and GNR interconnects for sub 16 nm node technology [53].

Table 3 shows the comparison between delay induced due to crosstalk in Cu and GNR interconnects. It was observed that when the switching of the input signals was in the same direction in both the aggressor and the victim nets, the interconnect delay was reduced due to a decrease in the signal transition time. However, when switching occurred in the opposite direction, the interconnect delay increased due to an increase in the signal transition time.

7.2 Propagation Delay in Cu/CNT Composite Interconnects

Gao et al. (2016) provided a comparison between the propagation delay of Cu, CNT, and Cu/CNT composite interconnects for various dimensions and CNT filling ratios. Table 4 shows the % Reduction in the propagation delay of interconnects when Cu and pure CNT are replaced by Cu-CNT composite material for different dimensions as calculated by Gao et al. [54], Koo et al. [55]. The performance of the composite interconnect improves with the increase in length.

One of the main parameters that characterize an on-chip interconnect is bandwidth density. It is a measure of the rate of data transmitted through a unit cross-section of interconnect. The bandwidth density is generally determined by the wire pitch and is given by the bandwidth divided by the pitch (width of the wire). Thus, interconnect with a large pitch offers poor bandwidth density which is a downside of optical

Table 4 % reduction in the propagation delay of interconnects

When Cu is replaced by Cu/CNT composite	Interconnect lengths(μm) ($f_{cnt} = 20\%$, Aspect ratio $= 2$)	Widths (nm)	% reduction in propagation delay
	600	210	33
		180	36
		150	40
	1000	210	~ 35
		180	~ 35
		150	~ 35
When pure CNT is replaced by Cu/CNT composite	1000	100	30

interconnects. This can be alleviated by using wave division multiplexing (WDM). Manipatruni et al. (2010) demonstrated that in an optical interconnect with 25 WDM channels operating at 12.5 Gbit/s, a bandwidth density of ~ 200 Gbit/s. μm can be attained [56]. The bandwidth density of optical interconnects increases merely because of the higher bit rate via waveguides with a fixed pitch. The bandwidth density of an optical line is given by Manipatruni (2010)

$$\beta = \frac{kB}{p} \tag{25}$$

where $p = 0.12\log_e\left(\frac{56.6z}{\pi}\right)$ is the pitch (in microns), k is the no. of wavelength channels, B is the bit rate and z is the crosstalk distance (in microns) [55]. Comparisons between the bandwidth densities of different interconnects have been presented by Manipatruni et al. [56], Haurylau et al. [57]. The wire pitch of an electrical on-chip interconnect (Cu, CNT) is conventionally smaller (~0.1 μm) than that of an optical interconnect (~0.6 μm), thus leading to a better bandwidth density. For global interconnects, significant improvement in the bandwidth density was observed by Naeemi and Meindl (2008) using SWCNT and MWCNT, without deteriorating the delay or energy per bit [58]. Due to its high mean free path, SWCNTs can provide a better bandwidth density than that of a Cu interconnect. Also, the density improves with an increase in the metallic fraction of the SWCNTs. There are several materials and technologies are proposed in the semiconductor industry. Such as, high-speed interconnect, high-speed semiconductor devices by Maity et al. (2020), Maity et al. (2019) solar cell by Muchahary et al. (2020), Narzary et al. (2020), Maity et al. (2019), Muchahary et al. (2019), photo-detector by Baruah et al. (2020), Maity et al. (2019) etc. [59–69]. In this chapter, the detailed simulation, fabrication, and characterization of interconnection have been discussed.

8 Conclusion

This chapter presented a review of different interconnect materials. First, the usage of traditional material like Cu as interconnect material was reviewed followed by carbon nanomaterials such as graphene nanoribbons (GNRs) and CNTs. Although GNRs are more convenient to fabricate than CNTs because of their planar structure, monolayer GNR were observed to have extremely high resistance due to skin effects. Therefore, multi-layered GNR with dopant layer inserted between each graphene ribbon was taken. Hence, for graphene and GNR interconnects specific control over the defects such as edge scattering and edge roughness are some of the crucial challenges to overcome. Theoretically, CNT (both SWCNT and MWCNT) is observed to show better performance than Cu interconnects. However, practically their performance is entirely dependent on the fabrication process. Growing extremely dense CNT with a high metallic fraction is a challenging task. The performance of individual SWCNTs was observed to wane due to their high imperfect contact resistance, hence bundles of SWCNT were investigated and the propagation delay was perceived to be better than that of MWCNT. However, the area occupied by MWCNT was minimized along with the electrostatic capacitance between the shells in comparison to that of the bundled CNT structure. In the next section, composite interconnect materials are reviewed. The performance characteristics of the composite materials were found to be practically better than that of pure CNT. The necessary criteria for pure CNT to provide better performance or one comparable to Cu were to have a high filling ratio or density and controllable chirality which was difficult to achieve via fabrication. Thus, it is seen that composite materials provide a better interconnect performance than traditional materials and can be a pragmatic solution to high-speed interconnects for VLSI application until the prerequisites of fabrication of pure CNT are attained.

References

1. Davis JA, Meindl JD, Fellow L (2000) Compact distributed RLC interconnect models—Part I: single line transient. Time Delay Overshoot Expressions 47(11):2068–2077
2. Coulibaly LM, Kadim HJ (2005) Analytical crosstalk noise and its induced-delay estimation for distributed RLC interconnects under ramp excitation. In: 2005 IEEE International symposium on circuits and systems, vol 2, pp 1254–1257. https://doi.org/10.1109/ISCAS.2005.1464822
3. Ullah MS, Chowdhury MH (2017) Analytical models of high-speed RLC interconnect delay for complex and real poles. IEEE Trans Very Large Scale Integr (VLSI) Sys 25(6):1831–1841. https://doi.org/10.1109/TVLSI.2017.2654921
4. Bohr MT (1995) Interconnect scaling—the real limiter to high performance ULSI. IEDM Tech Dig 241–244
5. Davis JA et al (2001) Interconnect limits on gigascale integration (GSI) in the 21st century. Proc IEEE 89(3):305–322
6. Ogawa ET, Lee KD, Blaschke VA, Ho PS (2002) Electromigration reliability issues in dual-damscene Cu interconnections. IEEE Trans Reliab 51(4):403–419
7. Havemann RH, Hutchby JA (2001) High-Performance interconnects: an integration overview. IEEE 89(5)

8. Maex K, Baklanov MR, Shamiryan D, Brongersma SH, and Yanovitskaya ZS (2003) Low dielectric constant materials for microelectronics. Appl Phys Rev Focused Rev 93(11)
9. Yang CC, Cohen S, Shaw T, Wang PC, Nogami T, Edelstein D (2010) Characterization of 'ultrathin-Cu'/Ru(Ta)/TaN liner stack for copper interconnects. IEEE Electron Device Lett 31(7):722–724
10. Ceyhan A, Naeemi A (2013) Cu interconnect limitations and opportunities for SWNT interconnects at the end of the roadmap. IEEE Trans Electron Devices 60(1):374–382
11. Sarkar D, Xu C, Li H, Banerjee K (2011) High-frequency behavior of graphene-based inter-connects—Part I: impedance modeling. IEEE Trans Electron Devices 58(3):843–852. https://doi.org/10.1109/TED.2010.2102031
12. Contino A, Ciofi I, Wu X, Asselberghs I, Celano U, Wilson CJ, Tökei Z, Groeseneken G, Sorée B (2018) Modeling of edge scattering in graphene interconnects. IEEE Electron Device Lett 39(7):1085–1088. https://doi.org/10.1109/LED.2018.2833633
13. Murali R, Brenner K, Yang Y, Beck T, Meindl JD (2009) Resistivity of graphene nanoribbon interconnects. IEEE Electron Device Lett 30(6):611–613. https://doi.org/10.1109/LED.2009.2020182
14. Li H, Xu C, Banerjee K (2010) Carbon nanomaterials: the ideal interconnect technology for next-generation ICs. IEEE Des Test Comput 27(4):20–31
15. Naeemi A, Sarvari R, Meindl JD (2005) Performance comparison between carbon nanotube and copper interconnects for gigascale integration (GSI). IEEE Electron Device Lett 26(2):84–86
16. Naeemi A, Meindl JD (2007) Design and performance modeling for single-walled carbon nanotubes as local, semiglobal and global interconnects in gigascale integrated systems. IEEE Trans Electron Devices 54(1):26–37
17. Srivastava N, Li H, Kreupl F, Banerjee K (2009) On the applicability of single-walled carbon nanotubes as VLSI interconnects IEEE Trans. Nanotechnol 8(4):542–559
18. Srivastava N, Banerjee K (2005) Performance analysis of carbon nanotube interconnects for VLSI applications. In: IEEE/ACM International conference on computer-aided design. Tech. Pap. ICCAD 2005, pp 383–390
19. Li H, Yin WY, Banerjee K, Mao JF (2008) Circuit modeling and performance analysis of multi-walled carbon nanotube interconnects. IEEE Trans Electron Devices 55(6):1328–1337
20. Majumder MK, Pandya ND, Kaushik BK, Manhas SK (2012) Analysis of MWCNT and bundled SWCNT interconnects: impact on crosstalk and area. IEEE Electron Device Lett 33(8):1180–1182
21. Haruehanroengra S, Wang W (2007) Analyzing conductance of Mixed Carbon-Nanotube Bundles for Interconnect Applications. IEEE Electron Device Lett 28(8):756–759. https://doi.org/10.1109/LED.2007.901584
22. Majumder MK, Kaushik BK, Manhas SK (2014) Analysis of delay and dynamic crosstalk in bundled carbon nanotube interconnects. IEEE Trans Electromagn Compat 56(6):1666–1673
23. Ilani S, Mceuen P (2010) Electron transport in carbon nanotubes. Ann Rev Condens Matter Phys 1(1):1–25
24. Jarosz P, Schauerman C, Alvarenga J, Moses B, Mastrangelo T, Raffaelle R, Ridgley R, Landi B (2011) Carbon nanotube wires and cables: near term applications and future perspectives. Nanoscale 3:4542–4553
25. Behabtu N et al (2013) Strong, light, multifunctional fibers of carbon nanotubes with Ultra-high conductivity. Science 339:182–186
26. Sule R, Olubambi PA, Abe BT, Johnson OT (2012) Microelectronics reliability synthesis and characterization of sub-micron sized copper–ruthenium–tantalum composites for interconnection application 52:1690–1698
27. Nogami T et al (2017) Cobalt/copper composite interconnects for line resistance reduction in both fine and wide lines. In: 2017 IEEE international interconnect technology conference (IITC), Hsinchu 8–10
28. Aryasomayajula L, Rieske R, Wolter KJ (2011) Application of copper–carbon nanotubes composite in packaging interconnect. In: Proceedings of 2011 34th international spring seminar on electronics technology, pp l 531–536

29. Sundaram R, Yamada T, Hata K, Sekiguchi A (2017) The influence of Cu electrodeposition parameters on fabricating structurally uniform CNT-Cu composite wires. Mater Today Commun 13:119–125
30. Liu P, Xu D, Li Z, Zhao B, Kong E, Zhang Y (2008) Fabrication of CNTs/Cu composite thin films for interconnects application. Microelectron Eng 85(10):1984–1987
31. Chai Y, Chan PCH, Fu Y, Chuang YC, Liu CY (2008) Electromigration studies of Cu/carbon nanotube composite interconnects using Blech structure. IEEE Electron Device Lett 29(9):1001–1003
32. Cheng ZH et al (2017) Investigation of copper-carbon nanotube composites as global VLSI Interconnects. IEEE Trans Nanotechnol 16(6):891–900
33. Sun S, Mu W, Edwards M, Mencarelli D, Pierantoni L, Fu Y et al (2016) Vertically aligned CNT-cu nano-composite material for stacked through- silicon-via interconnects. Nanotechnology 27:335705
34. Tu JP, Yang YZ, Wang LY, Ma XC, Zhang XB (2001) Tribological properties of carbon nanotube-reinforced copper composites. Tribol Lett 10:225–228
35. Zheng L, Sun J, Chen Q (2017) Carbon nanotubes reinforced copper composite with uniform CNT distribution and high yield of fabrication. Micro Nano Lett 12(10):722–725
36. Wang J, Guo J, Zhang Y, Pan Y, Guo J (2008) The resistivity of a new composite system: CNT-ceramic. In: 2008 3rd IEEE international conference on nano/micro engineered and molecular systems, Sanya, pp 820–823
37. Angelopoulos M (2001) Conducting polymers in microelectronics. IBM J Res Dev 45(1):57–75
38. Long Y et al (2004) Synthesis and electrical properties of carbon nanotube polyaniline composites. Appl Phys Lett 85:1796–1798
39. Alshehri A et al (2011) Electrical performance of carbon nanotube-polymer composites at frequencies up to 220 GHz. Appl Phys Lett 99
40. Seol SK, Chang W, Kim D, Jung S (2012) Carbon nanotube-conducting polymer composite wires formed by fountain pen growth (FPG) route. RSC Adv 2:8926–8928
41. Jurn YN, Malek MFBA, Rahim HA (2016) mathematical analysis and modeling of single-walled carbon nanotube composite material for antenna applications. Progress Electromagnet Res M 45:59–71
42. Feng Y, Burkett S (2015) Modeling a copper/carbon nanotube composite for applications in electronic packaging. Comput Mater Sci 97:1–5
43. Rao M (2017) Electrical modeling and characterization of copper/carbon nanotubes in tapered through silicon vias. In: 30th international conference on VLSI design and 16th international conference on embedded systems (VLSID), Hyderabad, pp 366–371
44. Rao M (2016) Electrical modeling of copper/carbon nanotubes for 3D integration in Nanotechnology (IEEE-NANO). In: 16th IEEE conference on August 2016, pp 763–766
45. Naeemi A, Mceuen PL (2009) Review of CNT Interconnect Technology 2002:18–33
46. Mido T, Asada K (1997) Crosstalk noise in high density and high-speed interconnections due to inductive coupling. In: Proceedings of ASP-DAC '97: Asia and South Pacific design automation conference, Chiba, Japan, pp 215–220
47. Sainarayanan KS, Ravindra JV, Srinivas MB (2006) A novel, coupling driven, low power bus coding technique for minimizing capacitive crosstalk in VLSI interconnects. In: 2006 IEEE International symposium on circuits and systems, pp 4155–4158. https://doi.org/10.1109/ISCAS.2006.1693544
48. Agarwal K, Sylvester D, Blaauw D (2006) Modeling and analysis of crosstalk noise in coupled RLC interconnects. IEEE Trans Comp Aided Des Integr Circuits Sys 25(5):892–901. https://doi.org/10.1109/TCAD.2005.855961
49. Ismall YI, Friedman EG (2003) Effects of inductance on the propagation delay and repeater insertion in VLSI circuits: a summary. IEEE Circ Syst 3(1):24–28
50. Sahoo M, Ghosal P, Rahaman H (2013) An ABCD parameter-based modeling and analysis of crosstalk induced effects in single-walled carbon nanotube bundle interconnects. In: Fifth Asia symposium on quality electronic design (ASQED 2013), Penang, pp 264–273

51. Khezeli MR, Moaiyeri MH, Jalali A (2017) Analysis of crosstalk effects for multiwalled carbon nanotube bundle interconnects in ternary logic and comparison with Cu interconnects. IEEE Trans Nanotechnol 16(1):107–117. https://doi.org/10.1109/TNANO.2016.2633460

52. Khezeli MR and Jalali A (2018) A comparative performance analysis of copper and MWCNT Bundle interconnects in ternary logic. In: Iranian Conference on Electr Eng (ICEE), pp 173–177

53. Bhattacharya S, Das D, and Rahaman H (2014) A novel GNR interconnect model to reduce crosstalk delay. In: 2014 fifth international symposium on electronic system design, Surathkal, pp 5–9

54. Gao X, Zheng J, Zhao W, Wang G (2016) Electrical modeling of on-chip copper-carbon nanotube composite interconnects. In: 2016 Asia-Pacific international symposium on electromagnetic compatibility (APEMC), Shenzhen, pp 229–231

55. Koo K, Cho H, Kapur P, Saraswat KC (2007) Performance comparisons between carbon nanotubes, optical, and Cu for future high-performance on-chip interconnect applications. IEEE Trans Electron Devices 54(12):3206–3215

56. Manipatruni S, Chen L, Lipson M (2010) Ultra high bandwidth WDM using silicon microring modulators. Opt Express 18:16858–16867

57. Haurylau H et al (2006) On-chip optical interconnect roadmap: challenges and critical directions. IEEE J Sel Top Quantum Electron 12(6):1699–1705

58. Naeemi A, Meindl JD (2008) Performance modeling for single- and multiwall carbon nanotubes as signal and power interconnects in gigascale Systems. IEEE Trans Electron Devices 55(10):2574–2582

59. Maity R, Shuvro S, Maity S, Maity NP (2020) Collapse voltage analysis of central annular ring metallized membrane-based MEMS micromachined ultrasonic transducer. Microsyst Technol 26:1001–1009

60. Maity NP, Maity R, Maity S, Baishya S (2019) A new surface potential and drain current model of dual material gate short channel metal oxide semiconductor field effect transistor in sub-threshold regime: application to high-k material HfO2. J Nanoelectron Optoelectron 14(9):868–876

61. Maity NP, Maity R, Maity S, Baishya S (2019) Comparative analysis of the quantum FinFET and trigate FinFET based on modeling and simulation. J Comput Electron 18:492–499

62. Muchahary D, Maity S, Metya SK, Basumatary B (2020) A simulation approach to improve photocurrent through a double-layer of the emitter in a-Si1-xCx/c-Si heterojunction solar cell. Superlattices Microstruct 146:106–651

63. Narzary R, Phukan P, Maity S, Sahu P (2020) Enhancement of power conversion efficiency of Al/ZnO/p-Si/Al heterojunction solar cell by modifying morphology of ZnO nanostructure. J Mater Sci Mater Electron 31:1–8

64. Maity S, Das B, Maity R, Maity NP, Guha K, Rao KS (2019) Improvement of quantum and power conversion efficiency through electron transport layer modification of ZnO/perovskite/PEDOT: PSS based organic heterojunction solar cell. Sol Energy 185:439–444

65. Maity S, Sahu PP (2019) Efficient Si-ZnO-ZnMgO heterojunction solar cell with alignment of grown hexagonal nanopillar. Thin Solid Films 674:107–111

66. Muchahary D, Maity S, Metya SK (2019) Modelling and analysis of temperature-dependent carrier lifetime and surface recombination velocity of Si–ZnO heterojunction thin film solar cell. IET Micro & Nano Letters 14:399–403

67. Baruah S, Bora J, Maity S (2020) Investigation and optimization of light trapping through hexagonal-shaped nanopillar (NP) array of indium gallium arsenide material-based photodetector. Opt Quant Electron 52:380

68. Baruah S, Bora J, Maity S (2020) High performance wide response GaAs based photo detector with nano texture on nanopillar arrays structure. Microsyst Technol 26:2651–2660

69. Maity S, Thomas T (2019) Hybrid-organic-photodetector containing chemically treated ZnMgO layer with promising and reliable detectivity, responsivity and low dark current. IEEE Trans Device Mater Reliab 19:193–200

Voltage-Programmed Pixel Circuit Design for AMOLED Displays

Kavindra Kandpal, Aryamick Singh, and Akriti Srivastava

Abstract Nowadays, the display industry is a driving force of consumer electronics, including various gadgets like mobile phones, laptops, and HD-TVs. This chapter focuses on thin film transistor (TFT)-based circuits, mainly used to drive an organic light-emitting diode (OLED) present in a pixel of an active-matrix organic light-emitting diode (AMOLED) display. TFTs are much cheaper transistors than silicon-based MOSFETs. However, they face low electron mobility, threshold voltage shift, low on-to-off ratio, and subthreshold slope. An increase in threshold voltage over time may reduce OLED current and can result in a dark pixel. However, the TFT offers a significant advantage in lower cost/cheaper electronics if the challenges can be addressed. TFTs do not require costly processes and costly wafers. One can make TFT-based circuits and devices on a plastic substrate using printable electronics. However, before the product goes to the market, the performance, yield, and lifetime of pixel circuits in particular or displays in general need to be substantially increased. Therefore, as a circuit designer, it becomes crucial to address the challenging issues of threshold voltage shift over time for the display's uniform brightness. This chapter will focus on design strategy and challenges for designing voltage-programmable pixel circuit.

Keywords ZnO-TFT · Voltage-programmed pixel circuits · AMOLED display

1 Introduction

An AMOLED display consists of a TFT backplane for superior control over the light emitted by the display panel. An AMOLED is driven by TFTs containing storage

K. Kandpal (✉) · A. Srivastava
Department of Electronics and Communication Engineering, Indian Institute of Information Technology, Uttar Pradesh, Allahabad, Prayagraj 211015, India
e-mail: kavindra@iiita.ac.in

A. Singh
Department of Electrical and Electronics Engineering, Birla Institute of Technology and Science Pilani, Pilani Campus, Rajasthan 333031, India

© The Author(s), under exclusive license to Springer Nature Singapore Pte Ltd. 2022
R. Goswami and R. Saha (eds.), *Contemporary Trends in Semiconductor Devices*,
Lecture Notes in Electrical Engineering 850,
https://doi.org/10.1007/978-981-16-9124-9_12

capacitors that hold the line pixel states stable, allowing for large size (and resolution) displays. AMOLEDs can be made much larger than passive-matrix OLEDs, and there are no limitations on their size or resolution. One of the most significant advantages of AMOLED displays is their ability to be made flexible. Flexible AMOLEDs can be used in smartphones and wearables and are becoming increasingly popular. More often mentioned advantages of AMOLED displays are that they produce vibrant colors, have superior blacks, consume less power, and are lighter and thinner than conventional LED models. In AMOLED, TFTs are mainly used to turn on and drive an individual pixel [1].

The operation of TFTs relies on the similar principle of field-effect transistor. This means that one can modulate a deposited semiconducting thin film's conductivity or surface potential by applying a voltage to the gate electrode. One can say TFTs are interesting and commercial applications of disordered semiconductors. We call these semiconductors disordered because the deposited channel material is polycrystalline or amorphous, unlike in CMOS technology, where substrate or channel is primarily crystalline [2]. The commonly used substrate materials for TFTs are corning glass, polyethylene terephthalate (PET), and polyethylene naphthalate (PEN) substrate [3]. These substrates are much cheaper than conventional single crystal silicon substrates used for CMOS microelectronics. The most commonly used channel materials for a TFT are amorphous silicon (a-Si), hydrogenated amorphous silicon a-Si:H, poly-crystalline silicon (poly-Si), oxide semiconductors such as ZnO, IGZO, SnO_2, and organic semiconductors such as pentacene. Figure 1 shows a schematic diagram of a bottom-gate TFT, which uses glass as a substrate material, HfO_2 as a gate dielectric, ZnO as a semiconducting channel layer, and indium-tin-oxide ITO as a source and drain contacts [4].

At the early stage of TFT development, a-Si:H or poly-Si-based TFTs were mostly fabricated and used in most displays. However, in recent years, the most commonly used semiconductors for TFT fabrication are either oxide semiconductors or organic semiconductors [5]. The deposition of both oxide and organic semiconductors can mainly be carried out at room temperature, thus enabling various cheaper, transparent, and flexible substrates for TFT fabrication. Though organic TFTs support printable electronics but lack in providing good electron mobility. However, the use of oxide semiconductors can overcome the problem of low electron mobility.

ZnO
ITO
HfO_2
Glass
Gate electrode (Pt)

Fig. 1 Schematic diagram of Bottom Gate ZnO TFT

As per the article published by Nikkei Electronic Asia [6], 'Transparent product soon a reality,' the future generation will see ZnO-based TFTs enabling high-performance integrated chips made over the transparent and flexible substrate [7–9]. ZnO is intrinsically an *n*-type wide bandgap semiconductor ($E_g = 3.4$ eV) [10, 11]. ZnO-based TFTs offer various advantages over their *a*-Si and *poly*-Si TFT counterparts, including low-temperature deposition, higher mobility, and higher on-to-off (I_{on}/I_{off}) ratio, steeper subthreshold slope, and optical transparency in the visible region of the spectrum [4]. This chapter shall use adapted ZnO TFTs SPICE (Simulation Program with Integrated Circuit Emphasis) models for the circuit simulation. However, the methods can be applied to all kinds of TFTs.

Figure 2 shows a typical transfer characteristic of a thin film transistor, where the drain current is shown on a logarithmic scale. From the transfer characteristic, one can extract various performance parameters of a TFT, which include subthreshold slope (SS), on-to-off ratio (I_{on}/I_{off}), the threshold voltage (V_T), and field-effect mobility (μ_{FE}). The transfer characteristic shows that the transfer characteristics are very similar to that of a MOSFET. Therefore, one can adapt the existing MOS SPICE models for the TFT circuit simulation [12, 13]. Moreover, one can also use a Verilog-A-based model to simulate TFT circuits. In terms of other available physical CAD compatible models, one model developed for a-Si and poly-Si TFTs by Rensselaer Polytechnic Institute (RPI) has been used in various CAD simulators, including HSPICE [14, 15]. For oxide or organic TFTs, however, no model is the standard. One can adapt the existing SPICE model and calibrate it with an oxide TFT's experimental characteristics.

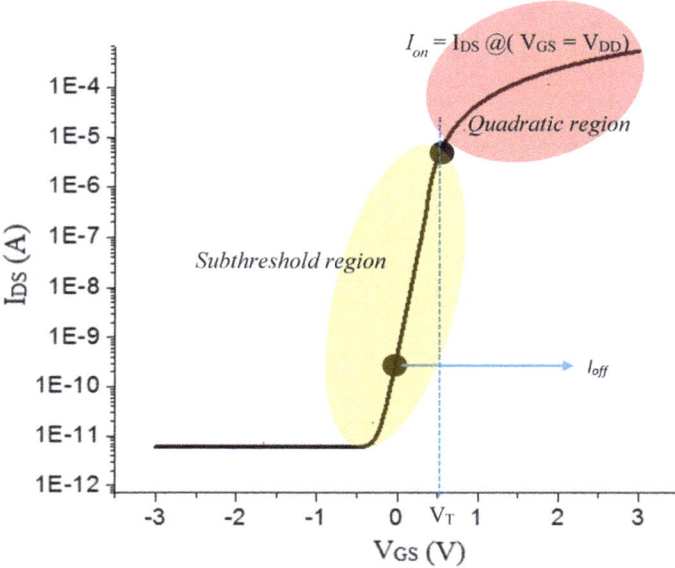

Fig. 2 Typical transfer characteristic of a TFT

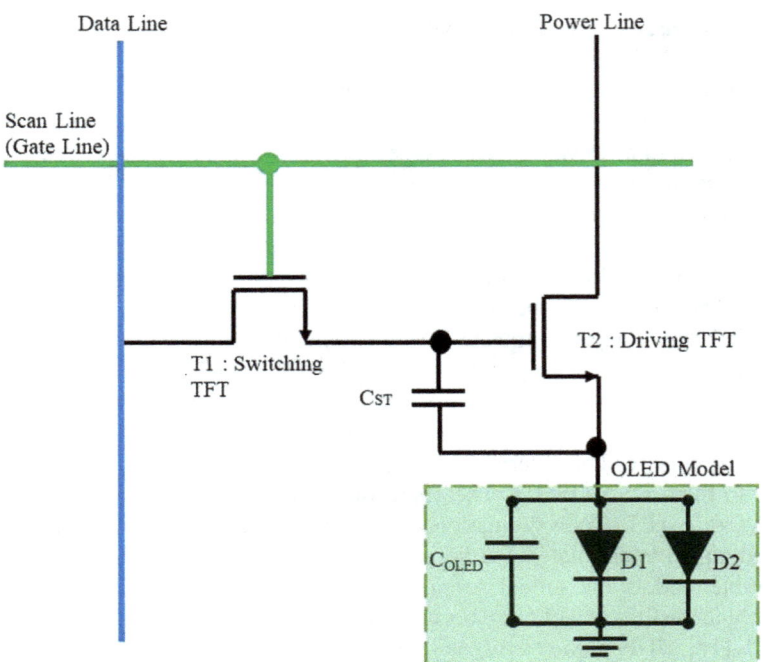

Fig. 3 Circuit diagram of a 2T1C pixel driver circuitry where OLED is shown using two-diode model

Figure 3 shows a typical 2T1C (2 TFT and 1 Capacitor) circuit used to drive an OLED. TFT T2 is used as a switching transistor where the scan line controls the switching action. TFT T1 is a driving transistor that provides the current to an OLED. Here, we have used a two-diode equivalent model to model OLED in SPICE. When TFT T1 is on, it is biased in the saturation region. In the saturation region, the current is a square function of overdrive voltage ($V_{ov} = V_{GS} - V_T$). If threshold voltage V_T is constant, a constant current can be provided to the OLED, making the brightness or luminescence consistent. However, in practice, the TFTs suffer from threshold voltage instability and threshold voltage shift over a prolonged gate bias stress. There could be various reasons for the threshold voltage instability, including interface trap charge, dielectric trap charges, grain boundary trap charges at the polycrystalline semiconductor, and dielectric interface. Charge trapping dominates at higher gate bias stress, while at lower gate bias stress, charge state creation seems to be overpowering. In general, NBTS or PBTS (negative or positive bias temperature stress) tests are carried out to investigate electrical stability. For an a-IGZO TFT, a shift in threshold voltage under gate bias stress can be approximated by a stretched exponential equation [16], as given below.

$$\Delta V_T = \Delta V_{T\infty}\{1 - \exp[-(t/\tau)^{\beta}]\} \tag{1}$$

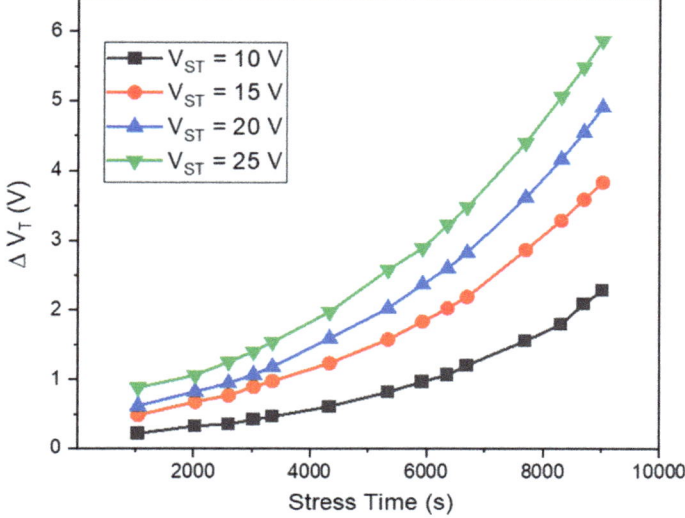

Fig. 4 Bias stress time vs. threshold voltage shift in a-IGZO TFT

where $\Delta V_{T\infty}$ is a shift in threshold voltage at infinite time, τ is the characteristic trapping time of the carriers, and β is an exponent. Figure 4 shows the dependency of threshold voltage over bias stress time for different magnitude of gate bias voltages (V_{ST}) as obtained from the stretched exponential equation given above using characteristic trapping time τ of 2×10^4 s and a stretched exponential exponent β of 0.42 [16]. As the magnitude of gate bias voltages is increased, the shift in threshold voltage also increases.

From the discussion so far, it is clear that the TFT technology's major problem is the threshold voltage shift over time. As in TFT, current in the saturation region is square of overdrive voltage; therefore, a change in threshold voltage directly impacts OLED current. The 2T1C circuit is shown in Fig. 3, as no threshold voltage compensation is involved; therefore, the OLED current directly depends upon threshold voltage. Any increase in threshold voltage results in a decrease in OLED current. Figure 5 shows the dependency of OLED current of a 2T1C circuit with stress time. As gate bias stress time increases, threshold voltage increases, thereby decreasing OLED current (Fig. 4).

2 Concept of Voltage Program Technique

A shift in TFT's threshold voltage poses significant reliability and lifetime concern in a display's lifetime. Moreover, an increase in threshold voltage also results in a degradation in pixel brightness. Various solutions have been proposed to address the

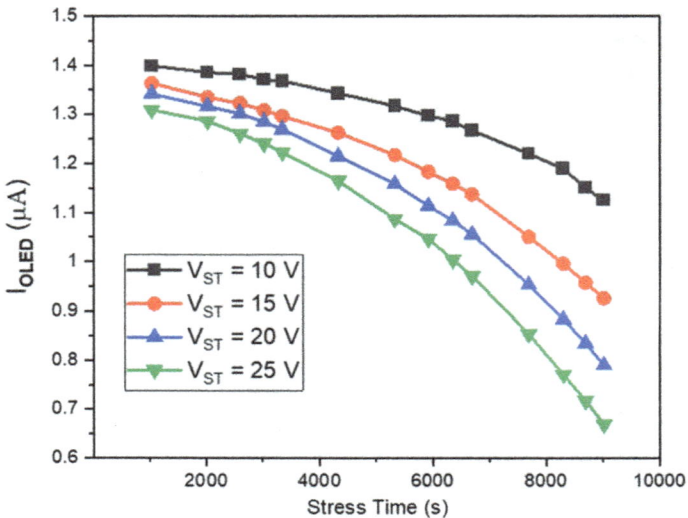

Fig. 5 Stress time versus OLED current in conventional 2T1C pixel driving circuit

issue of the threshold voltage shift. Different materials for TFTs are evolving at the device level, and their interface is being investigated with various gate dielectrics for reliable electrical performance. At the circuit level, given the TFT and its threshold voltage shift over time, a circuit designer attempts to devise the topologies where OLED current can be made independent of threshold voltage variation. Two primary compensation schemes are current-programming and voltage-programming-based pixel circuit design. Both approaches are widely used, but the voltage-programming-based approach has a faster settling time [17]. This chapter shall focus on the voltage-programming technique only.

In the voltage-programming technique, the idea is before the OLED emission phase, the gate-to-source voltage (V_{GS}) of a driving TFT must include the programming voltage and the threshold voltage of the driving TFT [18, 19]. This ensures that a TFT overdrive voltage is independent of threshold voltage in the emission phase, making OLED current independent of threshold voltage (V_T).

To achieve voltage-programmed compensation, this scheme can be divided into four phases.

(a) **Precharge Phase**: All of the voltage-programmed circuits use at least one storage capacitor, and its value is charged to a desired precharge value in this phase.

(b) **Compensation phase**: In the compensation phase, the storage capacitor C_{ST} is either discharged using a diode-connected TFT T1 to V_T as shown in Fig. 6 or charged to a value ($V_{reference} - V_T$) as shall be discussed in example 2. At the end of the compensation phase, the capacitor C_{ST} stores the threshold voltage of the diode-connected transistor (T1) and cannot be discharged/charged further

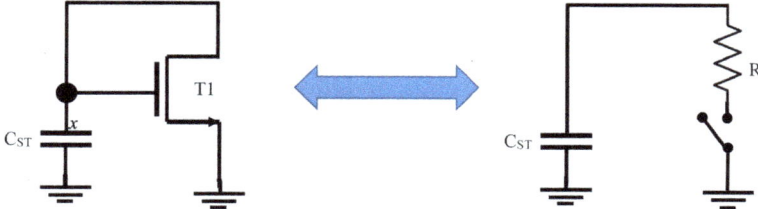

Fig. 6 Schematic of a diode-connected TFT and its equivalent RC delay model

as T1 goes in cutoff mode. However, in a realistic scenario, the finally stored value can further decay due to subthreshold conduction.

If we assume the V_T detection using C_{ST} discharging scenario, then the capacitor C_{ST} must be charged to V_P voltage during the precharge phase. In the compensation phase, its final value must be V_T. To ensure this, we need to discharge the capacitor C_{ST} to V_T. The net time taken for this discharge is known as compensation time. One can use the RC delay model to estimate compensation time (t_{comp}). If we assume the C_{ST} is the net value of capacitance appearing at node 'x', then the equivalent RC diagram is shown in Fig. 6, where R is the resistance of unity sized ($W/L = 1$) TFT. The capacitor C_{ST} shall discharge as

$$V_x = V_P e^{(-t/RC)} \tag{2}$$

The TFT T1 shall be off when V_x reaches V_T. Therefore, the compensation time (t_{comp}) can be given as:

$$t_{comp} = RC \ln\left(\frac{V_P}{V_T}\right) \tag{3}$$

As a designer, we need to make sure the compensation phase lasts for at least t_{comp} time. However, keeping higher than the required value of t_{comp} can degrade the voltage at V_x due to subthreshold leakage through transistor T1. This can similarly be analyzed when C_{ST} is charged through a diode-connected TFT, as discussed in example 2.

(c) **Data input phase**: In the data input phase, an additional data voltage V_{DATA} is added to the voltage across C_{ST}, which effectively makes the gate voltage of TFT T1 equal to ($V_T + V_{DATA}$). One can fix the source voltage of TFT to the desired value.

(d) **Emission Phase**: Once the required value of V_{GS}, which is equal to some constant value (V_{DATA}) plus threshold voltage (V_T), is obtained, we are ready, and the necessary setup can ensure OLED is ON only at this phase. In other phases, OLED must be turned off. In the emission phase, the OLED current I_{OLED}, therefore, can be given using the following equation

$$I_{\text{OLED}} = \mu_n C_{\text{ox}} \frac{W}{2L} (V_{\text{GS}} - V_{\text{T}})^2 \tag{4}$$

$$I_{\text{OLED}} = \mu_n C_{\text{ox}} \frac{W}{2L} (V_{\text{DATA}})^2 \tag{5}$$

Therefore, in the voltage-programmed technique, the idea is simple: to sample the threshold voltage and make it a part of V_{GS}, ensuring ($V_{\text{GS}} - V_{\text{T}}$) independent of V_{T}, making OLED current constant.

3 Example of Voltage-Programmed Techniques

To simulate and analyze the circuit performance explained earlier, one can use either a Verilog-A model or an RPI model or adapt any existing SPICE model after the calibration with experimental results. This chapter uses the adapted and calibrated SPICE Level-3 model for the a-IGZO TFT [12] and two-diode model for describing examples based upon a voltage-programmed technique for OLED [20].

Example 1 A five TFT, one capacitor (5T1C) voltage-programmed V_{T} compensation circuit and its timing waveforms are shown in Fig. 7. The TFT T2 acts as the driving

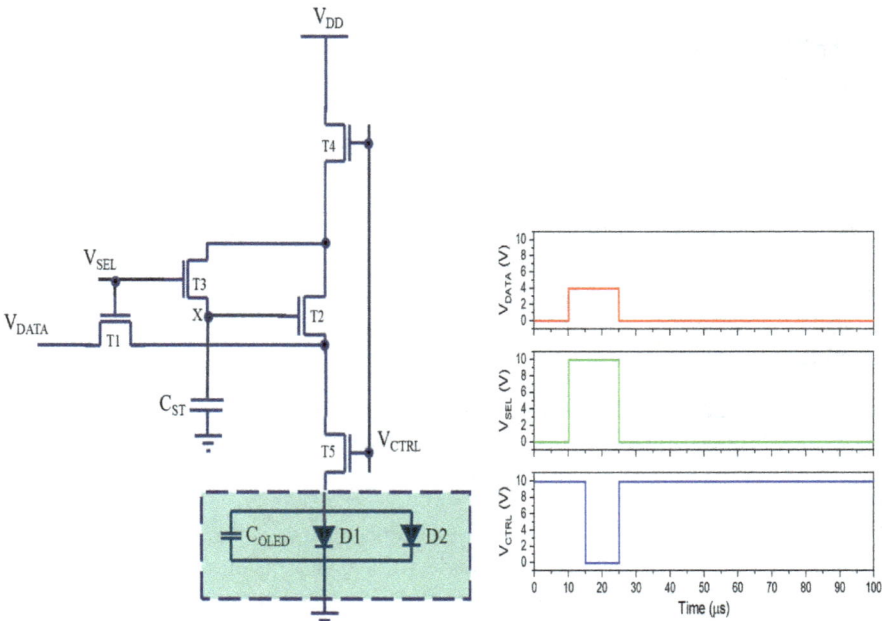

Fig. 7 Schematic of a 5T1C voltage-programmed pixel circuit with its timing waveform[21]

TFT, whereas the other TFTs, i.e., T1, T3, T4, and T5, serve as the switching TFTs. The functioning of the circuit is divided into three phases, which are described as follows:

(a) Precharge phase

During this phase, all the TFTs are kept on by making V_{SEL} and V_{CTRL} signals high. This phase acts as the reset phase, and no matter what the previously stored voltage on capacitor C_{ST} was, it would be charged to a precharge voltage V_P.

(b) Data input phase

In this phase, the V_{CTRL} signal is kept low, thereby turning off T4 and T5. V_{SEL} is kept high, and data input voltage V_{DATA} is supplied through the T1, working in a deep triode region, to the source of the driving TFT T2. The TFT T3 is in a deep triode region, making driving TFT T2 in a diode connection. Because of this diode connection capacitor C_{ST} discharges until it reaches the value of $(V_{T2} + V_{DATA})$ by the end of this phase.

(c) Emission phase

V_{SEL} is kept low during this phase, and V_{CTRL} signal is high, thereby turning on T4 and T5. The OLED is driven by the driving TFT T2 whose gate voltage is the same as the voltage across the capacitor C_{ST}, i.e., $(V_{T2} + V_{DATA})$. Thus, the current through the OLED is given as

$$I_{OLED} = \mu_n C_{ox} \frac{W_2}{2L_2} (V_{GS2} - V_{T2})^2 \tag{6}$$

$$I_{OLED} = \mu_n C_{ox} \frac{W_2}{2L_2} (V_{DATA} + V_{T2} - V_{T2})^2 \tag{7}$$

$$I_{OLED} = \mu_n C_{ox} \frac{W_2}{2L_2} (V_{DATA})^2 \tag{8}$$

Thus, the OLED current comes out to be independent of the threshold voltage V_{T2} of the driving TFT T2. Figure 8 shows the OLED current's square characteristics with the V_{DATA} independent of threshold voltage shifts (ΔV_{TH}) as derived in Eq. (8).

The compensation can be more clearly understood through the transient simulation of the pixel circuit. Figure 9 shows the transient analysis of voltage across the capacitor C_{ST}, i.e., V_X. It can be seen from the graph that V_T is successfully compensated and is captured on node X as $(V_{T2} + V_{DATA})$. During the compensation, capacitor C_{ST} discharges through the driving TFT T2, which turns off when its gate-to-source voltage V_{GS2} becomes equal to its threshold voltage V_{T2}, causing the capacitor to discharge to a voltage $(V_{T2} + V_{DATA})$. The compensation time (t_{comp}) of the circuit can be thus be given by the RC delay model as

$$t_{comp} = RC_{ST} \ln\left(\frac{V_P}{V_{DATA} + V_{T2}}\right) \tag{9}$$

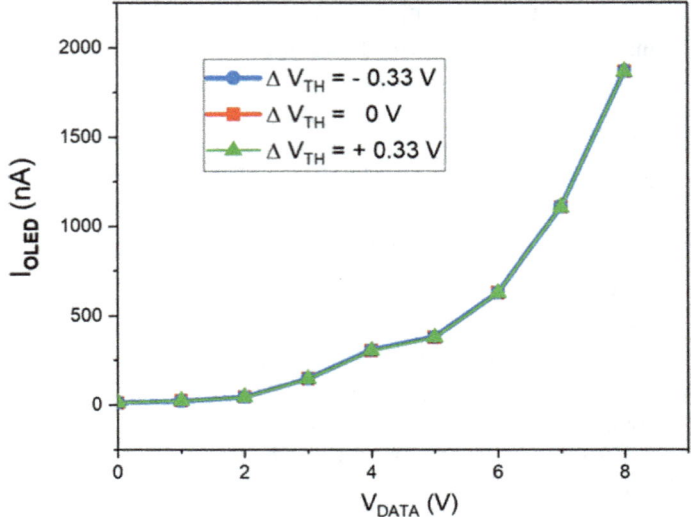

Fig. 8 OLED current as a function of V_{DATA} during the emission phase for threshold voltage shifts of -0.33 V, 0 V, and 0.33 V

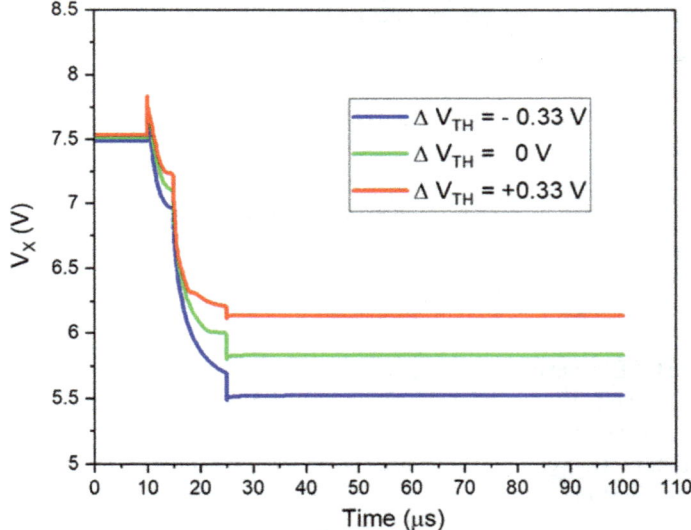

Fig. 9 Variation of voltage at node X with time for threshold voltage shifts of -0.33 V, 0 V, and 0.33 V

where V_P is the initial voltage on the capacitor C_{ST}, and R is the unit-sized TFT T2 resistance. The performance of the circuit depends much on the compensation time. In practicality, the voltage reached by the capacitor during the compensation is always accompanied by some error and is not exactly equal to the voltage (V_{T2} + V_{DATA}). This error in voltage across the capacitor is determined by writing the discharging current equation of the capacitor during the compensation, which will be

$$C_{ST}\frac{dV_X}{dt} + \mu_n C_{ox}\frac{W_2}{2L_2}(V_X - V_{DATA} - V_{T2})^2 = 0 \tag{10}$$

Solving the above, we get the value of node voltage V_X as

$$V_X = (V_{DATA} + V_{T2}) + \frac{(V_P - V_{DATA} - V_{Ts})}{\frac{\mu_n C_{ox} W_2 t_{comp}(V_P - V_{DATA} - V_{T2})}{2L_2 C_{ST}} + 1} \tag{11}$$

As obtained in Eq. (11), the voltage V_X consists of some additional terms other than (V_{T2} + V_{DATA}), which is the erroneous part. It can be seen from the equation that the error in voltage at node X is inversely proportional to the compensation time.

Figure 10 shows the voltage at the anode of the OLED, which comes out to be independent of V_T during the emission phase and thus provides a constant stable current through the OLED.

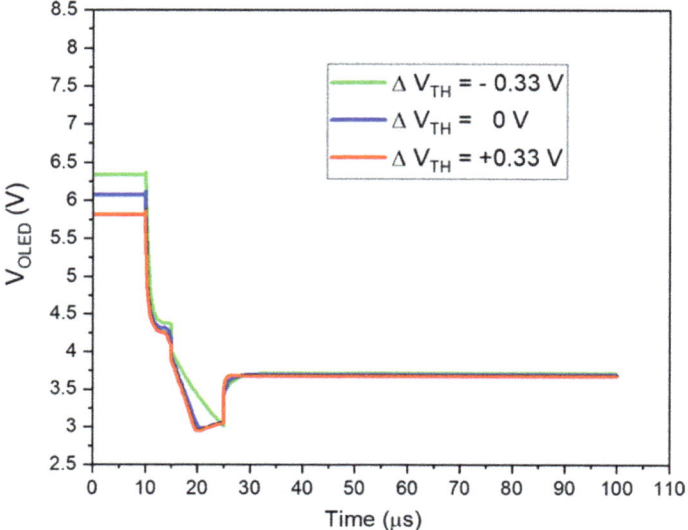

Fig. 10 Variation of voltage at the anode of the OLED with time for threshold voltage shifts of − 0.33 V, 0 V, and 0.33 V

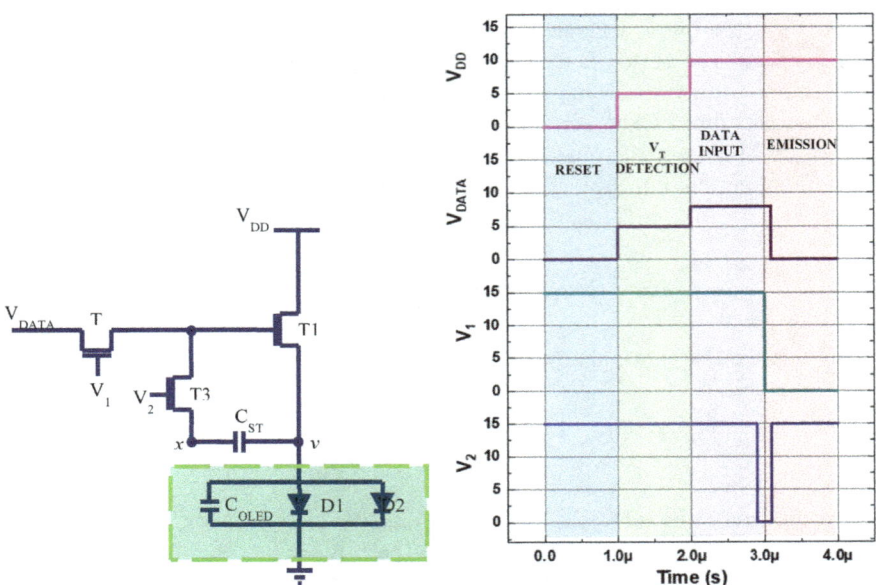

Fig. 11 Schematic of a 3 T-1C voltage-programmed pixel circuit with programming voltages V_1, V_2, V_{DATA}, and V_{DD} [22]

Example 2 Figure 11 shows a three TFT, one capacitor (3T1C) circuit that performs V_T compensation along with the associated signal waveforms. This circuit captures in a very simplistic manner the core concepts involved in voltage programming. This circuit's functioning can be understood from two aspects—firstly, the core principles of operation and secondly, the error introductions that prevent ideal compensation from happening.

(a) **Core operating principles**:

It can be observed that during the emission phase, T2 is turned off (T3 is on). This means C_{ST} holds the V_{GS} for the driving TFT T1 during the emission phase. To ensure V_T compensation (refer Eq. 2), voltage on C_{ST} should be of the form $[\alpha V_{DATA} + \beta] + V_T$ in the emission phase (α, β are arbitrary constant which depends only on the circuit design). C_{ST} is charged to the desired voltage with the help of the V_T detection phase and data input phase.

During the V_T detection phase, T2 is in a deep triode region, and V_{DATA} and V_{DD} each are at 5 V. Therefore, both gate and drain of T1 are at 5 V ensuring T1 to be in saturation, thereby charging node 'y' to the value (5 V $-$ V_T) during the V_T detection phase. Further, during the data input phase, the voltage at V_{DATA} is changed to the desired data input voltage. As a result, a component of data input voltage equal to $[C_{ST}/(C_{ST} + C_{OLED})] * V_{DATA}$ is added to node y due to the capacitive divider between C_{ST} and C_{OLED}. Hence, the final voltage at node 'y' becomes voltage at the end of V_T detection phase plus voltage added during data input phase, i.e., $[(5 - V_T)$

$+ \{C_{ST}/(C_{ST} + C_{OLED})\} * V_{DATA}]$. Also, as T2 and T3 are in the deep triode region, the voltage at node 'x' is V_{DATA}. Therefore, the voltage across C_{ST} at the end of the data input phase becomes ($V_X - V_Y$) equal to [$\{C_{OLED}/(C_{ST} + C_{OLED})\} * V_{DATA} + (-5) + V_T$] as desired to achieve compensation.

Figure 12 shows the simulated transient waveform for the V_T detection performed by the 3T1C circuit. It can be seen that by the emission phase, C_{ST} is charged to a value $f(V_{DATA}) + V_T$, ensuring ($V_{GS} - V_T$) is independent of the threshold of the driving TFT.

(b)   **Error introduced during V_T detection and data input phase**:

The core operating principle leads us to believe that V_T's compensation is perfect and no dependence on V_T is observed in I_{OLED} during the emission phase. However, this is far from real, and Fig. 13 helps highlight the errors introduced during the V_T detection and data input phase using simple RC abstractions.

The charging of node 'y' during the V_T detection phase can be modeled as shown in Fig. 9a. Although R_{OLED} can be easily neglected as the voltage at node 'y' is smaller than the OLED cut-in voltage, the charging of node 'y' follows an exponential (5 − V_T)*(1 − e$^{-t/\tau}$), requiring infinite time to settle at the steady-state value. Thus, the finite duration of the V_T detection phase does not allow for the accurate detection of the threshold voltage of the driving TFT T1, and some error is introduced.

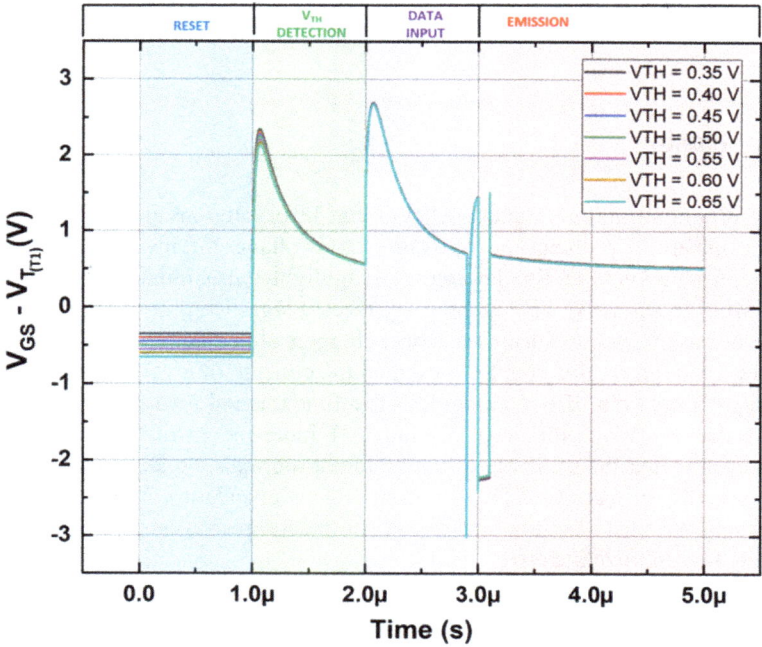

Fig. 12 Simulation result for the V_T compensation performed by the 3T1C circuit [22]

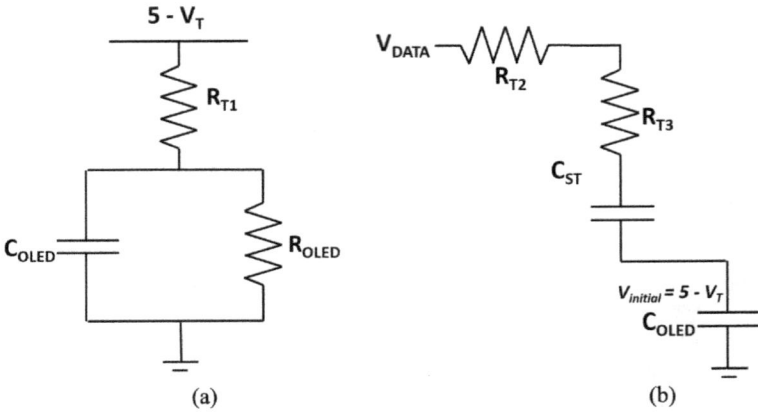

Fig. 13 RC models for error identification in **a** V_T detection phase and **b** data input phase

Further, the error introduced during the data input phase can be understood from Fig. 13b. It is assumed that the system's response is simply (forced response + initial conditions at C_{OLED}). However, in reality, the initial conditions will result in a natural response that will degrade the detected threshold voltage at node 'y'.

This example highlights in a simplified way the application of the core principles of voltage programming to the actual design of pixel circuits as well as highlights some key challenges that prevent ideal compensation from happening.

4 Conclusion

Voltage-programming techniques are key circuit-level solutions to compensate TFTs' electrical instability problem caused by threshold voltage shift in displays. We started with an introduction to TFT technology and highlighted its similarity and differences with CMOS technology. TFT technology faces a significant problem of threshold voltage instability due to various reasons: charge state creation, interface trapping, and aging. Therefore, this chapter presented the concept of a voltage-programming scheme with the help of two examples. The fundamental idea in this approach is to make the overdrive voltage of driving TFT independent of VT. The examples in Sect. 3 reflected the principle of the voltage-programming technique. Though it is not possible to make OLED current strictly independent of V_T due to various implicit reasons such as subthreshold conduction, compensation time can be chosen optimally to minimize the error.

References

1. Bagheri M, Cheng X, Zhang J, Lee S, Ashtiani S, Nathan A (2016) Threshold voltage compensation error in voltage programmed AMOLED displays. J Disp Technol 12:658–664
2. Chen B-T, Tai Y-H, Kuo Y-J, Tsai C-C, Cheng H-C (2006) New pixel circuits for driving active matrix organic light emitting diodes. Solid State Electron 50:272–275
3. Chen C, Kanicki J, Abe K, Kumomi H (2009) P-14: AM-OLED pixel circuits based on a-InGaZnOThin film transistors. In: SID symposium digest of technical papers. Wiley Online Library, pp 1128–1131
4. Cheng X, Lee S, Yao G, Nathan A (2016) TFT Compact modeling. J Disp Technol 12:898–906. https://doi.org/10.1109/JDT.2016.2556980
5. Fortunato EMC, Barquinha PMC, Pimentel AC, Gonçalves AMF, Marques AJS, Martins RFP, Pereira LMN (2004) Wide-bandgap high-mobility ZnO thin-film transistors produced at room temperature. Appl Phys Lett 85:2541–2543
6. Griffin PB, Plummer JD, Deal MD (2000) Silicon VLSI technology: fundamentals, practice, and modeling. 1a edição. Prentice Hall Inc
7. Hossain FM, Nishii J, Takagi S, Sugihara T, Ohtomo A, Fukumura T, Koinuma H, Ohno H, Kawasaki M (2004) Modeling of grain boundary barrier modulation in ZnO invisible thin film transistors. Phys E Low-dimensional Syst Nanostruct 21:911–915
8. Jiang A, Yuan Y, Liu N, Han L, Xiong M, Sheng Y, Ye Z, Liu Y (2019) Transparent capacitive-type fingerprint sensing based on zinc oxide thin-film transistors. IEEE Electron Device Lett 40:403–406
9. Kandpal K, Gupta N (2018) Perspective of zinc oxide based thin film transistors: a comprehensive review. Microelectron Int 35:52–63. https://doi.org/10.1108/MI-10-2016-0066
10. Kandpal K, Gupta N (2019) Adaptation of a compact SPICE level 3 model for oxide thin—film transistors. J Comput Electron 18:1037–1044. https://doi.org/10.1007/s10825-019-01344-0
11. Kandpal K, Gupta N, Singh J, Shekhar C (2020) On the threshold voltage and performance of ZnO-based thin-film transistors with a ZrO_2 gate dielectric. J Electron Mater 1–9
12. Kwon J-Y, Lee D-J, Kim K-B (2011) Review paper: transparent amorphous oxide semiconductor thin film transistor. Electron Mater Lett 7:1–11. https://doi.org/10.1007/s13391-011-0301-x
13. Lee JM, Cho IT, Lee JH, Kwon HI (2008) Bias-stress-induced stretched-exponential time dependence of threshold voltage shift in InGaZnO thin film transistors. Appl Phys Lett 93. https://doi.org/10.1063/1.2977865
14. Lee S, Jeon S, Chaji R, Nathan A (2015) Transparent semiconducting oxide technology for touch free interactive flexible displays. Proc IEEE 103:644–664
15. Nomura K, Ohta H, Takagi A, Kamiya T, Hirano M, Hosono H (2004) Room-temperature fabrication of transparent flexible thin-film transistors using amorphous oxide semiconductors. Nature 432:488–492
16. Nozawa T (2007) Transparent electronic products soon a reality. Nikkei Electron Asia 1024–1030
17. Perumal C, Ishida K, Shabanpour R, Boroujeni BK, Petti L, Munzenrieder NS, Salvatore GA, Carta C, Troster G, Ellinger F (2013) A compact a-IGZO TFT model based on MOSFET SPICE Level=3 template for analog/RF circuit designs. IEEE Electron Device Lett 34:1391–1393. https://doi.org/10.1109/LED.2013.2279940
18. Singh A, Goswami M, Kandpal K (2020) Design of a voltage-programmed V_{TH} compensating pixel circuit for AMOLED displays using diode-connected a-IGZO TFT. IET Circ Devices Sys, Sep 29, 14(6):876–880
19. Singh A, Kandpal K (2020) Design of a threshold voltage insensitive 3t1c pixel circuit using a-IGZO TFT for AMOLED displays. In: 2020 24th international symposium on VLSI design and test (VDAT). IEEE, pp 1–5
20. Sodhani A, Kandpal K (2020) Design of threshold voltage insensitive pixel driver circuitry using a-IGZO TFT for AMOLED displays. Microelectronics J 104819

21. Torricelli F, Meijboom JR, Smits E, Tripathi AK, Ferroni M, Federici S, Gelinck GH, Colalongo L, Kovacs-Vajna ZM, de Leeuw D (2011) Transport physics and device modeling of zinc oxide thin-film transistors part I: long-channel devices. IEEE Trans Electron Devices 58:2610–2619

22. Zeumault A, Subramanian V (2015) Anomalous process temperature scaling behavior of sol-gel ZrOxgate dielectrics: Mobility enhancement in ZnO TFTs. In: Device research conference—Conf. Dig. DRC 2015-August, pp 203–204. https://doi.org/10.1109/DRC.2015.7175634

Reprogrammable Optical Logic Circuit Using Opto-Electro-Mechanical Device

Arighna Deb

Abstract The advances in silicon photonics allow the applications of optical devices in on-chip functional computations. Several works on the development of synthesis and design methods for optical logic circuits have been performed, where, the circuits are composed of optical devices such as semiconductor optical amplifier based Mach–Zehnder Interferometer, electrically controlled Mach–Zehnder Interferometer. These optical devices may not allow compact and efficient realizations of large-scale optical circuits. Besides that, all the existing approaches suffer from unsatisfactory realizations and restricted flexibility. In fact, large number of splitting of optical signals, waveguide crossing and bending are the reasons for which the outputs of the optical circuits degrade. To solve these issues, one solution is an optical circuit structure which can be reprogrammable to realize arbitrary Boolean functions, while at the same time, making the physical realizations practical. To this end, one can utilize a recently developed optical device based on opto-electro-mechanical effect which enables smaller footprint, low optical loss and rapid switching. Using this device, a reprogrammable optical circuit structure can be designed which is a $(2^n \times 1)$ optical multiplexer composed of several (2×1) optical multiplexers. Experimental evaluations confirm the efficacy of the proposed approach over an existing approach.

Keywords Optical logic · Opto-electro-mechanical device · MZI · Optical multiplexer

1 Introduction

Deep learning [1] has received significant attention from the researchers across the globe because of its applications in different areas such as image processing [2, 3], video processing [3], natural language processing [4]. Various deep learning networks [5–8] perform massive amount of multiplication and addition operations (MAC) leading to the requirements of large memory size, high throughput.

A. Deb (✉)
School of Electronics Engineering, KIIT Deemed To Be University, Bhubaneswar, India

© The Author(s), under exclusive license to Springer Nature Singapore Pte Ltd. 2022
R. Goswami and R. Saha (eds.), *Contemporary Trends in Semiconductor Devices*,
Lecture Notes in Electrical Engineering 850,
https://doi.org/10.1007/978-981-16-9124-9_13

Graphic processing units (GPUs) [9], Tensor processing units (TPUs) [10] are used as hardware to implement the multiplication and addition operations of the deep learning networks. Very fast processing speed, smaller footprint are the major requirements in order to provide the advantages of deep learning networks in mobile and IoT devices/applications, which however, may not be possible to provide with the state-of-the-art GPUs or TPUs.

Although the miniaturization of devices i.e., downscaling the size of the transistors results in smaller footprint, however, achieving ultra-high speed to perform millions of multiplication and addition operations of deep learning networks in real-time is the current bottleneck of the conventional transistor technology. Several architectural designs have been proposed [11–14] to accelerate the computations of multiplication and addition operations, nevertheless, the processing speed still remains an issue.

To achieve extremely high speed of processing, researchers have proposed photonic-based solution, an alternative to the current transistor based technology. Thanks to the advances in silicon photonics [15, 16], optical devices and circuits can now be fabricated on silicon wafers. As a result, optical technology is considered to be a potential candidate for the development of ultra-fast circuits realizing deep learning networks. Motivated by this, optical logic circuits realizing important arithmetic operations, for e.g., addition, multiplication are proposed in the literature [17–20].

While the design approaches introduced in [17–20] allow the optical realizations of only fixed Boolean functions i.e., addition, multiplication, several alternative approaches are proposed which can generate optical logic circuits realizing arbitrary Boolean functions [21–25]. Typically, the existing logic design and synthesis approaches for optical circuits use either Semiconductor Optical Amplifier based Mach–Zehnder Interferometer (SOA-MZI) device or electrically controlled Mach–Zehnder Interferometer device [26, 27], leading to the all-optical or electro-optic realizations of the complex Boolean functions. The utilizations of SOA-MZI, electrically controlled MZI devices as basic building blocks of the optical circuits fail to satisfy all the primary requirements (such as compact footprint, low driving voltage, low optical power loss, short switching time) necessary to realize the physical optical circuits [28].

Opto-electro-mechanical (OEM) devices satisfy the above primary requirements since the devices consume negligible power and area, require small driving voltage and short switching time [28]. As a result, OEM devices provide an alternative to the SOA-MZI devices and electrically controlled MZI devices. Such advantages make the physical realizations of the optical circuits composed of OEM devices more pragmatic. Since the logic synthesis or design is an initial step in the design flow, therefore, the optical logic design approaches are re-investigated in order to generate optical circuits consisting of OEM devices.

To this end, the state-of-the-art logic synthesis approaches [21–25] for optical circuits can easily be adapted to generate the optical circuits consisting of OEM devices. While this seems an obvious solution, but the resulting optical circuits may not be practically realizable. This is mainly because, the optical logic synthesis is treated like a conventional logic design approach [29].

In particular, the conventional approach focuses on the realizations of logic circuits with minimal number of logic gates. Minimization of the number of logic gates, which can be achieved by sharing the sub-functions, leads to the reduction in overall circuit area. Sharing is typically realized by fan-out of the gate realizing the sub-functions.

Similarly, in optical logic design, the sub-functions sharing has been considered which reduces the number of optical gates; thereby reducing the circuit area [21–23]. However, the sub-function sharing requires the utilizations of splitters which reduce the overall optical signal power (i.e., result in optical output of poor signal strength). To overcome this, alternative synthesis approaches are proposed which allow the redundant realizations of the shared sub-functions [24, 25]. Consequently, the number of optical logic elements increases in those optical circuits (obtained from the approaches introduced in [24, 25]) which ultimately increases the overall circuit area.

Moreover, the existing optical design approaches have only considered the gate count and splitting effect as the metrics to evaluate the complexity and quality of the generated circuits, while have ignored the other metrics such as waveguide crossing [30, 31], waveguide bending [32, 33]. Since waveguide crossing and bending result in the optical loss, optical circuits are required to be evaluated in terms of these metrics.

In addition to that, the existing approaches generate optical circuits which are specific to the given Boolean functions. That is, given a Boolean function, any existing approach generates an optical circuit realizing the corresponding function only. This makes the generated circuits function-specific. In other words, the circuit structure changes with the considered function.

In the domain of optical computing, such design approach may not be suitable as the generated optical circuits suffer from high optical losses due to the splitting, waveguide crossing, waveguide bending. Because of this, an alternative design scheme is necessary which can overcome the shortcomings of the existing design approaches. For this purpose, an alternative design scheme is proposed which generates a reprogrammable optical circuit structure.

More specifically, this chapter introduces an optical circuit structure composed of an array of (2×1) optical multiplexers. Herein, each (2×1) optical multiplexer is realized by OEM gates and beam combiners making the proposed structure physically realizable. The proposed optical circuit structure is reprogrammable which allows the realizations of any Boolean functions without making any changes to the structure. Experimental evaluations confirm the effectiveness of the proposed approach over a similar such approach composed of SOA-MZI devices [34].

In the following, the proposed approach is introduced as follows. Section 2 reviews the applied optical gate libraries and cost metrics. Section 3 discusses the motivation and the general idea behind the proposed approach, which afterward, is described in detail in Sect. 4. The method of realizing any Boolean functions using the proposed approach is discussed in Sect. 5. Finally, Sect. 6 summarizes the obtained experimental results and Sect. 7 concludes the chapter.

2 Optical Logic Models

This section provides a brief overview of optical devices, optical logic models, gate libraries and cost metrics that can be used to design optical logic circuits. To design digital circuits using optical logic, logic models, gate libraries and cost metrics are abstracted from detailed physical and optical issues but discretely reflect the major constraints.

2.1 MZI Gate Library

Figure 1 shows a schematic of an optical device which is constructed using two *Semiconductor Optical Amplifiers (SOAs)* and two couplers. The resulting all-optical devices is known as the *Semiconductor Optical Amplifier based Mach–Zehnder Interferometer* or simply, the *SOA-MZI* device. A coupler in this device acts as a passive optical component that either splits a signal or combines multiple signals, whichever is required. In particular, the coupler c-1 at the input side of the device splits an input signal p into two optical signals which then propagate through SOAs to the coupler c-2 at the output side of the device, where the signals are merged. A control signal q is applied at the second input port of the device to change the phase of the signals. The output ports of the device are called a *bar port* and a *cross port*.

In the logic domain, the resulting structure is abstracted to a so-called *MZI gate*. Each MZI gate has two input ports and two output ports. The inputs can either be sourced by light (representing binary 1) or darkness (representing binary 0). Logically, an MZI gate is defined as follows:

Definition 1 An MZI gate realizes a Boolean function $\mathbb{B}^2 \rightarrow \mathbb{B}^2$ consisting of two optical inputs p and q and two optical outputs f and g. When both the inputs signals are 1, the outputs f and g become 1 and 0, respectively. When the input signal p is set to logic 1 and another input signal q is set to logic 0, the outputs f and g become 0 and 1, respectively. For rest of the input combinations, both the outputs become logic 0. The input–output relation can be defined as the Boolean functions $f = pq$ and $g = p\bar{q}$.

A functional block representing an MZI gate is depicted in Fig. 2a.

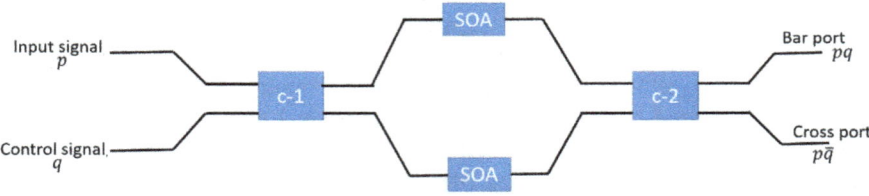

Fig. 1 SOA-MZI optical device

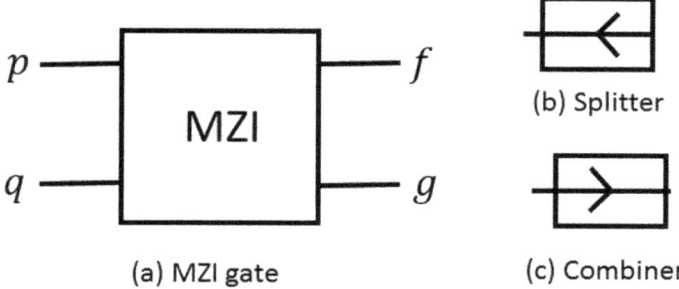

Fig. 2 MZI gate library

Definition 2 A splitter splits a single optical signal into two or more optical signals, whereas a combiner combines two or more optical signals into a single optical signal.

The basic operation of a splitter results in multiple output signals, where, each signal strength is reduced by a factor of the number of outputs. In case of a combiner, the basic operation inherently realizes the OR function.

Figure 2b, c show the graphical representations of splitter and combiner, respectively.

Using the MZI gate, splitter and combiner, an optical circuit library called *MZI gate library* is composed (as depicted in Fig. 2) that can be used to realize any Boolean function.

Example 1 Figure 3 shows an optical circuit based on the MZI gate library which realizes the Boolean function $F = x_0\overline{x_1} + \overline{x_0}x_2$.

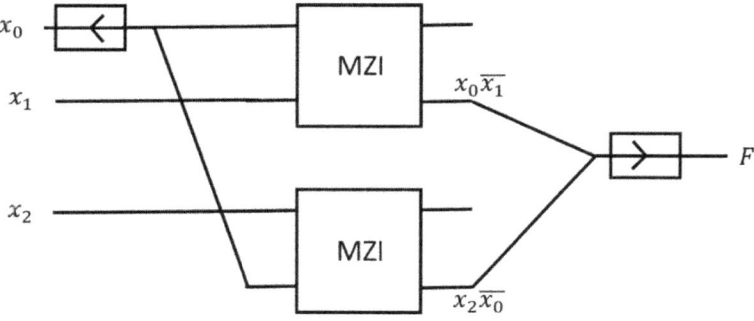

Fig. 3 An optical circuit using MZI gates

2.2 Crossbar Gate Library

Figure 4 shows an alternative optical device which employs an electrically controlled Mach–Zehnder Interferometer (MZI). Like Fig. 1, the device has two inputs and two outputs, where, the waveguides having refractive index of say, n connect inputs to the outputs. The device also uses two 3 dB couplers, C-1 and C-2. The coupler C-1 splits an input signal between the upper and lower waveguides. A phase shift of $\frac{\pi}{2}$ in the lower waveguide is introduced by the coupler C-1.

An external electrical input x is applied at the center of the upper waveguide which alters the refractive index of the upper waveguide by Δn. Such modification in refractive index causes a phase difference between the signals of upper and lower waveguides. At coupler C-2, the signals from upper and lower waveguides are combined together which will interfere constructively at one output and destructively at the other one. Because of this, the device produces output signals having a phase difference of 0 or π. Since the phase difference of the outputs is controlled by an electrical signal x, hence the device is called a *controlled MZI* device.

In the optical logic domain, the operation of the controlled MZI device can be defined as a *crossbar gate*. More formally,

Definition 3 A crossbar gate realizes a Boolean function $\mathbb{B}^3 \to \mathbb{B}^2$ when an electrical input x is set to $0(1)$ resulting in the propagations of two optical inputs p and q, to the two optical outputs $f(g)$ and $g(f)$, respectively.

Figure 5 depicts the functional block of a crossbar gate.

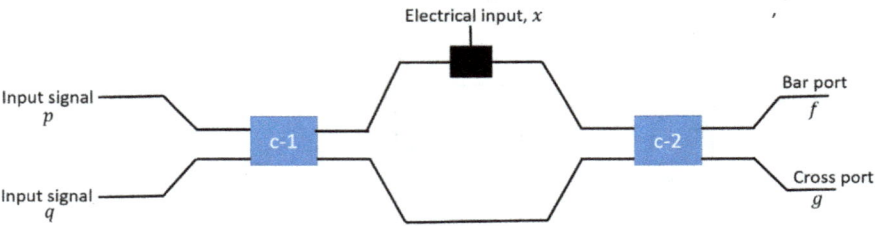

Fig. 4 Electrically controlled MZI optical device

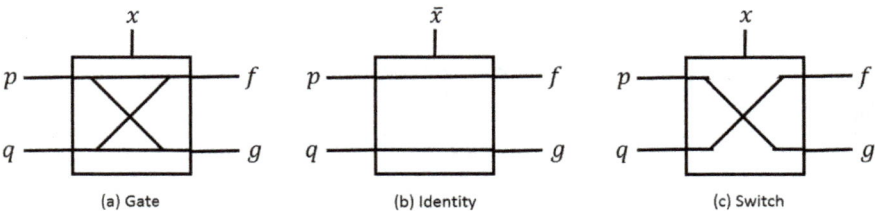

Fig. 5 Crossbar gate

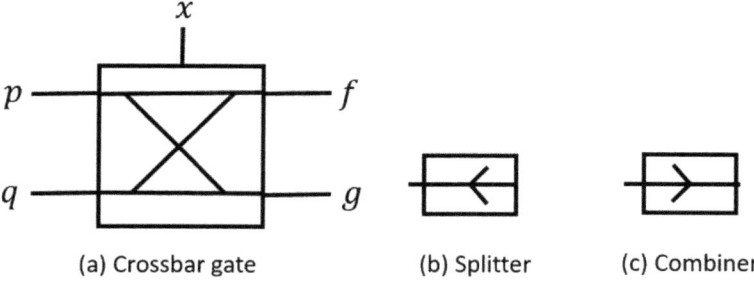

(a) Crossbar gate   (b) Splitter   (c) Combiner

Fig. 6 Crossbar gate library

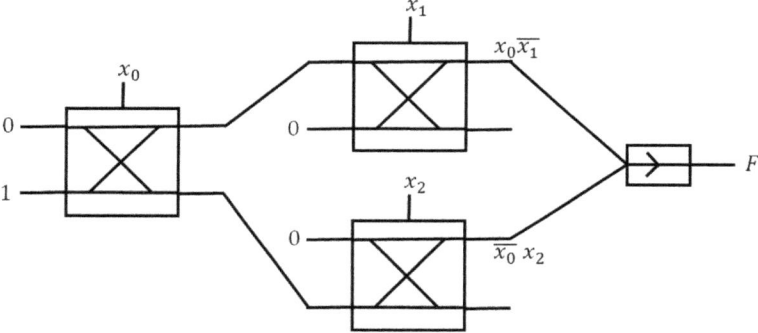

Fig. 7 Optical circuit consisting of crossbar gates and a combiner

A crossbar gate with a splitter and a combiner form a *crossbar gate library* as depicted in Fig. 6 which realizes any Boolean function as an optical circuit.

Example 2 An optical circuit consisting of crossbar gate library to realize the Boolean function $F = x_0\overline{x_1} + \overline{x_0}x_2$ is depicted in Fig. 7.

2.3 Opto-Electro-Mechanical (OEM) Gate Library

An alternative to all-optical and electro-optic devices, the opto-electro-mechanical (OEM) switch employs a thin gold disk which is kept suspended above a layer of silicon disk as shown in Fig. 8. The contacts are created over the gold disk and silicon layer, enabling one to apply a control voltage to the disks. Depending on the voltage applied, the gold disk moves up or down resulting in the propagation of light either in the through port or in the drop port. More specifically, a light wave is launched at the input end of the through port and the light propagates easily to the output end of the through port when $1V$ is applied as control voltage to the gold disk keeping

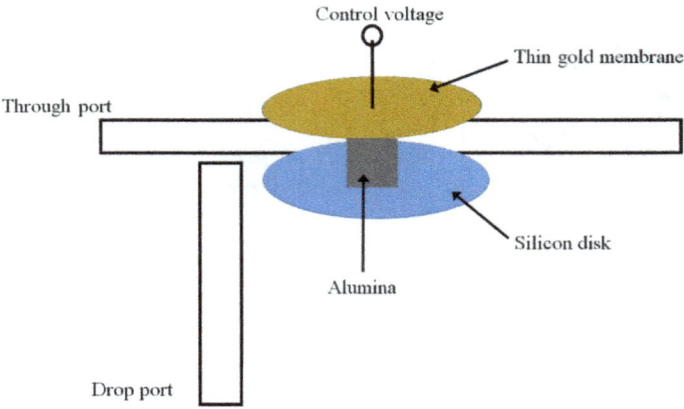

Fig. 8 Opto-electro-mechanical device [28]

the disk at the up position as shown in Fig. 9a. . However, when $0V$ is applied as control voltage, the gold disk moves down and the light leaks from the input side of the through port into the silicon disk. When another waveguide (drop port) is placed close to the silicon disk, the light moves from the silicon disk to the drop port as depicted in Fig. 9b. . This means, by varying the control voltage, the propagation of the light can be controlled or altered. Since the path of the light is controlled by applying a voltage and mechanically changing the position of the gold disk, hence the device is called an *opto-electro-mechanical (OEM) switch* [28].

In the logic domain, the operation of the OEM switch can be modeled as an optical logic gate, called an *OEM gate*, which has two inputs—one optical and one electrical and two optical output ports, *through and drop*. The optical input is routed between two output ports depending on the electrical input. The presence and absence of light in input and output ports are represented by binary 1 and binary 0, respectively. Logically, an OEM gate is defined as follows:

Definition 4 An OEM gate realizes a Boolean function $\mathbb{B}^2 \to \mathbb{B}^2$ composed of one optical input p, one electrical input x and two optical outputs t and d. The optical input p appears at output t when the electrical input x becomes logic 1. No light appears at the output d making the corresponding output logic 0. When the electrical input x is set to logic 0, the optical input p appears at output d and no input appears at output t. That is, $t \equiv p$ for x and $d \equiv p$ for $\bar{x}$ are realized.

Figure 10 provides the functional schematic of an OEM gate.

An OEM gate together with a beam combiner and a splitter form an *OEM gate library* as shown in Fig. 11 which can realize any Boolean function.

Example 3 Figure 12 shows the optical circuit realizing the Boolean function $F = x_0\overline{x_1} + \overline{x_0}x_2$, where, the circuit is designed using the OEM gate library.

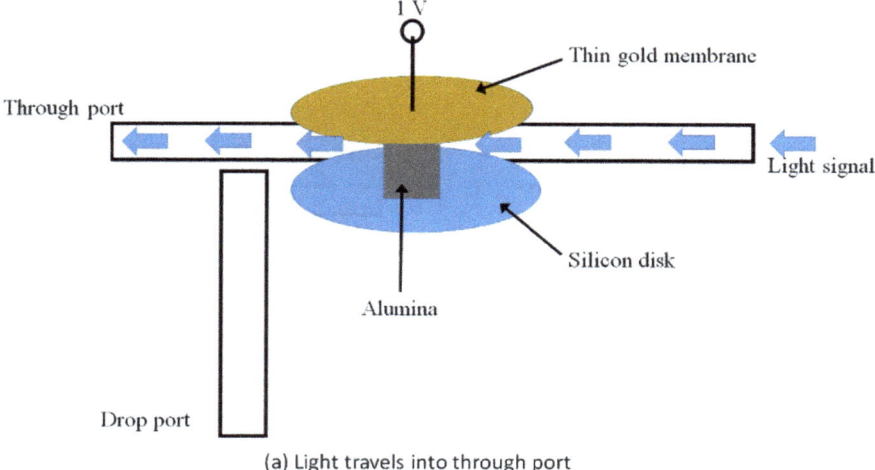

(a) Light travels into through port

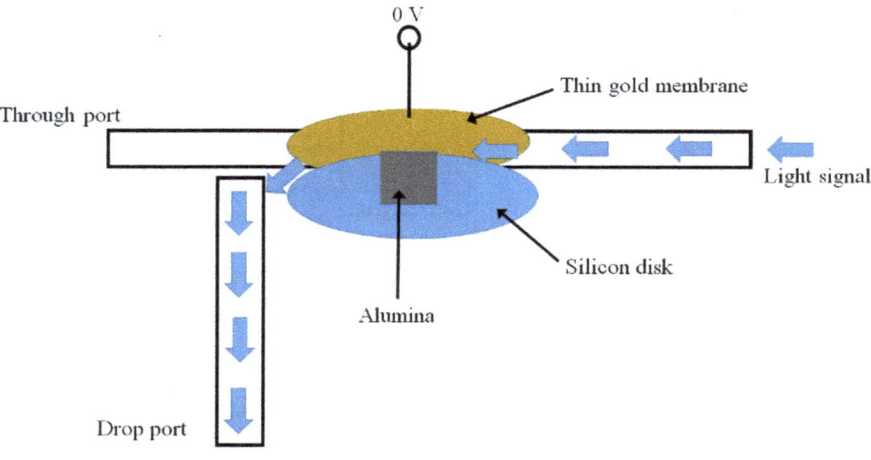

(b) Light travels into drop port

Fig. 9 Working principle of opto-electro-mechanical device [28]

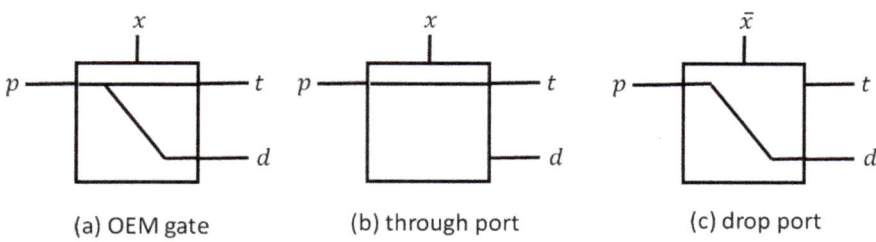

(a) OEM gate   (b) through port   (c) drop port

Fig. 10 Opto-electro-mechanical (OEM) gate

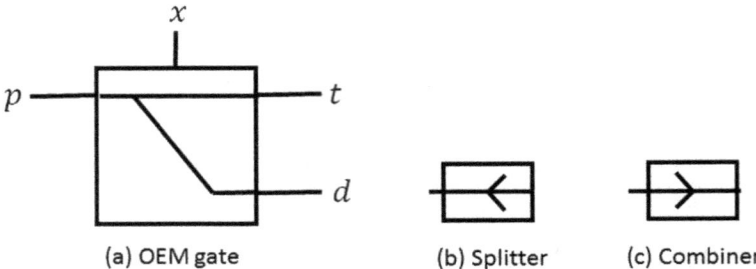

Fig. 11 OEM gate library

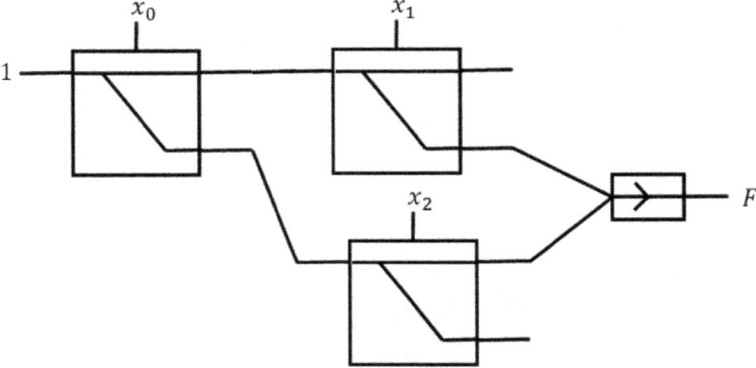

Fig. 12 Optical circuit consisting of OEM gates and a combiner

2.4 *Optical Circuit Complexity*

The optical circuit complexity is measured using several cost metrics. In particular, the following cost metrics are considered.

- The number of gates (or gate count): This metric is considered due to the fact that each gate (crossbar or MZI or OEM) will be physically realized. The number of splitters and the number of combiners are also considered. The total number of optical elements in any optical circuit is estimated as the sum of the number of optical gates, splitters and combiners.
- The worst case fraction: This metric is considered because splitters in the circuit decrease the strength of the optical signal. If each splitter with n outputs produces an output signal of strength $\frac{1}{n}$, then the worst case fraction of the optical circuit is determined by traversing all paths from primary inputs to the primary outputs and multiplying all signal strengths produced by the splitters visited on the current path. Finally, the minimum value of all paths is considered as the worst case fraction of the circuit [21, 25].

- Number of waveguide crossings: When two waveguides cross each other, a loss occurs resulting in the degradation of signal power in both the crossing waveguides. Each crossing causes a signal loss of the order of 0.1–0.2db [30, 31]. This motivates to count the number of waveguide crossings in the optical circuits since more the number of crossings, higher will be the signal loss. As a result, the number of waveguide crossings is considered a metric to evaluate the quality of any optical circuit.
- Number of waveguide bending: Bending a waveguide may result in signal loss in optical circuits [32, 33], which is why, the number of waveguide bends is estimated in any optical circuits. Since large the number of bends, higher will be the loss, because of which, number of waveguide bends is used as a metric to evaluate the quality of the optical circuit.

Although these metrics are just approximations that may not perfectly correspond to the physical realization, it provides a better idea of the resulting optical circuit complexity and the signal strength. Besides that, these metrics can easily be used to measure the quality and complexity of the optical circuits composed of any optical gate library (i.e., either MZI or crossbar or OEM gate library). For example, Fig. 3 shows an optical circuit based on the MZI gate library, in which, the total number of optical elements is 4, the worst case fraction is 1/2 (since the splitter splits the single input signal into two output signals), the number of waveguide crossings is 2 and the number of waveguide bending is 4. Similarly, the complexity of the optical circuit shown in Fig. 7 can also be estimated. Herein, the circuit consists of 3 crossbar gates, 1 combiner and the worst case fraction is 1 as there is no splitter in the circuit. Also, the circuit has 4 waveguide bending but zero waveguide crossings. All the metrics in the case of the optical circuit shown in Fig. 12 remain same as that of the circuit shown in Fig. 7 except the number of crossbar gates, in place of which, the number of OEM gates is to be considered.

3 Motivation

In this section, the state-of-the-art process of realizing optical logic circuits is briefly reviewed. Also, the physical constraints which are the potential obstacles in realizing large-scale optical circuits are discussed. This motivates the investigations toward alternative design approaches for optical logic circuits using the efficient optical logic gates.

3.1 Existing Optical Logic Synthesis Approaches

Logic synthesis for optical circuits helps to generate large-scale optical logic circuits for the Boolean functions. Thus far, several synthesis approaches for optical circuits have been introduced which synthesize arbitrary Boolean functions into optical circuits consisting of crossbar or MZI gates, combiners and splitters. These existing approaches utilize different Boolean function representations, for e.g., two-level descriptions, Binary Decision Diagrams (BDDs) [35], AND Inverter Graphs (AIGs) [36], OR Inverter Graphs (OIGs) [24], which are mapped to optical circuits realizing the desired Boolean functionalities.

More precisely, the synthesis methods for optical circuits considering two-level function representations are introduced in [25], where, the two most commonly used two-level representations of Boolean functions, Sum-of-Products (SoPs) and Exclusive-Sum-of-Products (ESoPs) have been utilized to generate optical circuits. Two different optical gate libraries i.e., a crossbar gate library and a MZI gate library were used to map the function descriptions into corresponding optical circuits. This approach generates optical circuits with large number of optical gates. To overcome this, synthesis methods using Binary Decision Diagrams (BDDs) are proposed in [21, 23]. The general idea of BDD-based synthesis approaches is: given a BDD of the Boolean function, the synthesis algorithm traverses the entire BDD and inserts optical gates for each node of the BDD into the circuit. A shared BDD node (i.e., a node with multiple predecessors) is realized by a splitter which is inserted at the output of the optical gates realizing that node. The gate count of the resulting optical circuits is smaller than the gate count of optical circuits generated from two-level descriptions, however, the optical circuits generated from BDDs contain large number of splitters which results in the outputs of poor signal quality.

Synthesis approach for optical circuits relying on *AND Inverter Graph* (AIG) representation of Boolean functions is proposed in [22]. Each node of an AIG represents a 2-input AND function which directly corresponds a single MZI gate in optical gate library. This one-to-one mapping of an AIG node to a MZI gate is utilized to develop the AIG-based synthesis approach for optical circuits. The experimental results reported in [22] show that the optical circuits generated from AIG-based scheme require less number of optical gates as compared to the optical circuits derived from the BDD-based and two-level description based approaches.

Shared product terms (in case of two-level descriptions such as SoPs and ESoPs) and shared nodes (in case of BDDs and AIGs) represent those sub-functions which are common to the multiple Boolean functions. The concept of shared product terms or nodes allows the synthesis approaches [21, 23] to generate optical circuits with reduced gate complexity. However, the reduction in gate count comes at the cost of increasing worst case fraction since each shared product term or node is realized by a splitter which reduces the signal strength. To optimize the number and usage of splitters in optical logic circuits, alternative synthesis approaches are developed [25], which, either partially or completely ignore the shared product terms/nodes by enabling the redundant realizations of common sub-functions. Several approaches employing for e.g., traversing BDD in reverse direction [37], optimizing the number of shared nodes in BDD [21], duplicating the shared product terms [25], generate optical circuits having reduced worst case fraction or splitting effect (and thereby, improving the output signal quality). However, the reduction of worst case fraction leads to an increase in the gate count. This means, the gate count and the worst case fraction are complementary optimization objectives, where, reducing the one objective increases the other one and vice-versa. This fact is experimentally confirmed in [25]. Motivated by the complementary nature of the two metrics (gate count and worst case fraction), a new function representation, called *OR Inverter Graphs (OIGs)* is introduced in [24], which is an alternative to the AIG-based synthesis scheme [22]. As compared to the AIG-based synthesis scheme, the optical circuits obtained from OIGs have smaller worst case fractions but contain relatively large number of gates.

3.2 Physical Constraints

The primary objective of designing optical circuits is to provide design automation to generate and integrate large-scale circuits realizing the desired complex functionalities. The existing synthesis approaches for optical logic circuits achieve this initial objective by automatically generating the large-scale circuits. The resulting optical logic circuits are then considered for physical design process which involves several steps such as floorplan, placement, routing, after which, the layouts of the optical circuits are generated. The layouts are then fabricated into actual physical circuits.

However, there are some physical constraints such as high driving voltage, high optical loss, large device length, which may not allow the realizations of low power, miniaturized final physical optical circuits [28]. In fact, the basic optical gates such as MZI and crossbar have these physical constraints. Since the existing logic synthesis approaches generate optical circuits composed of either MZI gates or crossbar gates, the final physical realizations may not guarantee smaller footprints and low power dissipations due to the physical restrictions of the considered optical logic gates.

3.3 Potential Alternatives

Since the MZI and the crossbar gates have some physical constraints which adversely impact the realizations of final physical optical circuits composed of the corresponding gates, the opto-electro-mechanical (OEM) gate [28] can be a suitable alternative basic optical element utilizing which any optical circuits can be designed. More importantly, the OEM gate overcomes the physical constraints that the MZI and crossbar gates have. In fact, the OEM gate consumes negligible amount of energy, requires low drive voltages, allows rapid switching and requires small footprint which are the typical advantages over the existing optical gates. Recently, an opto-electro-mechanical (OEM) device is demonstrated which has very low optical loss (0.1 dB) and requires driving voltage of 1 V [28]. These potential capabilities of OEM gates motivate a more promising and pragmatic physical realizations of the optical circuits. Consequently, the design of optical logic circuits is investigated where the circuits are composed of OEM gates, splitters and combiners i.e., OEM gate library, since the optical logic design is an initial step toward the development of physical optical circuits.

The optical circuits generated from the existing logic synthesis approaches are composed of large number of optical elements where the elements are interconnected using waveguides. Generally, the optical logic elements in the circuits are arranged in such a fashion that the waveguides crossing and bending occur very frequently leading to the optical losses. While the waveguide crossing and bending can be optimized during the placement and routing steps of the physical design process, the same can also be considered as an optimization objective during the logic design step. To the best of our knowledge, no optical logic synthesis approaches have considered the waveguide crossing and bending as the cost metrics for optical circuits. In other words, logic synthesis of optical circuits has thus far considered only two metrics— gate count and the worst case fraction, while ignored the other important metrics such as waveguide crossings and waveguide bending to evaluate the circuit complexities.

Additionally, the optical circuits generated from the existing logic synthesis approaches cannot be reprogrammed to realize any functionality other than the implemented one. This is because, the resulting optical circuits are function or application-specific, thereby allowing no flexibility for reprogramming.

An alternative logic design approach is, therefore, necessary for designing large-scale optical circuits which, in one hand, will allow efficient physical realizations of the circuits (i.e., circuits having compact footprint, low optical loss, rapid circuit switching) and in other hand, will make the optical circuits reprogrammable (i.e., no additional changes in the circuit structure will be required). For this purpose, an optical logic circuit structure is introduced in the next section, in which, the circuit consisting of OEM gates is designed in such a manner that it allows to achieve both the objectives—reprogrammable and efficient physical realization.

4 Proposed Reprogrammable Optical Logic Circuit

In this section, the reprogrammable optical logic circuit structure is introduced which can realize any function without making any changes in the structure. As a matter of fact, the proposed reprogrammable optical circuit structure is a generalized $(2^n \times 1)$ multiplexer consisting of an array of (2×1) multiplexers. Herein, each (2×1) multiplexer is realized using opto-electro-mechanical (OEM) gates and combiners only. In the following subsections, the optical realization of a (2×1) multiplexer is first discussed and afterward, a multiplexer tree realizing a generalized $(2^n \times 1)$ optical multiplexer i.e., the proposed reprogrammable optical circuit is outlined.

4.1 Design of (2 × 1) Optical Multiplexer

A (2×1) multiplexer realizes a function $F = I_0 \overline{S} + I_1 S$, where, I_0 and I_1 are the input lines and S is a select line. If $S = 1$, then output F follows the input I_1. Otherwise, the output F follows the input I_0. Figure 13 shows the schematic diagram of the multiplexer.

The optical circuit realizing the (2×1) multiplexer is shown in Fig. 14, which consists of three OEM gates and a combiner. The select line S, the inputs I_1, I_0 are applied to the electrical inputs of the OEM 1, OEM 2 and OEM 3 gates respectively, as depicted in Fig. 14a. . The through port of OEM 1 is connected to the optical input of OEM 2 and the drop port of OEM 1 is connected to the optical input of OEM 3, whereas, an optical signal (indicated by 1) is applied to the optical input of OEM 1. This arrangement of OEM gates realizes the sub-functions $I_1 S$ and $I_0 \overline{S}$ at the through ports of OEM 2 and OEM 3 respectively. The resulting sub-functions are then combined together using a combiner which realizes the desired functionality $F = I_0 \overline{S} + I_1 S$. In the following, an optical realization of the (2×1) multiplexer is represented by a functional block as depicted in Fig. 14b.

The OEM gates in the proposed optical circuit are arranged in such a manner that the resulting circuit realizing the (2×1) multiplexer avoids any waveguide crossing

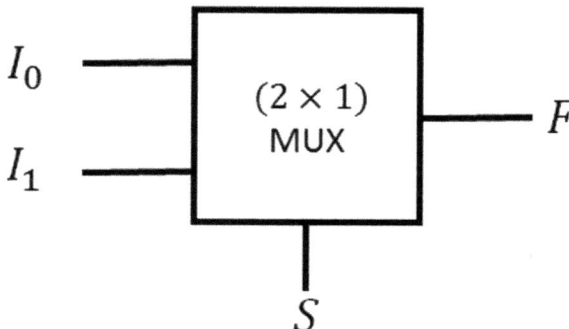

Fig. 13 Schematic of (2×1) multiplexer

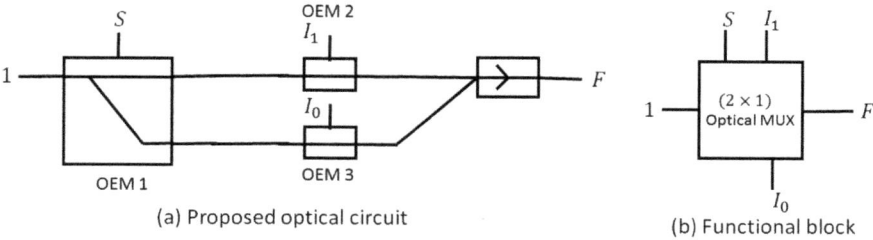

(a) Proposed optical circuit   (b) Functional block

Fig. 14 Proposed optical realizations of (2×1) multiplexer

as shown in Fig. 14a. However, there are 2 occasions in which the waveguides are bent i.e., the number of waveguide bending in the proposed circuit is 2. Note that, the OEM gates 2 and 3 are represented by smaller blocks (with no drop ports shown) as compared to the OEM gate 1 since the drop ports of OEM gates 2 and 3 are not necessary and hence, can be logically/physically omitted.

4.2 Design of Proposed Optical Circuit Structure

The proposed optical circuit structure is designed as a higher-order multiplexer which allows the scope of reprogramming the structure based on the function to be realized. The method of designing the higher-order multiplexer, i.e., multiplexer of larger size using the multiplexers of smaller input sizes is illustrated in detail. This is a standard design approach which is followed in the conventional logic design. In a similar fashion, the optical circuits realizing a $(2^n \times 1)$ multiplexer, where, $n > 11$, can be constructed using lower-order i.e., $(2^{n-1} \times 1)$ optical multiplexers.

More precisely, an optical realization of a $(2^n \times 1)$ multiplexer, where, $n(n > 1)$ denotes the number of select lines $S_0, S_1, \ldots, S_{n-1}$ and $I_0, I_1, \ldots, I_{N-2}, I_{N-1}$ denote the $N = 2^n$ input lines, is designed using two $(2^{n-1} \times 1)$ optical multiplexers, two OEM gates and a combiner as shown in Fig. 15. The $(2^{n-1} \times 1)$ optical multiplexers at the top realizes the sub-function $F_{11} = I_0 \overline{S_{n-2}} \ldots \overline{S_1} \overline{S_0} + I_1 \overline{S_{n-2}} \ldots \overline{S_1} S_0 + \ldots + I_{M-1} \overline{S_{n-2}} \ldots S_1 S_0$, where, $I_0, I_1, \ldots, I_{M-2}, I_{M-1}$ are the first half of the inputs with $M = \frac{N}{2}$ and $S_0, S_1, \ldots, S_{n-2}$ are the $(n-1)$ select lines of $(2^{n-1} \times 1)$ optical

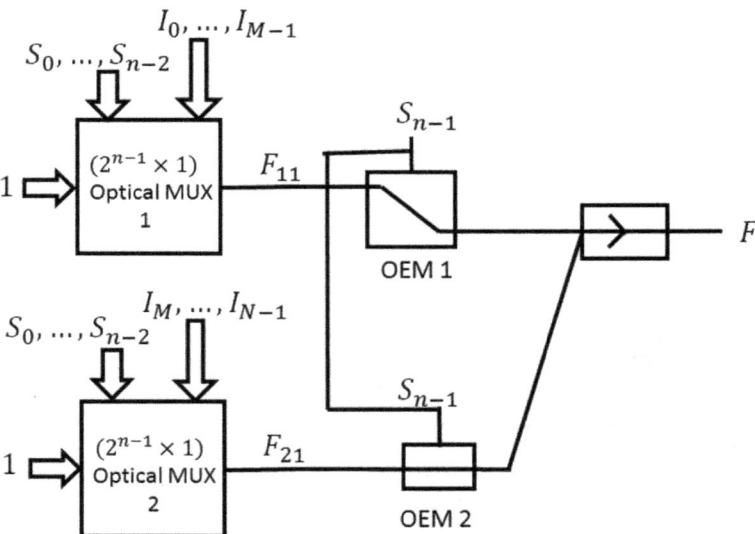

Fig. 15 Proposed general circuit structure

multiplexer. Similarly, the $(2^{n-1} \times 1)$ optical multiplexers at the bottom realizes the sub-function $F_{21} = I_M S_{n-2} \dots \overline{S_1} \, \overline{S_0} + I_{M+1} S_{n-2} \dots \overline{S_1} S_0 + \dots + I_{N-1} S_{n-2} \dots S_1 S_0$, where, $I_M, I_{M+1}, \dots, I_{N-2}, I_{N-1}$ are the next half of the input lines with $M = \frac{N}{2}$ and $S_0, S_1, \dots, S_{n-2}$ are the $(n-1)$ select lines of $(2^{n-1} \times 1)$ optical multiplexer. Now, the outputs of the $(2^{n-1} \times 1)$ optical multiplexers are applied to the optical input port of two opto-electro-mechanical gates, OEM 1 and OEM 2, whereas, the select input S_{n-1} is applied to the electrical inputs of both the OEM gates. Because of this circuit arrangement, the OEM 1 realizes a sub-function $F_{11} \overline{S_{n-1}}$ at its drop port and the OEM 2 realizes a sub-function $F_{21} S_{n-1}$ at its through port, where, F_{11} and F_{21} are the outputs of the $(2^{n-1} \times 1)$ optical multiplexers located at the top and bottom (of Fig. 15) respectively. Finally, the beam combiner combines the drop port of OEM 1 and through port of OEM 2, thereby resulting in a final optical output realizing the desired functionality of the $(2^n \times 1)$ multiplexer.

Corollary 1 An optical realization of the $(2^n \times 1)$ multiplexer requires $G(n) = (2^{n-1} + 2^{n+1} - 2)$ OEM gates.

Proof For $n = 1$, $G(1) = 3$ OEM gates are required. For $n = 2$ (i.e., (4×1) optical multiplexer), two (2×1) optical multiplexers and 2 OEM gates are used. As a result, for $n = 2$, the circuit contains $G(2) = 2.G(1) + 2 = 2.3 + 2 = 8$ $+2 = 4.G(1) + 4 + 2 = 2^2 G(1) + 2^2 + 2$ OEM gates. Similarly, for $n = 3$, the circuit contains $G(3) = 2.G(2) + 2 = 2.[2.G(1) + 2]$ OEM gates. Therefore, one can determine the number of OEM gates $G(n)$ using a recurrence relation in the following manner.

$$\begin{aligned} G(n) &= 2.G(n-1) + 2 \\ &= 2^{n-1}.G(1) + 2^{n-1} + 2^{n-2} + \dots + 2 \\ &= 2^{n-1}.3 + 2^{n-1} + 2^{n-2} + \dots + 2 \\ &= 2^{n-1} + 2^{n+1} - 2. \end{aligned}$$

Hence the proof. ∎

Corollary 2 An optical realization of the $(2^n \times 1)$ multiplexer requires $C(n) = (2^n - 1)$ combiners.

Proof For $n = 1$, the circuit requires $C(1) = 1$ combiner. For $n = 2$ (i.e., (4×1) optical multiplexer), the circuit requires two (2×1) optical multiplexers and 2 OEM gates. As a result, for $n = 2$, the circuit contains $C(2) = 2.C(1) + 1 = 2.1 + 2 = 3$ combiners. Similarly, for $n = 3$, the circuit contains $C(3) = 2.C(2) + 1 = 2.[2.C(1) + 1] + 1 = 4.C(1) + 2 + 1 = 2^2 C(1) + 2^1 + 2^0$ combiners. Therefore, the number of combiners $C(n)$ can be expressed using a recurrence relation in the following manner.

$$\begin{aligned} C(n) &= 2.C(n-1) + 1 \\ &= 2^{n-1}.C(1) + 2^{n-2} + \dots + 2^1 + 2^0 \end{aligned}$$

$$= 2^{n-1}.1 + 2^{n-2} + \ldots + 2^1 + 2^0$$
$$= 2^{n-1} + 2^{n-2} + \ldots + 2^1 + 2^0$$
$$= 2^n - 1$$

Hence the proof.   ∎

Corollary 3 An optical realization of the $(2^n \times 1)$ multiplexer has $(2^{n+1} - 2)$ waveguide bending.

Proof At each combiner input, the waveguide bending occurs twice. It follows from Corollary 2 that the number of combiners in the proposed structure is $(2^n - 1)$, as a result, the number of waveguide bending in the proposed circuit structure is $2(2^n - 1) = 2^{n+1} - 2$. Hence, the proof.   ∎

Note that, the proposed structure requires no splitters, because of which the effect of splitting in the circuit is zero i.e., the worst case fraction is 1. Furthermore, the OEM gates are arranged in the circuit in such a manner that no waveguide crossing occurs leading to an optical circuit free of waveguide crossings.

The general optical structure realizing the $(2^n \times 1)$ multiplexer can further be illustrated with the following examples.

Example 4 The proposed optical realization of a (4×1) multiplexer is shown in Fig. 16. The structure uses two (2×1) optical multiplexers and two OEM gates, in which, the outputs from both the multiplexers are connected to the optical input ports of the OEM gates. The final output of the (4×1) multiplexer is realized by combining

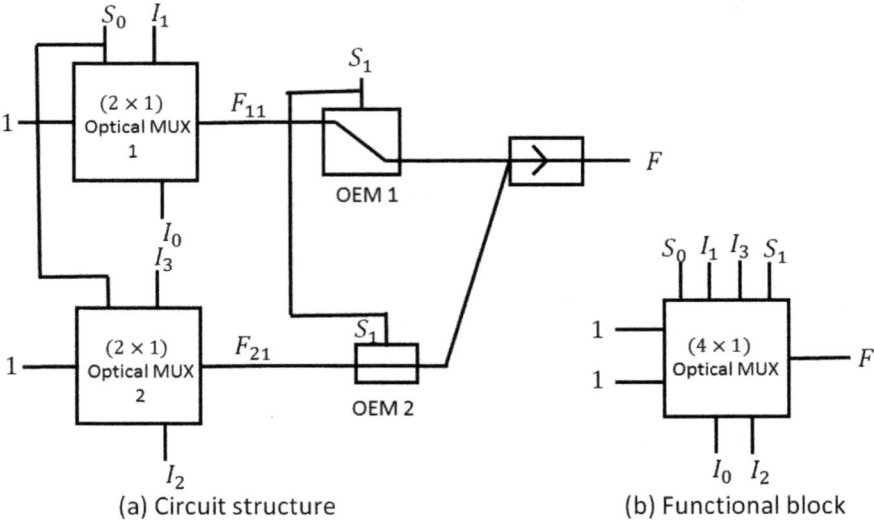

(a) Circuit structure   (b) Functional block

Fig. 16 Proposed optical realizations of (4×1) multiplexer

the drop port of OEM 1 and through port of OEM 2. In this circuit structure, the select input S_0 is common to both the (2×1) optical multiplexers, whereas, another select input S_1 is applied as electrical inputs to both the OEM gates. The four input lines I_0, I_1, I_2, I_3 are applied to the two (2×1) optical multiplexers as shown in Fig. 16a. The proposed circuit structure for (4×1) optical multiplexer is represented by a functional block shown in Fig. 16b. Since each (2×1) optical multiplexer contains 3 OEM gates and a single combiner, overall the proposed circuit contains 8 OEM gates and 2 combiners. Besides that, the circuit has 6 waveguide bending in total (each (2×1) optical multiplexer has 2 bending and another 2 bending occurs during the realization of the final output), while, no waveguide crossing occurs in the circuit.

5 Mapping of Boolean Functions to the Proposed Circuit Structure

In this section, the method of realizing arbitrary Boolean functions using the proposed optical circuit structure is described. Realization of arbitrary Boolean functions using the proposed circuit structure involves—(1) the selection of a suitable optical multiplexer based on the number of variables of the functions to be realized and (2) the appropriate mapping of the input variables of the considered functions to the select and input lines of the optical multiplexer.

More precisely, a $(2^{n-1} \times 1)$ optical multiplexer to realize any Boolean function of n variables is utilized. Once the size of the multiplexer is determined, then the mapping of the input variables $(x_0, x_1, \ldots, x_{n-1})$ of the function to the proposed optical multiplexer is carried out in the following manner:

- The $(n-1)$ variables $(x_0, x_1, \ldots, x_{n-2})$ of the function to be realized are mapped to the $(n-1)$ select lines $(S_{n-2}, \ldots, S_1, S_0)$ of the $(2^{n-1} \times 1)$ optical multiplexer. That is, input variable x_0 is mapped to the select input S_{n-2}, input variable x_1 is mapped to the select input S_{n-3} and so on.
- The input lines $(I_0, I_1, \ldots, I_{M-1})$ of the proposed multiplexer are connected to the fixed values (i.e. $0\ V$ and/or $1\ V$) or the nth input variable x_{n-1} as x_{n-1} and/or $\overline{x_{n-1}}$ whenever the output of the function equals to logic 0 or logic 1 or follows the input variable x_{n-1} or its cplement.

The above idea of mapping the inputs of the functions to the multiplexer structure allows implementation of arbitrary Boolean functions using the proposed optical circuit structure without requiring additional hardware or making any changes in the structure. This makes the proposed optical circuit structure a reprogrammable structure.

The above idea is further illustrated with the following example.

Example 5 Consider the Boolean function F to be realized is shown in Fig. 17a. A (4×1) optical multiplexer is employed to alize the function since it has three variables, x_0, x_1, x_2. The first two variables x_0 and x_1 are mapped to the two select lines

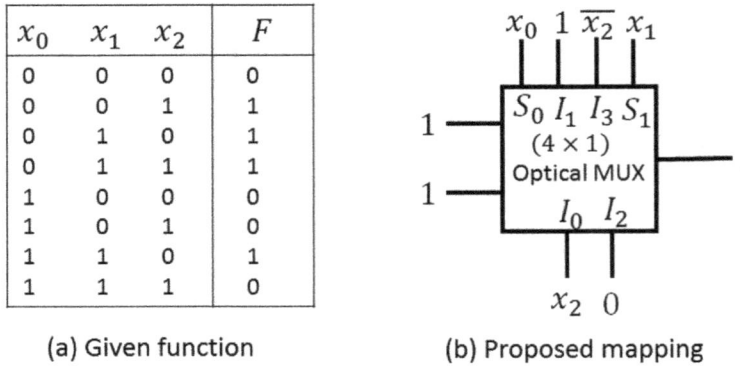

(a) Given function   (b) Proposed mapping

Fig. 17 Proposed realization of arbitrary Boolean function

S_1 and S_0 respectively. Depending on the output values for the input combinations, the input lines I_0, I_1, I_2, I_3 are connected to x_2, $1V$, $0V$, $\overline{x_2}$ respectively as shown in Fig. 17b. Such mapping realizes the output as $F = x_2$ or $\overline{x_2}$ for combinations 00 or 11 at $x_0 x_1$ and $F = 1$ or 0 for combinations 01 or 10 at $x_0 x_1$; thereby leading to the desired Boolean function.

6 Experimental Results

In this section, the results obtained by the proposed approach are evaluated and compared to the scheme previously proposed in [34] in order to verify whether the proposed approach is indeed beneficial and provides solution to the concerns discussed in Sect. 3.2. Note that, the results are obtained based on theoretical and functional analysis of the proposed approach. Once the theoretical results confirm the efficacy of the proposed approach over the existing approaches, physical realizations of optical circuits can then be considered as the next steps toward actual optical circuits. However, realizing the physical optical circuit is currently beyond the scope of this chapter since physical realizations of optical circuits involve several design steps, for e.g., circuit partitioning, floorplanning, placement and routing.

For results and discussions, the (8×1), (16×1), (32×1), (64×1) optical multiplexers composed of OEM gate library are designed and compared with the all-optical realizations consisting of MZI gate library proposed in [34]. The results are summarized in Table 1 and Table 2. The first column of Table 1 provides the different sizes of the multiplexers. In the second column of Table 1, the number of MZI gates, splitters, combiners and the total number of optical elements (which is computed as the sum of number of MZI gates, splitters and combiners in the circuit) are provided for each optical realization of the multiplexer based on the existing approach [34]. The final column provides the number of OEM gates, combiners and the total number

Table 1 Comparisons in terms of number of optical gates

Multiplexer size	Existing approach [34]				Proposed approach		
	MZI	Splitter	Combiner	Total Gates	OEM	Combiner	Total Gates
8 × 1	14	11	7	32	18	7	25
16 × 1	30	26	15	71	38	15	53
32 × 1	62	57	31	150	78	31	109
64 × 1	126	120	63	309	158	63	221

Table 2 Comparisons in terms of worst case fraction, waveguide crossings and bending

Multiplexer size	Existing approach [34]			Proposed approach	
	Fraction	Crossing	Bending	Fraction	Bending
8 × 1	$1/8$	8	52	1	14
1 16 × 1	$1/16$	22	114	1	30
32 × 1	$1/32$	52	241	1	62
64 × 1	$1/64$	114	495	1	126

of optical elements (which is obtained by summing the number of OEM gates and combiners) that are required to realize the considered multiplexers using the proposed scheme. Note that, the proposed approach generates optical circuits consisting of zero splitters, which is why, the number of splitters is not explicitly shown in Table 1.

Table 1 clearly shows that, the optical circuits generated from the proposed approach are much more compact as compared to the existing approach [34]. In fact, significant reductions in the total number of optical elements can be achieved using the proposed approach as shown in Fig. 18. More precisely, a reduction up to 28.5% in total optical elements can be obtained in comparison to the existing work [34]. Interestingly, as the size of the multiplexer increases, the percentage of reduction in total optical elements increases, which makes the proposed scheme a suitable choice for designing large-scale optical circuits.

Like Table 1, the first column of Table 2 provides the different sizes of the multiplexers. In the next column of Table 2, the worst case fraction (*Fraction*), waveguide crossing and bending are reported for each optical realization of the considered multiplexer based on the existing work [34]. The final column reports the worst case fraction and the number of bending obtained from the proposed realizations. Since no waveguide crossings occur in the proposed optical realizations, because of which, the number of crossings are not explicitly reported in case of proposed approach in Table 2.

Table 2 clearly demonstrates the quality of the optical circuit that can be generated from the proposed scheme. The results show that the optical circuits generated from the existing approach [34] have high worst case fraction (i.e., high effect of splitting), large number of waveguide crossing and bending, all of which result in poor optical signal. In contrast, the optical circuits obtained from the proposed approach have

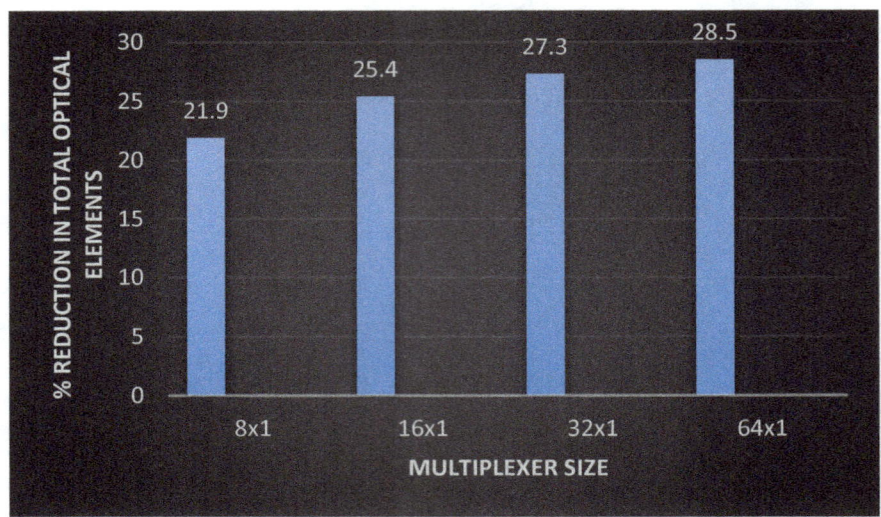

Fig. 18 Percentage reduction in total optical elements in comparison to [34]

no splitting effect (leading to a fraction of 1) and zero waveguide crossing. Besides that, the number of waveguide bending are much less in the proposed realizations as compared to the optical realizations obtained from the existing work [34]. In fact, reduction of up to 74.5% in the number of bending can be achieved as shown in Fig. 19.

In addition, the optical realizations of some benchmark functions [37] are considered. The results are summarized in Table 3. The first column provides the details

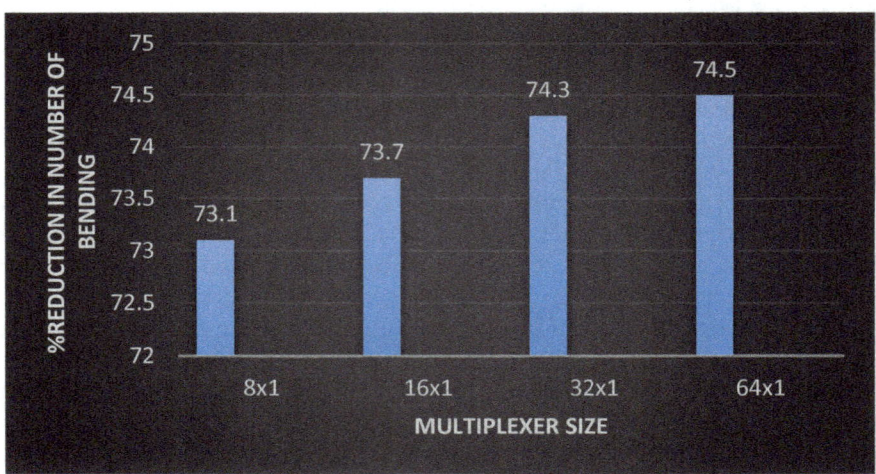

Fig. 19 Percentage reduction in the number of bending in comparison to [34]

Table 3 Comparisons on benchmark functions

Benchmark		Existing approach [34]				Proposed approach	
Name	PI/PO	Total	Fraction	Crossing	Bending Elements	Total	Bending elements
4gt5	4/1	32	$1/8$	8	52	25	14
4gt12	4/1	32	$1/8$	8	52	25	14
4mod5	4/1	32	$1/8$	8	52	25	14
ex2	5/1	71	$1/16$	22	114	53	30
alu	5/1	71	$1/16$	22	114	53	30
majority	5/1	71	$1/16$	22	114	53	30
Sym6	6/1	150	$1/32$	52	241	109	62

of the benchmark i.e., name and number of primary inputs and outputs (*PI/PO*) of each benchmark. The second column lists the total number of optical elements, worst case fraction (*Fraction*), crossing and bending for the realizations obtained from the existing approach [34]. In the final column, the number of optical elements and the number of bending obtained from the proposed circuits are reported. It is clearly observed that the significant reductions for the benchmarks can be obtained by the proposed approach, thereby making the optical realizations way cheaper.

7 Conclusion

In this chapter, a reprogrammable optical circuit structure is introduced which can realize any Boolean function without making any changes in the structure. The proposed structure is a $(2^n \times 1)$ multiplexer which is composed of several (2×1) multiplexers, where, each (2×1) multiplexer is realized by recently developed opto-electro-mechanical devices (which are referred as OEM gates in the chapter). The OEM gates have several advantages over the other optical devices, which makes the optical circuits composed of OEM gates physically realizable. Such potential advantages of the OEM gates are utilized in the proposed circuit structure, besides which, the proposed structure is designed in such a manner that the resulting optical circuits are much more compact and efficient as compared to the previously proposed approaches. The effectiveness of the proposed approach is further confirmed by the experimental evaluations which show zero waveguide crossing, zero effect of splitting and significant reductions in optical elements and number of waveguide bending.

References

1. Goodfellow I, Bengio Y, Courville A, Bengio Y (2016) Deep learning. MIT press, Cambridge
2. Pitas I (2000) Digital image processing algorithms and applications. Wiley
3. Bovik AC (2010) Handbook of image and video processing. Academic press
4. Jurafsky D (2000) Speech and language processing. Pearson Education India
5. Krizhevsky A, Sutskever I, Hinton GE (2012) Imagenet classification with deep convolutional neural networks. In: Pereira F, Burges CJC, Bottou L, Weinberger KQ (eds) NIPS' 12: Advances in neural information processing systems, Lake Tahoe, December, 2012
6. Dong C, Loy CC, He K, Tang X (2015) Image super-resolution using deep convolutional networks. IEEE Trans Pattern Anal Mach Intell 38(2):295–307
7. Szegedy C, Liu W, Jia Y, Sermanet P, Reed S, Anguelov D, Rabinovich A (2015) Going deeper with convolutions. In: Proceedings of 28th IEEE conference on computer vision and pattern recognition, Boston, USA, 7–12 June 2015
8. Simonyan K, Zisserman A (2014) Very deep convolutional networks for large-scale image recognition. In: arXiv Available. https://arxiv.org/abs/1409.1556
9. Nam BG, Yoo HJ (2014) Graphics processing unit: algorithm, architecture, and power. CRC Press Inc.
10. Jouppi N, Young C, Patil N, Patterson D (2018) Motivation for and evaluation of the first tensor processing unit. IEEE Micro 38(3):10–19
11. Wang C, Gong L, Yu Q, Li X, Xie Y, Zhou X (2016) DLAU: a scalable deep learning accelerator unit on FPGA. IEEE Trans Comput Aided Des Integr Circuits Syst 36(3):513–517
12. Shawahna A, Sait SM, El-Maleh A (2018) FPGA-based accelerators of deep learning networks for learning and classification: a review. IEEE Access 7:7823–7859
13. Ma Y, Suda N, Cao Y, Vrudhula S, Seo JS (2018) ALAMO: FPGA acceleration of deep learning algorithms with a modularized RTL compiler. Integration 62:14–23
14. Guo K, Han S, Yao S, Wang Y, Xie Y, Yang H (2017) Software-hardware codesign for efficient neural network acceleration. IEEE Micro 37(2):18–25
15. Reed GT, Knights AP (2004) Silicon photonics: an introduction. Wiley
16. Vivien L, Pavesi L (eds) (2016) Handbook of silicon photonics. Taylor and Francis
17. Datta K, Chattopadhyay T, Sengupta I (2015) All optical design of binary adders using semiconductor optical amplifier assisted Mach-Zehnder interferometer. Microelectron J 46(9):839–847
18. Singh P, Singh AK, Arun V, Dixit HK (2016) Design and analysis of all-optical half-adder, half-subtractor and 4-bit decoder based on SOA-MZI configuration. Opt Quant Electron 48(2):1–14
19. Sharma S, Chakrabarty K, Roy S (2018) On designing all-optical multipliers using mach-zender interferometers. In: Proceedings of 21st IEEE Euromicro conference on digital system design, Prague, Czech Republic, 29–31 August 2018
20. Kumar S, Bisht A, Singh G, Amphawan A (2015) Implementation of 2-bit multiplier based on electro-optic effect in Mach-Zehnder interferometers. Opt Quant Electron 47(12):3667–3688
21. Deb A, Wille R, Keszöcze O, Shirinzadeh S, Drechsler R (2017) Synthesis of optical circuits using binary decision diagrams. Integr VLSI J 59:42–51
22. Deb A, Wille R, Drechsler R (2017) Dedicated synthesis for MZI-based optical circuits based on AND-inverter graphs. In: Proceedings of 36th IEEE/ACM international conference on computer-aided design, Irvine, USA, 13–16 November 2017
23. Schönborn E, Datta K, Wille R, Sengupta I, Rahaman H, Drechsler R (2015) BDD-based synthesis for all-optical Mach-Zehnder interferometer circuits. In: Proceedings of 28th international conference on VLSI design, Bangalore, India, 3–7 January 2015
24. Deb A, Wille R, Drechsler R (2017) OR-inverter graphs for the synthesis of optical circuits. In: Proceedings of IEEE 47th international symposium on multiple-valued logic, Novi Sad, Serbia, 22–24 May 2017
25. Deb A, Wille R, Keszöcze O, Hillmich S, Drechsler R (2016) Gates vs. splitters: contradictory optimization objectives in the synthesis of optical circuits. ACM J Emerging Technol Comput Syst 13(1):1–13

26. Liao L, Samara-Rubio D, Morse M, Liu A, Hodge D, Rubin D, Franck T (2005) High speed silicon Mach-Zehnder modulator. Opt Express 13(8):3129–3135
27. Ding M, Wonfor A, Cheng Q, Penty RV, White IH (2017) Hybrid MZI-SOA InGaAs/InP photonic integrated switches. IEEE J Sel Top Quantum Electron 24(1):1–8
28. Haffner C, Joerg A, Doderer M, Mayor F, Chelladurai D, Fedoryshyn Y, Aksyuk VA (2019) Nano–opto-electro-mechanical switches operated at CMOS-level voltages. Science 366(6467):860–864
29. Vahid F (2010) Digital design with RTL design, VHDL, and Verilog. Wiley
30. Bogaerts W, Dumon P, Van Thourhout D, Baets R (2007) Low-loss, low-cross-talk crossings for silicon-on-insulator nanophotonic waveguides. Opt Lett 32(19):2801–2803
31. Xu F, Poon AW (2008) Silicon cross-connect filters using microring resonator coupled multimode-interference-based waveguide crossings. Opt Express 16(12):8649–8657
32. Cardenas J, Poitras CB, Robinson JT, Preston K, Chen L, Lipson M (2009) Low loss etchless silicon photonic waveguides. Opt Express 17(6):4752–4757
33. Vlasov YA, McNab SJ (2004) Losses in single-mode silicon-on-insulator strip waveguides and bends. Opt Express 12(8):1622–1631
34. Datta K, Sengupta I (2014) All optical reversible multiplexer design using Mach-Zehnder interferometer. In: Proceedings of 27th international conference on VLSI design, Mumbai, India, 5–10 January 2014
35. Akers SB (1978) Binary decision diagrams. IEEE Trans Comput 6:509–516
36. Kuehlmann A, Paruthi V, Krohm F, Ganai MK (2002) Robust Boolean reasoning for equivalence checking and functional property verification. IEEE Trans Comput Aided Des Integr Circuits Syst 21(12):1377–1394
37. Wille R, Keszocze O, Hopfmuller C, Drechsler R (2015) Reverse BDD-based synthesis for splitter-free optical circuits. In: Proceedings of 20th IEEE Asia and South Pacific design automation conference, Tokyo, Japan, 19–22 January 2015

SELBOX TFET and DTD TFET for DC and RF/Analog Applications

Puja Ghosh and Brinda Bhowmick

Abstract This chapter presents the comparison of electrical and RF parameters of selective buried oxide (SELBOX) TFET with dual tunnel diode (DTD) TFET. The main objective is to analyze the kink reduction in both the TFET structures. The major disadvantage of SOI device is floating body effect (FBE). Kink appears as an instantaneous proliferation in the drain current (I_D). FBE reduces the output resistance, minimizes the power gain and leads to fluctuation of threshold voltage. Hence, mitigation of the FBE is crucial. The presence of undesirable kink influences the electrical parameters of the device. The electric field crowding caused by high band-to band generation rate increases the number of charge carriers near the tunnel junction and leads to the aggregation of holes. The holes cumulated in the floating body rise the potential of the body and minimize the threshold voltage. This elevates I_D which is defined as kink effect. The release of cumulated holes by using a small buried oxide gap in the SELBOX device and L-patterned trench in the DTD TFET attenuates kink. The DTD device becomes more economic and reliable due to its small device area compared to body contact-based devices. The simulation results manifest that the proposed structures significantly deteriorate kink by postponing it to higher drain voltage while still maintaining the prime benefits of conventional SOI device. The RF parameters of SELBOX TFET are compared with DTD TFET at various temperatures. The reliability issues of different TFET structures are investigated. The dependency of RF parameters like cut-off frequency (f_T), transconductance (g_m), and gate capacitance (C_{GG}) on temperature has been analyzed. Moreover, linearity parameters of both the structures are compared and 1-dB compression point has been studied for a range of temperature.

Keywords SELBOX · DTD · FBE · SS · BTBT · TGF · TFP · Linearity

P. Ghosh (✉)
Department of Electronics and Communication Engineering, Indian Institute of Information Technology Ranchi, Ranchi 834010, India

B. Bhowmick
Department of Electronics and Communication Engineering, National Institute of Technology Silchar, Silchar 788010, India

1 Introduction

The power density can be mitigated by minimizing the supply voltage with scaling of MOSFETs. The mitigation of subthreshold swing (SS) has emanated as an essential technological issue. To overcome the demerits of MOSFETs, different research groups came forward with various alternative devices by inducting gate engineering, material engineering, structural engineering and gate oxide engineering. One of the alternatives to MOSFET, the tunneling FET (TFET) is a p-i-n-diode which is reverse biased and operates on the concept of interband tunnelling with a gate to control the tunnelling phenomenon. It has evolved as a potential candidate for applications related to low power. Under the section Process Integration of 2013's ITRS (now morphed to IRDS) edition, for future low power applications, tunnel FET has been identified as a potential device [1, 2]. There is penetration of carriers through the barrier by the tunneling process instead of crossing of carriers over the barrier like MOSFET. It has the potential to overcome the issues of high I_{OFF}, SCEs, and SS >60 mV/decade. Higher ON current can be obtained by using group III-V compound semiconductors with direct bandgap. TFET turns on at low operating voltages. In TFET, the channel is intrinsic type with p-type (heavily doped) source. With the scaling down of supply voltage to 1 V, there is no deterioration in the device performance due to the different mode of operation of TFET from MOSFET. Increasing the gate voltage (V_G) more than the threshold voltage (V_T), results in the movement of electrons from the p-type source (valence band) to the intrinsic channel (conduction band) and thus, making the channel n-type. These carriers get collected at the drain terminal leading to current conduction. The holes move in opposite direction to that of the electrons and get collected at the source side from the channel region [3]. For complementary circuit applications, ambipolar current create a problem. Owing to BTBT operation, tuneling FETs (TFETs) can achieve SS below 60 mV/decade [4]. By controlling the channel region band bending, using the gate bias, it is possible to switch on and off abruptly the interband tunneling. BTBT current depends on transmission probability [5].

Various TFET geometries have been reported in literatures such as double-gate TFET [6–8], dual-material gate TFET [9], silicon-on-insulator (SOI) TFET [10–12], heterojunction TFET [13, 14], and circular-gate (CG) TFET [15, 16]. SOI devices have emerged with different benefits over bulk geometries for instance reduced short channel effects, minimized leakage current, low drain body capacitance resulting in excellent switching speed. However, few undesirable effects like FBE [17–20] are associated with partially depleted SOI devices. Owing to the development of neutral channel region, FBE occurs resulting in rapid elevation in the conducting current, which is defined as kink. Threshold voltage fluctuates due to FBE. In SOI devices concentration of holes within the floating body is caused by high BTBT rate which leads to congestion of electric field around the source corner close to the channel and source junction. The voltage of the body enhances and the threshold voltage minimizes due to the accumulation of holes. This leads to kink effect.

Abrupt source doping leads to kink effect. In the L-patterned TFET device, for investigating the kink effect, BTB generation rates have been explored by analyzing the 2-dimensional contour plots for different front gate biases [21]. Various structures such as body-tied-source and T-gate have been proposed to suppress FBE [22–24]. These devices have large area and low effective device width. In circuits, for device realization, an enumerated analysis of the analog parameters is requisite. For a wide temperature range, the reliability and applicability of any device can be assessed by comprehensively analyzing the temperature affectability. The affectability of temperature variation on different electrical parameters have been studied in various literatures [25].

Although a lot of works have been reported on TFET but still there are some domains left to be fully explored. New TFET structures that have electrical properties in accordance with ITRS have to be proposed [6, 13]. The electrical properties of the devices like I_{ON}/I_{OFF} ratio, threshold voltage and subthreshold swing (SS) have to be investigated rigorously. In the literatures, various TFET geometries have been proposed to enhance the I_{ON}/I_{OFF} ratio and minimize the SS. The proposed geometry comprises of a gap created on buried oxide substrate with an SiGe layer at the tunnel junction and it is termed as selective buried oxide (SELBOX) TFET. The gap present within the buried oxide of the SELBOX TFET can be used to eliminate kink. A solution to alleviate kink effect has been discussed herein. One more geometry of TFET known as dual tunnel diode (DTD) TFET, very highly doped with p^{++} type L-patterned region has been proposed. The generation of dual tunnel diode in the proposed geometry reduces kink effect. This chapter presents the dependency of RF parameters on temperature of SELBOX TFET and DTD TFET. A comparison of DC parameters of both the proposed structures is also shown. For low voltage applications in nanoscale range, the proposed device proves to be a potential candidate. The rest of the chapter is systemized as: The doping specifications with the simulation methodology of the proposed device have been reported in Sect. 2. Section 3 discusses the results with summarization of the chapter in Sect. 4.

2 Simulation Methodology

Sentaurus TCAD is utilized perform the simulation analysis of the device [26]. The industrial TCAD device simulators are equipped with a number of models based on different physics principles. Within the tool, these models are represented by sets of parameters which mathematically provide identity to the simulation models. A number of TFET designs have been fabricated based on the structure modifications and exploring new materials. For making the TCAD models compatible with the experimental results of fabricated TFETs, the values of these simulation model parameters must be modified. This process of calibration is usually carried out by observing the parameters of the prime model that describes the operation of the device, and altering their values to bring about a satisfactory match [27, 28].

A TFET with δp^+ $Si_{1-x}Ge_x$ layer at the tunnel junction amidst the channel and the source has been configured on selective buried oxide substrate in Fig. 1a. A layer of material (silicon–germanium) having low bandgap is incorporated amidst the source and channel to amplify the ON current. For further improving the ON current a high-k material of thickness 2 nm is utilized as oxide of gate [29]. The entire structure is 100 nm long with 35 nm of drain and source regions. A 3 nm thick δp^+ $Si_{1-x}Ge_x$ layer is used with total channel length of 30 nm. The proposed device is called selective buried oxide (SELBOX) as a thin gap of 2 nm length is developed (nearby the tunneling junction) in the 10 nm thick buried oxide. The numerous doping concentrations are: source (1×10^{20} cm^{-3}), δp^+ layer (1×10^{18} cm^{-3}), drain (5×10^{18} cm^{-3}), and channel (1×10^{16} cm^{-3}). The presence of buried oxide gap elevates the availability of states density in the tunnel junction where the carriers get confined. This reduces the carrier concentration to be drawn by the drain voltage. The OFF current is diminished due to the existence of gap in comparison with fully depleted (FD) SOI TFET, where floating body effects lead to high OFF current [17, 30]. In

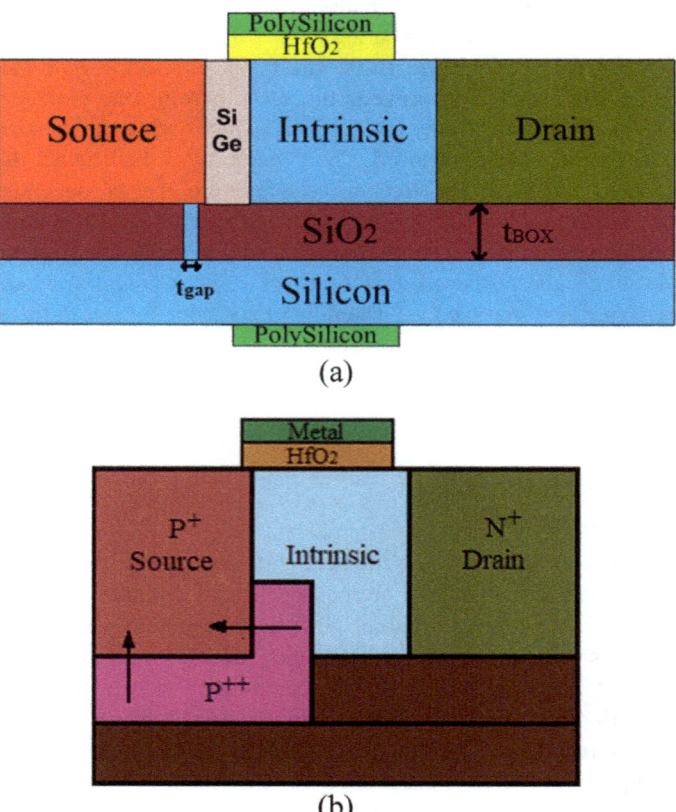

Fig. 1 Simulated **a** SELBOX TFET **b** DTD TFET with metal work function of 4.57 eV

the ON state, the effect of gap on the ON current is insignificant due to very large concentration of electrons inserted into the channel from the source compared to the available density of states in the gap. The presence of gap adjacent to the source reduces the OFF current. This boosts the I_{ON}/I_{OFF} ratio.

The proposed DTD device design is portrayed in Fig. 1b. Between the channel and source region and also below some portion of channel and source regions, heavily doped p^{++} L-shaped layer is inserted. The channel is 40 nm of 100 nm long device. The specifications of doping of various regions are source (1×10^{20} cm^{-3}), drain (1×10^{18} cm^{-3}), channel (1×10^{16} cm^{-3}), and L-shaped trench (1×10^{21} cm^{-3}). Two tunnel diodes are formed amidst the p^{++} type region and the source where the p^{++} type region and the source are considered as anode and cathode of the device, respectively. Therefore, the proposed design is termed as DTD TFET.

The I_D of the device is determined by employing the BTBT model. Relying upon the BTBT model which defines the device generation rate, various designed structures result in distinct sets of parameter values. Schenk's BTBT model based on Phonon-assisted tunneling provides simulation convergence. Fermi Dirac Statistics has been taken into account owing to different doping concentrations in different regions of the device. Narrow bandgap resulting from high concentration of doping causes tunneling of carriers in TFETs. Hence, Bandgap model of narrowing has been activated. Different regions have different doping concentrations and for considering the affectability of doping specification on charge carrier mobility, the mobility model of doping dependent has been utilized. SRH model of recombination has also been employed. For improving the accuracy, enormal model is used. The affectability of temperature alteration on device operation is considered by activating hydro dynamic model. To activate generation/recombination process, trap assisted tunneling (TAT) model is utilized. The simulation models are validated by calibrating with respect to the experimental results of [31] in Fig. 2. The fabrication process flows of the proposed devices are: With proper wafer bonding of normal wafer of silicon with oxidized wafer of silicon, etching process is carried out for the formation of SOI substrates. Using photolithography with appropriate masking, the gap position is

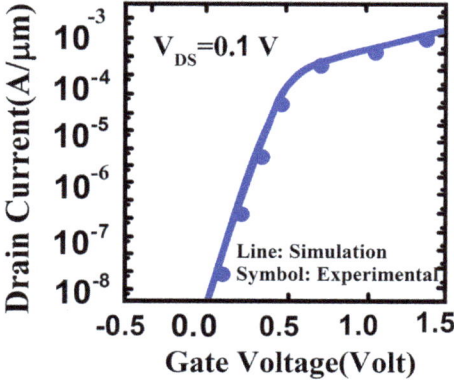

Fig. 2 Plot of simulation result calibrated with experimental SOI TFET device

defined in the oxide layer in case of SELBOX device. After etching out the oxides which are not required, the gap is epitaxially filled up with silicon. The processed wafer and the handle wafer are then bonded together. For the DTD-SOI device, similar steps are followed for the formation of SOI substrates. Suitable masking is used to deposit p++ doped silicon region in the patterned source and intrinsic region. Using similar photolithography process, high-k dielectric is deposited and metallization is used for developing the metal contacts.

3   Results

3.1   DC Performance of SELBOX TFET

ON current is elevated by the SiGe layer existing at the tunnel junction and the gap closer to the source reduces the OFF current. This increases the I_{ON}/I_{OFF} ratio of the device. Figure 3a portraits the alteration of I_D of the device for several positions of

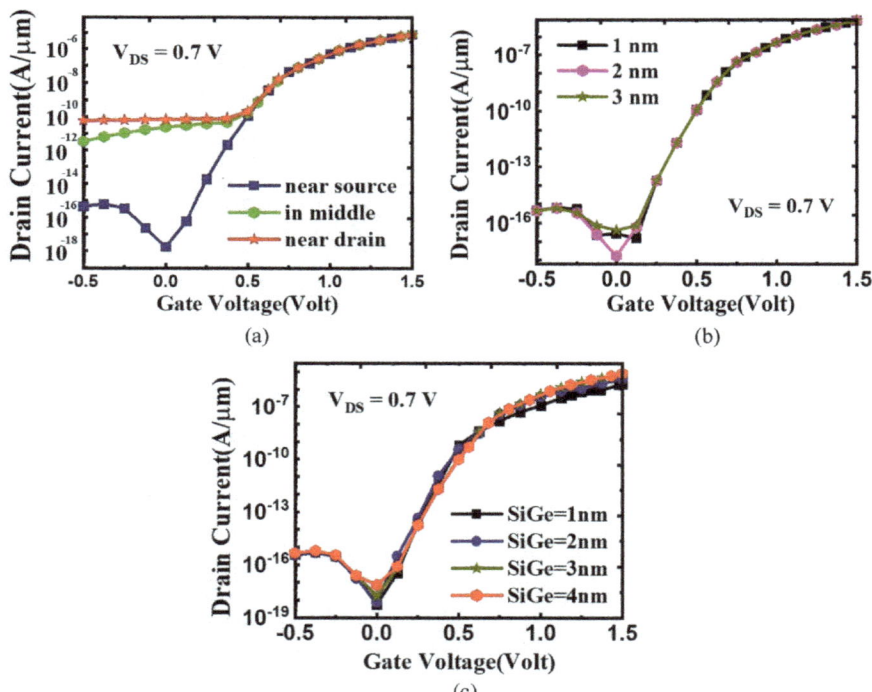

Fig. 3 Plot of transfer characteristics for various **a** position of gap **b** thickness of gap near the source, and **c** thickness of SiGe layer

buried oxide gap. The unfilled states density in the tunneling junction increases when the gap is situated close to the source. The gap situated on some other position will not effect the tunnel junction states density and therefore, I_D will persist to be high. Elevated I_{ON} of 8.3 × 10^{-6} A/μm and minimum OFF current of 1.7 × 10^{-18} A/μm have been acquired with high I_{ON}/I_{OFF} ratio of 4.62 × 10^{12} as depicted in Fig. 3a. The device provides low SS of 53 mV/decade. The operation of TFET is mainly dependent on BTBT as it results in bending of bands at the tunneling junction due to variation in doping concentration. In the SELBOX structure, the presence of gap adjacent to the source (31–33 nm) yields minimal OFF current due to the occupancy of highly available states density in the tunnel junction by the electrons [30]. Hence, the amount of carriers to be extracted by the drain bias alleviates as the unfilled states density confine most of the electrons. An elevation in the OFF current is observed as the gap relocates toward the drain. OFF current slightly increases when the gap is located far from the tunnel junction in the source due to the less influence of drain voltage. Moreover, the thickness of the buried oxide gap is optimized to 2 nm. As the gap thickness enhances, the characteristics of the device tends toward bulk TFET and the OFF current enhances. Diminished OFF current is achieved for 2 nm gap thickness as portrayed in Fig. 3b. The device behaves as FDSOI TFET with reduced thickness of gap and therefore, the OFF current increases. The OFF current of SOI TFET is lower than bulk TFET. The gap with thickness 2 nm situated close to the tunneling junction provides the highest I_{ON}/I_{OFF} ratio.

The low band gap material present adjacent to the tunneling junction amidst the source and the channel enhances the tunneling current from (1) and (2) [32]. ON current amplifies with the increase of SiGe layer thickness from 1 to 3 nm and remains comparable for 3 nm and 4 nm thickness as depicted in Fig. 3c. This is because low band gap material located only nearer to the tunneling junction effects the ON current and the tunneling junction extends till 3 nm in SiGe layer [33]. Although the ON current remains comparable but the OFF current of 4 nm thickness is more than 3 nm thickness. Therefore, the best output is achieved for 3 nm thick SiGe layer.

$$G_{BTB} = A \frac{|E|^2}{\sqrt{E_g}} \exp\left(-B \frac{E_g^{1.5}}{|E|}\right) \tag{1}$$

where G_{BTB} is the BTB generation rate, E_g is the energy bandgap, the tunnel junction electric field is represented by E, and A and B are constants.

$$I_D = q \int G_{BTB} dV \tag{2}$$

The variation of I_D-V_{DS} characteristics with SELBOX thickness keeping gap thickness fixed to 0.5 nm is shown in Fig. 4a. Kink varies with the SELBOX thickness as perceived in Fig. 4a. There is a dependence of gap resistance on SELBOX thickness where SELBOX thickness (t_{BOX}) denotes the resistance length and gap thickness

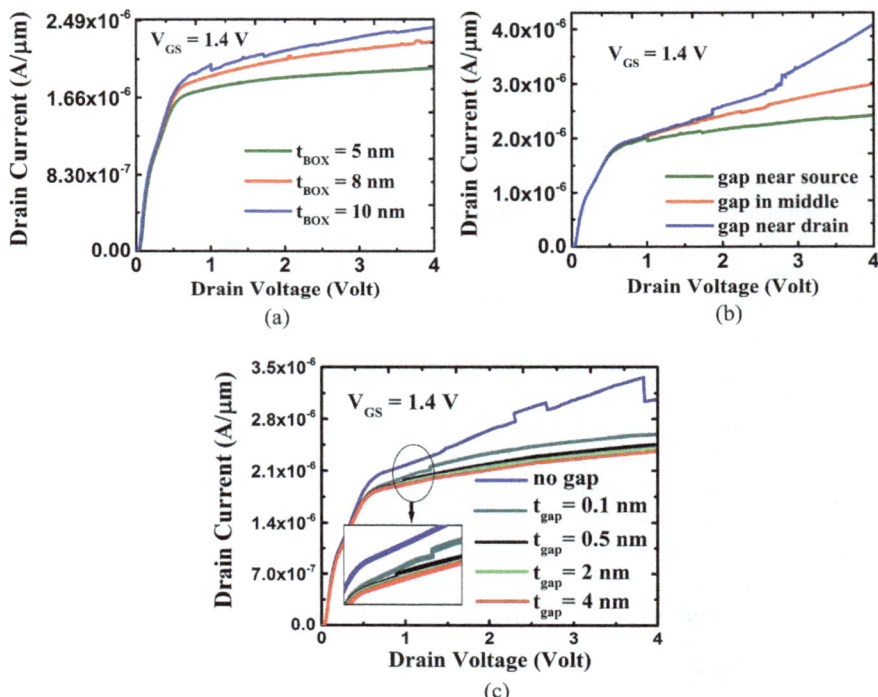

Fig. 4 Plot of output characteristics of SELBOX device for various **a** BOX thickness **b** gap position, and **c** gap thickness near source

(t_{gap}) denotes the resistance width. Gap resistance increases with the increase of t_{BOX} which rises the body potential leading to occurrence of kink at minimized drain voltage. Kink effect is maximum for 15 nm t_{BOX} and it decreases with the minimization of t_{BOX}.

The dependency of kink on different gap positions in the SELBOX with fixed 0.5 nm thick gap is shown in Fig. 4b. Kink effect is significant for the gap situated close to the drain and it is nearly insignificant when gap is located close to the source. n-TFET, with p-type source operates by the mechanism of BTB tunneling of carriers between the source and the channel. The holes present close to the tunnel junction can smoothly move through the small gap located nearer to the source in comparison with drain. More kink is deduced for the gap located away from the tunneling junction (25–27 nm) in the source in comparison with the gap located near the tunneling junction (31–33 nm) in the source. This is ascribed to the easy movement of holes in the latter in comparison with the former. Thus, highest resistance and sharp kink in the I_D–V_{DS} characteristics is obtained for gap location near the drain. For the holes a path of conduction between the tunneling junction and the substrate is provided by the gap within the SELBOX. With the expansion in gap thickness, more number of holes generated adjacent to the tunnel junction can move to the substrate leading to

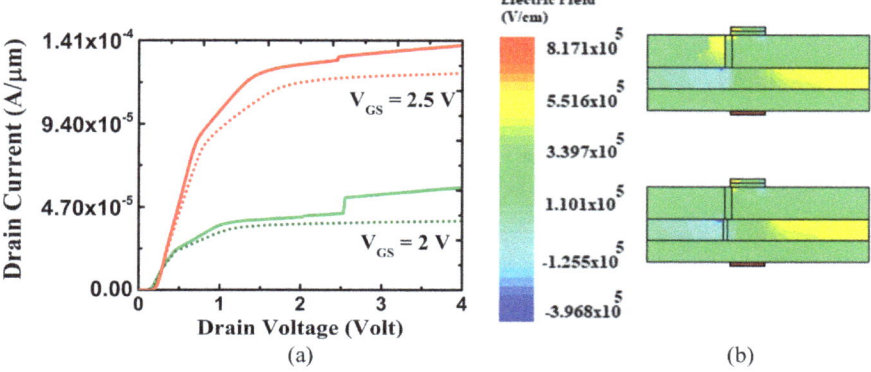

Fig. 5 **a** Output curves of C1-SOI TFET (with solid lines) and SELBOX TFET (with dotted lines) with variation in gate voltages **b** Electric Field of C1-SOI TFET and SELBOX TFET

reduction in the voltage drop across the gap and enhancement in the corresponding V_T. Thus, kink effect reduces with the increase of gap thickness. The SELBOX device with no gap acts as a fully depleted SOI device. It is noticeable from Fig. 4c that the device with 2 and 4 nm gap thickness provides minimum kink effect. For 4 nm gap thickness, drain current decreases as the device with larger gap thickness behaves as bulk TFET. The reduction of drain current for 4 nm gap thickness is due to the increase of voltage drop between drain and body as the body potential reduces. This leads to the early occurrence of pinch-off and drain current saturates at lower voltage. Thus, the device with 2 nm gap thickness provides the best result. The drain current upgrades owing to the alleviation of threshold voltage and rise of body potential by the gathered holes within the floating body for the device with no buried oxide gap.

Figure 5a depicts the output plots of the SELBOX TFET with 2 nm buried oxide gap and conventional1 SOI (C1-SOI) TFET with similar device parameters as SELBOX TFET without a buried oxide gap. SOI TFET shows kink effect whereas it is not present for SELBOX TFET. At various gate biases, SELBOX device with 2 nm gap has no kink effect as depicted by the dotted lines. The kink voltage, i.e., the voltage at which the kink appears, enhances with the rise of gate to source voltage. Mostly kink induces by the congestion of electric field surrounding the abrupt source edge corner [34, 35] in C1-SOI TFET owing to maximized BTB rate. So, in TFET, the electric field congestion at the source corner near the source-channel junction is the primary cause of arising kink effect. Kink phenomenon is investigated by examining the 2-D contour plots of electric field for C1-SOI TFET and SELBOX TFET in Fig. 5b. High electric field is observed adjacent to the source edge in C1-SOI device resulting in more kink.

In the SELBOX across the gap region, hole current is estimated by using the hole current density. For high drain voltage, with the increase of electric field more holes are created which can easily flow across the barrier near junction between the drain

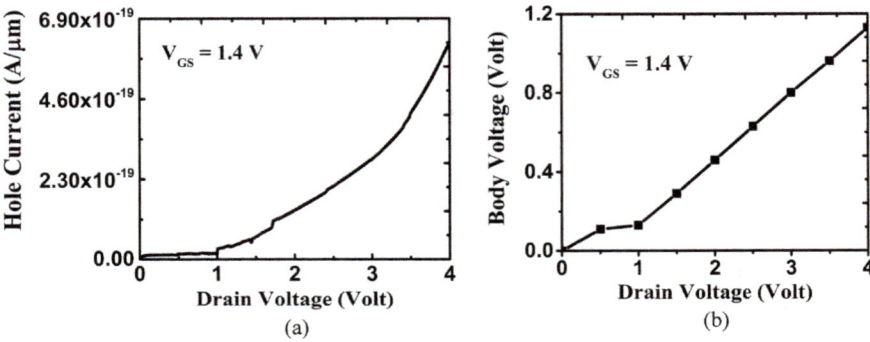

Fig. 6 Plot of **a** hole current **b** body voltage of SELBOX TFET

and the channel. Thus, drain bias increases the hole current as portrayed in Fig. 6a. The narrow gap region has high resistance in comparison with the body region. The movement of holes through this gap resistance develops a potential across the small gap in the SELBOX. This potential increases the body potential as portrayed in Fig. 6b..

3.2 DC Performance of DTD TFET

The output characteristics of DTD TFET is compared to conventional2 SOI (C2-SOI) TFET with similar device parameters as DTD TFET at $V_{GS} = 0.7$ V in Fig. 7. The L-patterned trench in the DTD device suppresses the kink effect contrary to the C2-SOI device. The high BTB generation rate in C2-SOI device rises the electric field gathering adjacent to the source edge at the tunneling junction. The BTBT generation rate of C2-SOI device is more in comparison with DTD device at $V_{DS} =$

Fig. 7 Plot of output curves of C2-SOI and DTD TFET

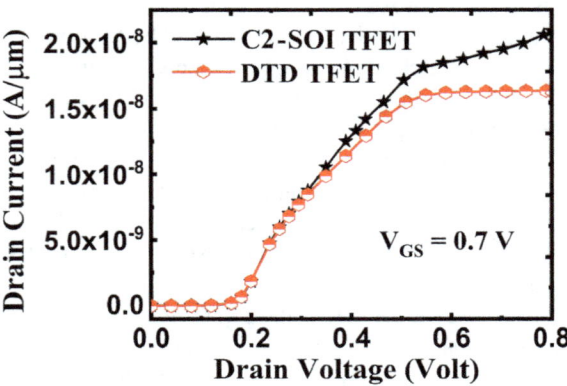

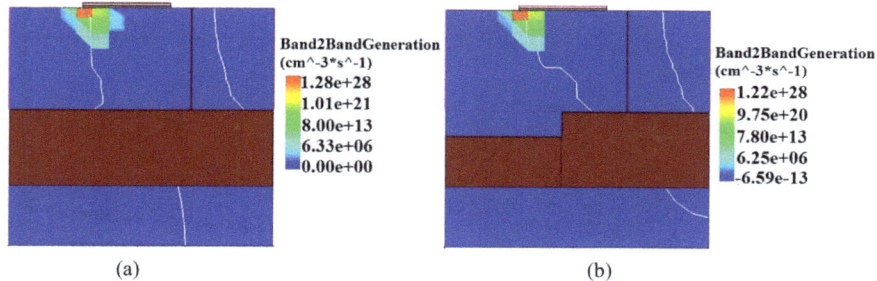

(a)   (b)

Fig. 8 Plot of BTBT Rate of **a** C2-SOI TFET **b** DTD TFET

0.8 V and $V_{GS} = 0.7$ V as depicted in Fig. 8a, b. Holes get accumulated owing to the gathering of electric field. These holes relocate to the floating body and increase the body voltage [22]. This leads to the reduction of threshold voltage and results in kink effect. Moreover, BTBT generation extends over a broad region close to the source at the tunneling junction in C2-SOI device. This expanded BTBT generation rate congregates the electric field which enhances the drain current.

In case of DTD device, the cumulated body holes can be easily extricated by the tunnel current of the two tunnel diodes [36]. This minimizes the rate of accumulated hole aggregation in the body and reduces the mitigation of V_T. Thus, the kink effect is delayed and shifted to elevated drain voltage due to the existence of L- patterned trench in DTD device. Thus, at a particular operating voltage floating body effect reduces. Drain voltage affects the tunneling current for short channels. With the increase of drain voltage the accumulation of holes increases due to high rate of BTBT generation at the tunneling junction. The enhanced aggregation of holes rises the body potential [17]. Thus, with the increase of drain voltage, body potential rises as depicted in Fig. 9. The release of aggregated holes by the two tunnel diodes using

Fig. 9 Plot of potential at the body of C2-SOI and DTD TFET

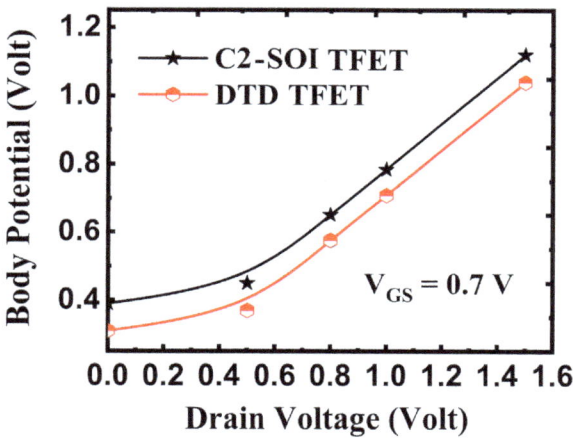

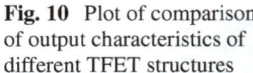

Fig. 10 Plot of comparison of output characteristics of different TFET structures

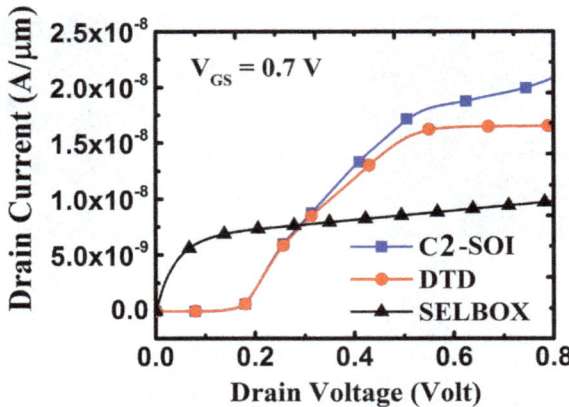

their tunnel currents lowers the body voltage of DTD TFET. DTD TFET shows low body potential compared to C2-SOI device signifying less mitigation of threshold voltage and minimum kink effect. For C2-SOI TFET, the accumulated hole density is more compared to DTD TFET indicating lower kink in DTD TFET.

3.3 Comparison of DC Performance of SELBOX and DTD TFET Structures

To compare the influence of kink in SELBOX TFET and DTD TFET, the channel length of both the structures is considered to be 35 nm with the similar doping concentrations, mentioned above. For simulating the output characteristics, it can be perceived from Fig. 10 that the kink effect is eradicated in both the structures, i.e., SELBOX TFET and DTD TFET with high drain current in the latter compared to the former. A high drain current of 2×10^{-8} A/μm is obtained by employing DTD TFET.

3.4 RF Analysis of SELBOX and DTD TFET Structures

The output characteristics of SELBOX TFET and DTD TFET for varying range of temperature is illustrated in Fig. 11a, b. With temperature a growing trend in drain current is deduced in SELBOX and C1-SOI devices in Fig. 11. The dependency of energy bandgap on temperature leads to the elevation of drain current with temperature. The minimized bandgap at elevated temperature enhances the drain current. Energy bandgap is dependent on temperature as [37]

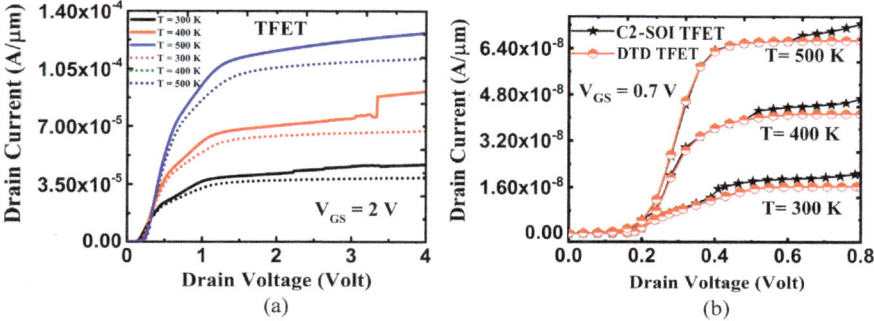

Fig. 11 Plot of temperature affectability on output characteristics of **a** SELBOX TFET (with dotted lines) and C1-SOI TFET (with solid lines) **b** DTD TFET

$$E_{bg}(T) = E_{bg}(300) - \frac{\beta T^2}{T + \gamma} \qquad (3)$$

where $\beta = 4.73 \times 10^{-4}$ eV/K and $\gamma = 636$ K and $E_{bg} = 1.08$ eV for silicon at 300 K.

Figure 11a compares the output characteristics of SELBOX TFET with C1-SOI TFET considering similar device parameters. It can be observed that there is increment in kink voltage with the hike in temperature, i.e., at minimal 300 K temperature, kink is obtained at low drain bias, whereas an opposite behavior is seen at higher temperature where kink is obtained at high drain bias. Kink voltage gets shifted to higher values of drain bias. There is no kink perceived in the output characteristics of SELBOX TFET even at higher temperatures. In case of DTD TFET in Fig. 11b I_D enhances with temperature due to the similar reasons given for SELBOX TFET. The kink in the output plots owing to the accretion of holes in the floating body is also observed at higher temperatures in C2-SOI TFET with dimensions similar to DTD TFET. In case of C1-SOI device kink is observed at very high gate voltage but C2-SOI TFET shows kink at low gate bias. It is perceived from Fig. 11 that with the rise in temperature kink relocates toward higher drain voltage in C2-SOI TFET whereas DTD TFET has no kink effect even at elevated temperature. Therefore, the proposed SELBOX and DTD devices show improved performance even at elevated temperature.

Figure 12 depicts the alteration of total gate capacitance, C_{GG} with temperature for SELBOX and DTD TFET. It can be observed that the C_{GG} of SELBOX TFET is less than DTD TFET. C_{GG} of SELBOX TFET is three times lesser than DTD TFET. The high drain current of DTD TFET indicates the presence of more inversion charge carriers at the channel and high gate capacitance balances these carriers. Considering fixed drain bias, C_{GG} shows a rising trend with gate bias. At low gate voltage, inversion layer starts creating nearer to the drain side and it further gets expanded toward the source with the elevation in gate voltage. Moreover, with the rise in temperature C_{GG} of the TFET devices elevates. The inversion charge carriers present at the semiconductor surface enhance due to the mitigation of energy barrier at

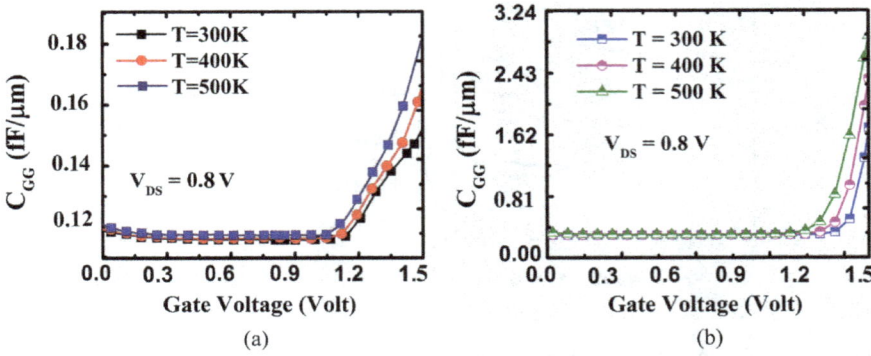

Fig. 12 Plot of temperature dependency on C_{GG} of **a** SELBOX TFET **b** DTD TFET

elevated temperature and these inversion charge carriers get balanced by the amplified gate charge resulting in high gate capacitance. The high value of total gate capacitance of DTD TFET indicates the dominance of gate over the channel is more in contrast with SELBOX TFET.

The comparison of transconductance ($g_m = \partial I_D / \partial V_{GS}$) which is an essential RF parameter of SELBOX TFET with DTD TFET is shown in Fig. 13. In both the TFET devices, g_m rises with gate bias. The total number of carriers that tunnel via the tunneling junction enhances owing to the upgraded electrostatic integration amidst the gate and the tunnel junction. The increase of drain current at high gate voltage boosts up g_m. There is a significant variation of transconductance in both the structures and the elevated drain current of DTD TFET leads to its enhanced g_m compared to SELBOX TFET. The increment of drain current as a result of energy barrier reduction enhances g_m with temperature and it is evident that both the devices shows enhancement of g_m with rise in temperature. The transconductance of DTD TFET is 7.7×10^{-6} S/μm, whereas SELBOX TFET has a transconductance of 7×10^{-6} S/μm at 300 K.

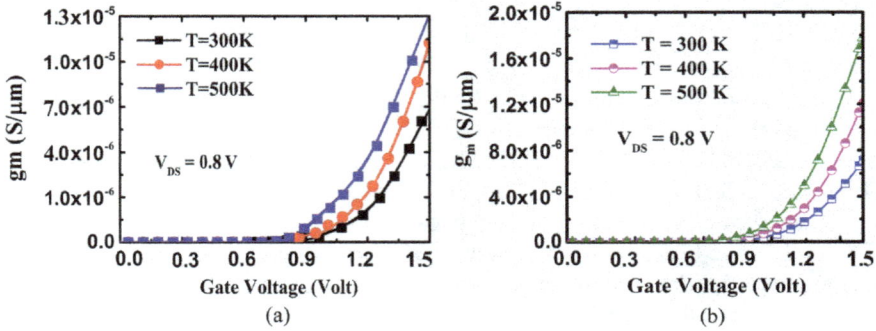

Fig. 13 Plot of temperature dependency on g_m of **a** SELBOX TFET **b** DTD TFET

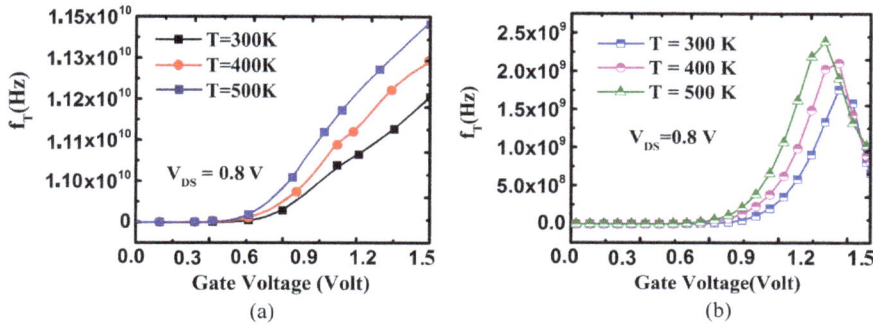

Fig. 14 Plot of affectability of temperature on f_T of **a** SELBOX TFET **b** DTD TFET

Figure 14 depicts the dependency of cut-off frequency (f_T), a vital figure of merit on gate bias of both the structures. It is specified as the frequency where the gain of the short circuited current sets off at unity. It is represented in terms of total gate capacitance (C_{GG}) and transconductance (g_m) as [38]

$$f_T = \frac{g_m}{2\pi C_{GG}} \tag{4}$$

It can be deduced from Fig. 14 that the cut-off frequency of DTD device is lower in comparison with the SELBOX TFET. This is accredited to the high value of C_{GG} of DTD TFET which is inversely proportional to f_T. The lower value of capacitance dominates over transconductance. The proposed SELBOX TFET shows a high cut-off frequency of 1.12×10^{10} Hz whereas for SELBOX TFET f_T is 1.76 GHz at 300 K. The reliance of f_T on temperature is portrayed in Fig. 14. Similar to transconductance, f_T also increases with the elevation of temperature. The upgraded BTBT rate of electrons at high temperature causes high g_m. This results in high f_T. Elevated frequency of GHz range is obtained by the proposed SELBOX device and thus, the device can be utilized for high frequency applications.

Figure 15 depicts the temperature affectability on transconductance generation factor (TGF = g_m/I_D) [39]. It is a vital parameter considering various circuits of analog design. TGF is also defined as device efficiency. In SELBOX TFET, there is a remarkable deviation in TGF with variation in temperature at low values of drain current with minimal variation at high values of drain current. The highest value of TGF at low temperature of 300 K is 120 V^{-1}. For DTD TFET, TGF mitigates at high temperature. This is ascribed to the enhancement of drain current. The rise of drain current dominates over gm resulting in low TGF at high temperature. TGF varies significantly with temperature at low values of drain current in DTD TFET. At moderate temperature, the proposed device has the potential to effectively change power into speed. However, with the increase of temperature, efficiency decreases.

The variation of the product of transconductance and frequency, (TFP = $g_m f_T / I_D$) with temperature is depicted in Fig. 16. Similar to g_m and f_T, TFP also

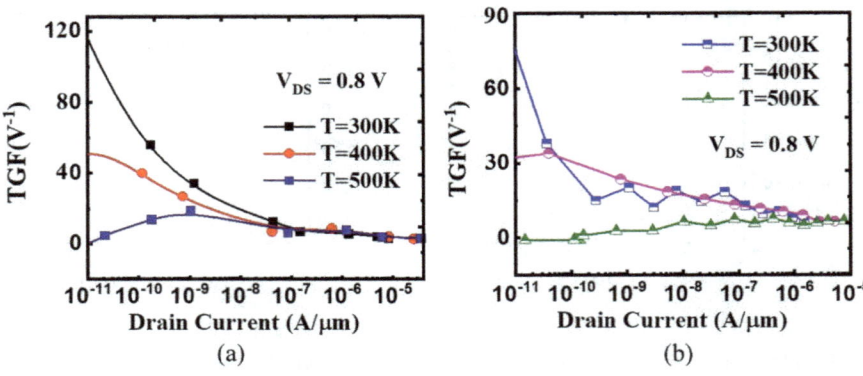

Fig. 15 Plot of TGF reliance on temperature of **a** SELBOX TFET **b** DTD TFET

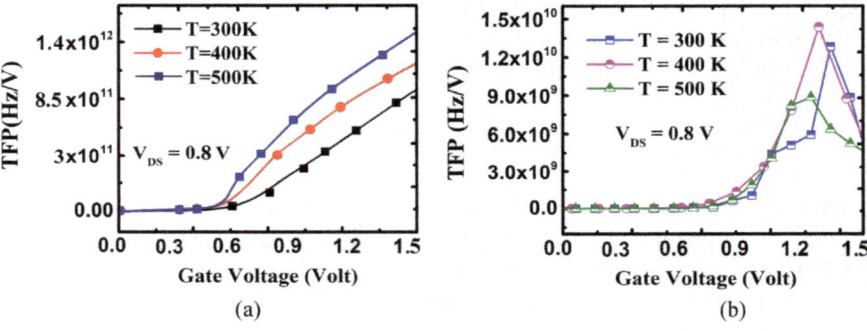

Fig. 16 Plot of temperature affectability on TFP of **a** SELBOX TFET **b** DTD TFET

shows positive temperature coefficient with temperature [40–42]. It is an essential RF parameter for applications related high speed. It enhances linearly and achieves a peak value prior to the formation of inversion region.

Another essential RF parameter is intrinsic delay (τ) and it is expressed with regard to C_{GG} and ON current (I_{ON}) as [38]

$$(\tau = ((C_{GG} V_{DD})/I_{on})) \tag{5}$$

where the drain voltage $V_{DD} = 0.8$ V.

Intrinsic delay is a crucial RF figure of merit that plays a pivotal role in digital logic applications. The dependency of temperature on intrinsic delay (τ) is portrayed in Fig. 17. It can be deduced that at minimal temperature of 300 K, τ exhibits higher value. τ and ON current are inversely related and as shown in Fig. 11, I_{ON} boosts up with the increase of temperature. The proposed SELBOX and DTD TFET show upgraded switching speed characteristics providing minimum intrinsic delay of the

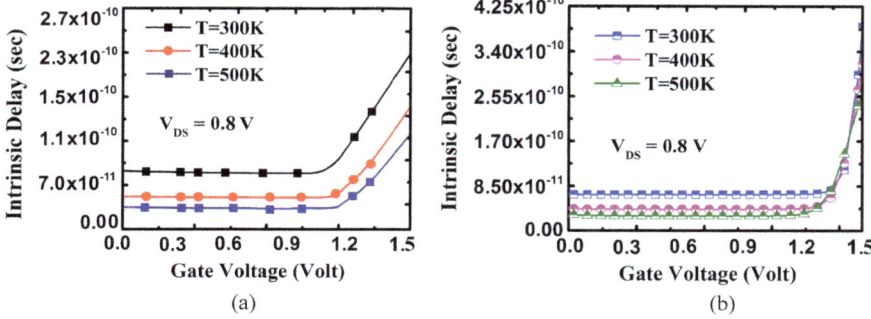

Fig. 17 Plot of temperature affectability on τ of **a** SELBOX TFET **b** DTD TFET

order of 10^{-10} s. The lower value of intrinsic delay of SELBOX TFET is caused by its low gate capacitance in comparison with DTD TFET.

Linearity analysis is the primary requisite to deduce the capability of the device. To compute the linearity performance of the proposed devices, 1-dB compression point (C-Point) has been studied by using the second and third order harmonics of I_D versus V_G [43, 44]. The temperature affectability on various linearity parameters such as C-Point has been explored in this work. Elevated values of linearity parameters specify that the output is less distorted. The input power which must be applied to shift the output power by 1-dB from linearity is defined as 1-dB compression point (C-Point). To evaluate the higher limit of linear operation, C-Point is a crucial parameter [43]. It depends on transconductance. Figure 18a shows the reliance of C-Point of SELBOX device on temperature. The high value of transconductance with lower signal distortion leads to amplification of C-Point with increment in temperature. The dependency of C-Point on temperature of DTD device is depicted in Fig. 18b. It can be deduced that the C-Point of the proposed DTD design is more at high temperature and it mitigates with the decrease of temperature. The

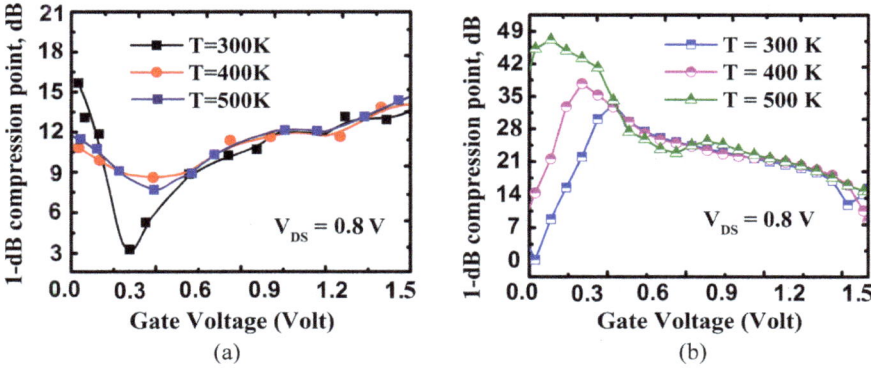

Fig. 18 Plot of temperature affectability on linearity parameter of **a** SELBOX TFET **b** DTD TFET

linearity performance of the design is superior at minimal temperature. In both the devices, the deviation in C-Point with temperature is more at minimal gate voltage with insignificant variation at high gate bias. DTD TFET shows higher values of C-Point compared to SELBOX TFET. The DTD device with high input power becomes suitable for amplifier applications.

4 Conclusion

The proposed SELBOX device provides minimum OFF current of 1.7×10^{-18} A/μm with gap of thickness 2 nm near the source. Thus, the device becomes worthy for low power applications. As the gap within the buried oxide moves away from the tunneling junction, OFF current increases. SiGe due to its low energy band gap provides high ON current. Drain current increases with temperature. A kink is glimpsed in the V-I characteristics at the output in conventional SOI TFET. Owing to the accretion of the floating body holes of the SOI device there in an instantaneous hike in the drain current and it is defined as kink. In C1-SOI TFET, the abrupt source doping intensifies kink effect as BTBT is the transmission mechanism in TFET. The irregularity perceived in the output curve can be eradicated with very small gap in the SELBOX layer. The divergence in the kink effect with different gap thickness and SELBOX thickness has been studied. Further, the effect of temperature has been examined. There is an elevation in kink voltage with the enhancement of temperature. The proposed SELBOX device provides reduced OFF current. The release of holes by DTD TFET device in the body minimizes the kink effect. In comparison with C2-SOI device, the voltage of the body of DTD TFET minimizes and the kink voltage elevates. Even at elevated temperature also there is no kink in the output characteristics of DTD TFET. Hence, in DTD TFET current increases in a controlled manner and can be utilized as a potential device. In both the SELBOX TFET and DTD TFET bandgap narrowing leads to rise in drain current with increase of temperature. It has been analyzed that the ON current of DTD TFET is higher than SELBOX TFET. The RF parameters of SELBOX TFET are compared with DTD TFET considering various temperatures. At high temperature, reduction of energy barrier leads to the increase of gate capacitance. The gate capacitance of DTD TFET is more than the SELBOX TFET that signifies remarkable controllability of the gate in DTD TFET. The high drain current enhances the transconductance of DTD TFET. However, the dominancy of capacitance over transconductance leads to high cut-off frequency in SELBOX TFET. The proposed device provides high cut-off frequency of 1.76 GHz and low intrinsic delay of the order of 10^{-10} s at 300 K making it worthy for high speed applications. SELBOX TFET proves to be better device in terms of efficiency due to its elevated values of TGF. Kink effect minimizes in both SELBOX TFET and DTD TFET but the latter device provides high drain current.

References

1. Gargini P (2000) The international technology roadmap for semiconductors (ITRS): "Past, present and future". In GaAs IC symposium. IEEE gallium arsenide integrated circuits symposium. 22nd annual technical digest 2000. (Cat. No. 00CH37084), pp 3–5
2. "International Technology Roadmap for Semiconductors", http://public.itrs.net/, 2013 Edition
3. Wang PF et al (2004) Complementary tunneling transistor for low power application. Solid-State Electron 48:2281–2286
4. Choi WY, Park BG, Lee JD, Liu TJ (2007) Tunneling field-effect transistors (TFETs) with subthreshold swing (SS) less than 60 mV/dec. IEEE Electron Device Lett 28(8):743–745
5. Knoch J, Appenzeller J (2005) A novel concept for field-effect transistors-the tunneling carbon nanotube FET. In 63rd Device Research Conference Digest, 2005. DRC'05 IEEE 153–156.
6. Boucart K, Ionescu AM (2007) Double-gate tunnel FET With High-k gate dielectric. IEEE Trans Electron Devices 54(7):1725–1733
7. Sivasankarana K, Mallickb PS (2013) A comparative study of radio frequency stability performance of double gate MOSFET and double gate tunnel FET. In2013 International conference on green computing, communication and conservation of energy (ICGCE) IEEE 220–224 Dec 2013
8. Pal A, Dutta AK (2016) Analytical drain current modeling of double-gate tunnel field-effect transistors. IEEE Trans Electron Devices 63(8):3213–3221
9. Balamurugan NB, Priya GL, Manikandan S, Srimathi G (2016) Analytical modeling of dual material gate all around stack architecture of tunnel FET. In2016 29th international conference on VLSI design and 2016 15th international conference on embedded systems (VLSID) IEEE 294–299 Jan 2016
10. Young KK (1989) Short-channel effect in fully depleted SOI MOSFETs. IEEE Trans Electron Devices 36(2):399–402
11. Chaudhry A, Kumar MJ (2004) Controlling short-channel effects in deep-submicron SOI MOSFETs for improved reliability: a review. IEEE Trans Device Mater Reliab 4(1):99–109
12. Bhushan B, Nayak K, Rao VR (2012) DC compact model for SOI tunnel field-effect transistors. IEEE Trans Electron Devices 59(10):2635–2642
13. Mitra SK, Goswami R, Bhowmick B (2016) A hetero-dielectric stack gate SOI-TFET with back gate and its application as a digital inverter. Superlattices Microstruct 92:37–51
14. Ahish S, Sharma D, Vasantha MH, Kumar YB (2016) Design and analysis of novel InSb/Si Heterojunction double gate tunnel field effect transistor. In 2016 IEEE computer society annual symposium on VLSI (ISVLSI) IEEE 105–109 July 2016
15. Goswami R, Bhowmick B (2016) Circular gate tunnel FET: Optimization and noise analysis. Procedia Computer Science 93:125–131
16. Goswami R, Bhowmick B, Baishya S (2015) Electrical noise in circular gate tunnel FET in presence of interface traps. Superlattices Microstruct 86:342–354
17. Narayanan M, Al-Nashash H, Mazhari B, Pal D, Chandra M (2012) Analysis of kink reduction in SOI MOSFET using selective back oxide structure. Active and Passive Electronic Components Article ID 565827:1–9
18. Wei A, Sherony MJ, Antoniadis DA (1998) Effect of floating-body charge on SOI MOSFET design. IEEE Trans Electron Devices 45(2):430–438
19. Casse M, Pretet J, Cristoloveanu S, Poiroux T, Fenouillet-Beranger C, Fruleux F, Raynaud C, Reimbold G (2004) Gate-induced floating-body effect in fully-depleted SOI MOSFETs with tunneling oxide and back-gate biasing. Solid-State Electron 48(7):1243–1247
20. Ren JZ, Salama CAT (2000) 1 V SOI NMOSFET with suppressed floating body effects. Solid-State Electron 44(11):1931–1937
21. Kim SW, Choi WY, Sun MC, Kim HW, Park BG (2012) Design guideline of Si-based L-shaped tunneling field-effect transistors. Japanese J Appl Phys 51(6S):06FE09
22. Koh YH, Choi JH, Nam MH, Yang JW (1997) Body-contacted SOI MOSFET structure with fully bulk CMOS compatible layout and process. IEEE Electron Device Lett 18(3):102–104

23. Min BW, Kang L, Wu D, Caffo D, Hayden J, Mendicino MA (2002) Reduction of hysteretic propagation delay with less performance degradation by novel body contact in PD SOI application. InProc. IEEE Int. SOI Conf 169–170 October 2002

24. Wu CL, Yu C, Shichijo H (2011) I-gate body-tied silicon-on-insulator MOSFETs with improved high-frequency performance. IEEE Electron Device Lett 32(4):443–445

25. Madan J, Chaujar R (2017) Temperature associated reliability issues of heterogeneous gate dielectric-Gate all around-Tunnel FET. IEEE Trans Nanotechnol 17(1):41–48

26. T.C.A.D. Synopsys, Manual, ver. E2010.12.

27. Kampen C, Burenkov A, Lorenz J (2011) Challenges in TCAD simulations of tunneling field effect transistors. In2011 Proceedings of the European Solid-State Device Research Conference (ESSDERC) IEEE 139–142 Sep 2011.

28. Hermle M, Letay G, Philipps SP, Bett AW (2008) Numerical simulation of tunnel diodes for multi-junction solar cells. Prog Photovoltaics Res Appl 16(5):409–418

29. Choi WY, Lee W (2010) Hetero-gate-dielectric tunneling field-effect transistors. IEEE Trans Electron Devices 57(9):2317–2319

30. Ghosh P, Bhowmick B (2020) Analysis of kink reduction and reliability issues in low-voltage DTD-based SOI TFET. Micro & Nano Letters 15(3):130–135

31. Kim SW, Kim JH, Liu T-JK, Choi WY, Park B-G (2016) Demonstration of L-shaped tunnel field-effect transistors. IEEE Trans Electron Devices 63(4):1774–1778

32. Goswami R, Bhowmick B, Baishya S (2016) Physics-based surface potential, electric field and drain current model of a $\delta p^+ Si_{1-x}Ge_x$ gate-drain underlap nanoscale n-TFET. Int J Electron 103(9):1566–1579

33. Goswami R, Bhowmick B (2014) Hetero-gate-dielectric gate-drain underlap nanoscale TFET with a $\delta p^+ Si_{1-x}Ge_x$ layer at source channel tunnel junction. In 2014 International conference on green computing communication and electrical engineering (ICGCCEE) IEEE 1–5 March 2014

34. Kim SW, Choi WY, Sun MC, Park BG (2013) Investigation on the corner effect of L-shaped tunneling field-effect transistors and their fabrication method. J Nanosci Nanotechnol 13(9):6376–6381

35. Fossum JG, Yang JW, Trivedi VP (2003) Suppression of corner effects in triple-gate MOSFETs. IEEE Electron Device Lett 24(12):745–747

36. Anvarifard MK, Orouji AA (2015) Enhanced critical electrical characteristics in a nanoscale low-voltage SOI MOSFET with dual tunnel diode. IEEE Trans Electron Devices 62(5):1672–1676

37. Ghosh P, Bhowmick B (2019) Reduction of the kink effect in a SELBOX tunnel FET and its RF/analog performance. J Comput Electron 18(4):1182–1191

38. Ghosh P, Bhowmick B (2020) Effect of temperature on reliability issues of ferroelectric dopant segregated schottky barrier tunnel field effect transistor (Fe DS-SBTFET). SILICON 12(5):1137–1144

39. Saha R, Bhowmick B, Baishya S (2018) Temperature effect on RF/analog and linearity parameters in DMG FinFET. Appl Phys A 124(9):642

40. Kranti A, Armstrong GA (2010) Nonclassical channel design in MOSFETs for improving OTA gain-bandwidth trade-off. IEEE Trans Circuits Syst I Regul Pap 57(12):3048–3054

41. Ghosh P, Bhowmick B (2020) Effect of temperature in selective buried oxide TFET in the presence of trap and its RF analysis. Int J RF Microwave Comput-Aided Eng 8:e22269 (1–9)

42. Gautam R, Saxena M, Gupta RS, Gupta M (2012) Effect of localised charges on nanoscale cylindrical surrounding gate MOSFET: Analog performance and linearity analysis. Microelectron Reliab 52(6):989–994

43. Kumar SP, Agrawal A, Chaujar R, Gupta RS, Gupta M (2011) Device linearity and intermodulation distortion comparison of dual material gate and conventional AlGaN/GaN high electron mobility transistor. Microelectron Reliab 51(3):587–596

44. Ghosh P, Haldar S, Gupta RS, Gupta M (2012) An investigation of linearity performance and intermodulation distortion of GME CGT MOSFET for RFIC design. IEEE Trans Electron Devices 59(12):3263–3268

CPSIA information can be obtained
at www.ICGtesting.com
Printed in the USA
BVHW051920220223
659019BV00005B/76